MTP International Review of Science

Volume 3
Spectroscopy

Edited by **D. A. Ramsay, F.R.S.C.**
National Research Council of Canada

Butterworths · London
University Park Press · Baltimore

THE BUTTERWORTH GROUP

ENGLAND
Butterworth & Co (Publishers) Ltd
London: 88 Kingsway, WC2B 6AB

AUSTRALIA
Butterworths Pty Ltd
Sydney: 586 Pacific Highway 2067
Melbourne: 343 Little Collins Street, 3000
Brisbane: 240 Queen Street, 4000

NEW ZEALAND
Butterworths of New Zealand Ltd
Wellington: 26–28 Waring Taylor Street, 1

SOUTH AFRICA
Butterworth & Co (South Africa) (Pty) Ltd
Durban: 152–154 Gale Street

ISBN 0 408 70264 8

UNIVERSITY PARK PRESS

U.S.A. and CANADA
University Park Press Inc
Chamber of Commerce Building
Baltimore, Maryland, 21202

Library of Congress Cataloging in Publication Data

Ramsay, Donald A., 1922–
  Spectroscopy.

  (Physical chemistry, series one, v. 3) (MTP
international review of science)
  1. Spectrum analysis—Addresses, essays, lectures.
I. Title.
QD453.2.P58 vol. 3 [QC451] 541'.3'08s [535'.84]
ISBN 0–8391–1017–0   72–4330

First Published 1972 and © 1972
MTP MEDICAL AND TECHNICAL PUBLISHING CO. LTD.
Seacourt Tower
West Way
Oxford, OX2 OJW
and
BUTTERWORTH & CO. (PUBLISHERS) LTD.

Filmset by Photoprint Plates Ltd., Rayleigh, Essex
Printed in England by Redwood Press Ltd., Trowbridge, Wilts
and bound by R. J. Acford Ltd., Chichester, Sussex

Consultant Editor's Note

The MTP International Review of Science is designed to provide a comprehensive, critical and continuing survey of progress in research. The difficult problem of keeping up with advances on a reasonably broad front makes the idea of the Review especially appealing, and I was grateful to be given the opportunity of helping to plan it.

This particular 13-volume section is concerned with Physical Chemistry, Chemical Crystallography and Analytical Chemistry. The subdivision of Physical Chemistry adopted is not completely conventional, but it has been designed to reflect current research trends and it is hoped that it will appeal to the reader. Each volume has been edited by a distinguished chemist and has been written by a team of authoritative scientists. Each author has assessed and interpreted research progress in a specialised topic in terms of his own experience. I believe that their efforts have produced very useful and timely accounts of progress in these branches of chemistry, and that the volumes will make a valuable contribution towards the solution of our problem of keeping abreast of progress in research.

It is my pleasure to thank all those who have collaborated in making this venture possible – the volume editors, the chapter authors and the publishers.

Cambridge   A. D. Buckingham

Preface

Twenty years ago, I was informed by a senior scientist that 'the trouble with spectroscopy is that it is dead and doesn't know it'. This remark is even more emphatic than the prophecy of Oliver Lodge in 1888 that 'the whole subject of electrical radiation seems to be working itself out splendidly'. It is now approximately 300 years since the early experiments of Newton, and yet the creativity of successive generations of research workers shows no signs of diminishing. Indeed, the Nobel Prize for Chemistry this year was awarded to a spectroscopist, Dr. Gerhard Herzberg, for his 'contributions to the knowledge of the electronic structure and geometry of molecules, especially free radicals'.

Another remark, to be taken more seriously, is that there are already far too many scientific and review journals. One can only hope that regulation will be provided by the law of supply and demand. Surveying the field of spectroscopy one finds that while broad areas of spectroscopy are reviewed from time to time in various journals, yet there is no current review journal in which topics of more detailed spectroscopic interest are discussed by experts active in these fields. It is hoped that the present series will fulfil this need.

The general policy has been to ask about ten authors to contribute a chapter of approximately 30 pages each, relating to their own particular fields of interest. Overall, an attempt has been made to keep a balance between electronic, vibrational and rotational spectroscopy and between smaller and larger molecules. No attempt has been made to change or edit the subject matter in any chapter. Indeed the individuality of approach of each author is a quality to be cherished and preserved. Also no attempt has been made to impose a regimented system of units. The choice of units has been left to the individual conscience of each author.

It is hoped that the present volume will commend itself to the reader. Criticisms and suggestions for future contributions will be most gratefully received.

Ottawa   D. A. Ramsay

Contents

1
Matrix Isolation Spectroscopy

DOLPHUS E. MILLIGAN and MARILYN E. JACOX
National Bureau of Standards, Washington, D.C.

1.1   SPECTRA OF MATRIX-ISOLATED MOLECULES

The matrix isolation technique was first utilised in the laboratory of Lewis[1] for studies of the emission spectrum of dilute solid solutions of fluorescein in a boric acid glass. Changes in the emission spectrum were related to the increasing rigidity of the boric acid as the temperature was lowered. These studies have formed the basis for a large volume of work on the photo-excitation of molecules trapped in an inert environment and have greatly furthered our knowledge of the properties of excited triplet species. In a subsequent study[2] the potential of the technique for studies of the spectra of free radicals and molecular ions formed upon irradiation of dilute rigid solid solutions with ultraviolet light was also recognised, and the feasibility of such experiments was demonstrated by the detection of electronic spectra of several large organic free radicals and molecular ions after irradiation of the parent molecule suspended in small concentration in EPA at 90 K.

Unlike the phosphorescence studies, the photodissociation studies were not immediately pursued. The ultraviolet absorption studies of Norman and Porter[200] demonstrated that a number of smaller, less stable free radicals, including CS, ClO and several simple aromatic radicals, could also be stabilised in rigid organic glasses at 77 K. These workers suggested that the technique might be extended to permit infrared observations of free radicals. This possibility had independently been recognised by Pimentel and co-workers[3], who had performed several experimental tests of the isolation efficiency of solid xenon, CO_2, CCl_4 and methylcyclohexane at various temperatures for such small molecules as NH_3, HN_3 and NO_2. Subsequently, the infrared studies of Becker and Pimentel[4] demonstrated that at 20 K xenon, nitrogen and argon are effective in isolating NO_2 and a number of small hydrogen-bonding molecules present in concentrations less than 1%. Further studies of the infrared spectra of methanol[5] and of water[6] trapped in a nitrogen matrix at 20 K led to the assignment of infrared absorptions to isolated molecules, to dimeric species and to simple polymeric aggregates. Because the importance of the cage effect in determining the photoprocesses which may occur in rigid matrices was incompletely realised, the early attempts at the *in situ* photoproduction of free radicals and other reactive species in concentration sufficient for direct spectroscopic observation encountered some difficulty. However, the understanding of the restrictions imposed by the cage effect has since proved invaluable for the design of experiments suitable for isolating reactive species. Infrared and ultraviolet spectra have by now been reported for more than 50 diatomic and small polyatomic free radicals stabilised upon ultraviolet or vacuum-ultraviolet photolysis of suitable precursors present in small concentration in inert solid matrices[7]. Electron spin resonance spectra have also been reported for an extremely large number of species with unpaired electrons produced *in situ* in various rigid solids. Because of the very great sensitivity of this technique, data have been obtained in many systems in which the concentration of the species of interest is insufficient for direct infrared or ultraviolet spectroscopic detection. With one important exception, such electron spin resonance studies will not receive further consideration.

The early matrix-isolation studies also include a number of experiments in which reactive molecules formed in gas-phase processes were rapidly frozen in an excess of an inert rigid matrix material. Robinson and McCarty[8-12] succeeded in detecting electronic absorptions of such reactive species as NH_2, NH, OH and HNO in deposits of the products of gas-phase electric discharges trapped in rare-gas matrices. The suggestion of Pimentel[13] that it should be feasible to study the spectra of molecules characteristic of high-temperature vapours by rapidly freezing them in a matrix environment was soon followed by the report of the isolation of lithium fluoride in rare-gas matrices by Linevsky[14]. A very large number of high-temperature species have since been studied using the matrix isolation technique, and a large fraction of the current research activity in the field of matrix isolation is concerned with the study of such high-temperature molecules.

1.1.1   Rotational and vibrational spectra

Upon isolation in a matrix, most molecules show sharp absorptions, with half-widths of only a few cm^{-1}; rotational structure does not appear. However, some of the low-J rotational lines have been observed in the spectra of a few simple hydride molecules isolated in rare-gas matrices.

1.1.1.1   Pure rotational spectra

There have been a number of recent studies of the pure rotational spectra of the various hydrogen halides isolated in rare-gas matrices. Katz, Ron and Schnepp[15] suggested the assignment of a band which appeared at 72 cm^{-1} in the spectra of HCl, DCl, HBr and DBr in an argon matrix, and between 45 and 95 cm^{-1} in other rare-gas matrices, to an activated phonon absorption. Evidence was also obtained for an absorption near 35 cm^{-1}, the low frequency limit of their observations, in matrix-isolated HCl and HBr samples. The interferometric studies of Barnes and co-workers[16] covering the 8–120 cm^{-1} spectral region have led to the assignment of an absorption at 18.6 cm^{-1} to the R(0) transition of HCl isolated in an argon matrix. Several other peaks, including one at 35 cm^{-1}, were assigned to activated phonon absorptions. Studies of the spectra of HCl in various rare-gas matrices out to 16 cm^{-1}, taken with a grating spectrometer, have recently been reported by Von Holle and Robinson[17], who confirmed the previous assignment of the R(0) absorption of HCl and obtained evidence for the R(1) absorption as well. These workers have emphasised the role of water impurity in accounting for discrepancies between their observations and those previously reported. Mason, Von Holle and Robinson[18] have also studied the vibration–rotation and pure rotation spectra of HF and DF in rare-gas matrices near 4 K. Except for the pure rotational absorption of DF in an argon matrix, the R(0) absorption appeared at successively greater frequencies as the mass of the rare-gas atom was increased. In all but the neon matrix observations, this peak was also split. In the vibration–rotation spectrum, the peak assigned to the R(0) absorption appeared at a frequency which increased steadily as the mass of the rare-gas atom was decreased, with the position in the neon matrix observations most closely approaching that of the gas-phase absorption. These workers also noted that, although its lattice remains rigid to a higher temperature than do the lattices of the lighter rare gases, xenon is relatively inefficient in isolating HF and DF, because of the relatively large lattice spacings. Their data were accounted for in some detail by considering the Lennard-Jones interaction between HF or DF and the neighbouring rare-gas atoms and the anisotropy of the rotational potential function.

1.1.1.2 Vibrational spectra

Vibrational absorptions characteristic of molecules isolated in rare-gas matrices generally appear within a few cm^{-1} of the gas-phase band centres. Usually a single sharp peak is observed, with a few weaker satellite absorptions which may be attributed to the trapping of molecules in different types of site in the rare-gas lattice. Thus, a molecule may be trapped in a substitutional or an interstitial site in the lattice or on an edge of a crystallite of the matrix deposit. Even at a relatively high dilution, there is a significant probability for the trapping of two molecules in adjacent sites in the matrix, and a site perturbation due to the interaction between pairs of molecules may result. The infrared spectra of a few molecules, including the hydrogen halides[19, 20] and HNC[21], are extraordinarily sensitive to the presence of traces of nitrogen impurity in the matrix. While such site splittings are often considered to be undesirable, typically their magnitude is small and frequently their presence is indicative of efficient isolation of the molecule. As the lattice is warmed irreversible shifts and broadenings in the peaks occur and often the splittings are lost.

A number of workers have proposed theories which attempt to provide detailed explanations of these site splittings. However, none of these theories is completely adequate and attempts to assign the individual peaks by changing the characteristics of the matrix are complicated by competing changes in the system. As the temperature is increased the rigidity of the matrix may decrease and molecules may begin to diffuse and to interact, counterbalancing changes resulting from the development of a more regular rare-gas lattice. An increase in the mass of the rare-gas matrix material leads to a more rigid lattice at a given temperature, but also to larger substitutional and interstitial sites. Weltner and co-workers[22, 23] have noted that the electronic spectrum of C_3 isolated in a neon matrix corresponds closely to the gas-phase spectrum and have suggested that the behaviour of molecules in a neon matrix might be expected to approximate most closely that of the gaseous species. As exemplified by the studies of HF in various rare-gas matrices[18], this generalisation has frequently been found to be valid. However, it is not without exception, as may be illustrated by the electronic absorption spectrum assigned to C_2^- in various matrices[24]. The origin of this transition is closest to the gas-phase value in the neon matrix experiments, but in a neon matrix each band is split into four peaks, whereas in an argon matrix each band is unsplit except for a partially resolved high-frequency satellite.

Like the rare gases, nitrogen cooled to cryogenic temperatures is very efficient for isolating reactive molecules. Rotational structure is completely suppressed even for the small hydride molecules isolated in a nitrogen matrix. Studies in a nitrogen matrix, as well as in rare-gas matrices, are frequently useful for determining whether peaks separated by a few cm^{-1} are contributed by two different fundamental absorptions or by site splittings. Absorptions associated with motions of a hydrogen atom which may hydrogen bond with the nitrogen matrix are shifted significantly; hydrogen stretching absorptions generally appear at lower frequencies and hydrogen deformation absorptions at higher frequencies than in the gas phase or in a rare-gas matrix. Vibrational absorptions of highly ionic species also are significantly shifted in a nitrogen matrix. The antisymmetric stretching fundamental of $FeCl_2$ is decreased by almost 50 cm^{-1}, or c. 10%, in going from an argon to a nitrogen matrix[25], and the symmetric stretching fundamental of CeO_2, derived by analysis of the emission spectrum, is decreased by about 10% in going from a neon to a nitrogen matrix[26].

Still other matrix materials have been used for studies of the infrared spectra of both stable and reactive molecules. CO and CO_2 form rigid matrices which are transparent over most of the infrared spectral region. A number of free radicals have been stabilised in sufficient concentration for direct spectroscopic study by the reaction of an atom or a free radical with these matrix materials, the so-called reactive matrix technique. Indeed, the first small polyatomic free radical to be identified in the infrared was HCO, produced by the reaction of H atoms with a CO matrix[27]. Of course, the possible occurrence of hydrogen bonding and the enhanced likelihood of polar interactions may result in greater matrix shifts in carbon monoxide and carbon dioxide matrices than in rate-gas matrices. Studies in a variety of other matrix materials have also been reported. However, most such matrix materials possess several infrared absorption regions, making them inherently unsuitable for detailed infrared spectroscopic observations. Furthermore, a variety of polar interactions may occur, resulting in significant deviations from the spectrum characteristic of the gas-phase material. Since the emphasis in this Review is on the properties of reactive species isolated in an environment in which their behaviour is closely similar to that of the gas-phase molecule, studies in large-molecule matrices will not be considered in detail.

The sharpness of the vibrational absorptions characteristic of matrix-isolated species often is a distinct advantage, even in studies of the infrared spectra of stable molecules. Thus, overlapping absorption bands may be resolved. However, the loss of the gas-phase band contour implies the loss of the information which it affords regarding the symmetry of the vibration. The utility of matrix isolation observations is illustrated by the recent identification of the v_3 absorption of H_2S and of D_2S in rare-gas matrices[28]. Because this absorption is relatively weak and lies close to the v_1 absorption, attempts at direct observation of it in the gas phase had been unsuccessful. The assignment of the vibrational fundamentals of monomeric acetic acid has also recently been clarified by argon matrix and nitrogen matrix studies[29].

Because of the sharpness of the absorptions of matrix-isolated molecules, frequently it is possible to obtain very accurate values for isotopic shifts and splittings. Although the v_3 absorption of $NiCl_2$ vapour is extremely broad, because of the important contributions of rotational structure and of vibrational 'hot bands' at the high temperature required for its observation, the corresponding absorption of $NiCl_2$ isolated in an argon matrix is so sharp that not only the individual chlorine isotopic contributions but also the nickel isotopic contributions have been resolved[30]. Such isotopic shifts and splittings have provided the key to the positive infrared spectroscopic identification of reactive species isolated in an inert matrix environment. Under favourable circumstances, a detailed analysis of the data for various isotopically substituted species has provided semi-quantitative data on the structure of the molecule. Several workers have attempted to extend the usefulness of such isotopic studies to quantitative determinations of molecular structure. Detailed comparisons between geometries derived from matrix data and from gas-phase studies have been possible for SO_2 [31, 32], for SeO_2 [32] and for SiF_2 [33]. In general, the geometries found for the matrix-isolated molecules were closely similar to those obtained from rotational analysis of gas-phase data. However, the analysis of vibrational data to obtain molecular geometries complicates very significantly if there are two or more vibrations of a given symmetry, and the resulting structure is susceptible to a large uncertainty as a result of small experimental errors in the determination of the positions of the various isotopic peaks.

It would be of considerable interest to supplement the infrared data for matrix-isolated species with Raman data. However, the size of the sample needed for conventional Raman observations has posed severe limitations on the use of the technique for matrix isolation studies, in which the amount of the species of interest typically ranges between ten and a hundred micromoles. The advent of laser Raman instrumentation has greatly eased this restriction and a number of laboratories are actively engaged in the design of sample cells suitable for laser Raman studies of matrix-isolated samples. The first successful report of such a study is that of Shirk and Claassen[34], who have succeeded in observing the three Raman-active fundamentals of SF_6 and all six vibrational fundamentals of $HCCl_3$ isolated in an argon matrix using both argon and krypton laser excitation.

(a) *Molecular aggregates* — Despite the considerable potential for the application of the matrix isolation technique to relatively stable molecules, its great power is, of course, that of making accessible spectroscopic data on species which cannot be studied using more conventional techniques. The first example of such an application was the observation of absorptions due to hydrogen-bonded dimers and simple polymers of methanol[5] and of water[6] in a nitrogen matrix. Further spectroscopic assignments for simple polymeric species of water[35] and of H_2S[36] in a nitrogen matrix have recently been offered. Argon matrix studies of aggregates of the hydrogen halides[37], of methanol[38] and of ethanol[39] also have recently been reported. Matrix isolation studies have proved useful in the assignment of the vibrational fundamentals of the cyclic dimers of acetic acid and of trifluoroacetic acid[40]. Absorptions have also been attributed to interacting pairs of pyridine molecules isolated in a nitrogen matrix[41]. The vibrational spectra of the *cis*- and *trans*-dimers of NO isolated in inert matrices, first studied by Fateley, Bent and Crawford[42], have recently been re-investigated by Guillory and Hunter[43], who also reported data for the ^{15}N- and ^{18}O-enriched species.

(b) *Reactive molecules* — The matrix isolation technique is extremely useful for the study of molecules which are difficult to handle in the gas phase because of their relatively great reactivity. The recent matrix isolation studies of the infrared spectra of H_2O_2 and of D_2O_2 [44] have indicated that the torsional motion in an argon matrix is quite similar to that of the gas-phase molecule, whereas that in a nitrogen matrix is much more hindered. The spectra of cis- and trans-HONO have been studied in the gas phase by a number of workers, but the relative instability of these molecules has resulted in spectral complications due to the absorptions of NO, NO_2 and H_2O in all of these studies. The infrared spectra of the two stereoisomers of HONO, stabilised in a nitrogen matrix by the reaction of photolytically produced NH with O_2, have been reported by Pimentel and co-workers[45, 46]. Recently, these species have been found to be formed in an argon matrix by the reaction of photolytically produced H atoms with NO_2 [47]. Spectra have been obtained for the ^{15}N- and ^{18}O-substituted species, but not for the two DONO species, and a detailed normal coordinate analysis has been offered for the cis-stereoisomer. A detailed study of the infrared spectra of SbF_5 and of AsF_5 isolated in argon and neon matrices[48] has led to the rather surprising result that SbF_5 possesses C_{4v} symmetry, whereas AsF_5 possesses D_{3h} symmetry. Neon, argon and nitrogen matrix observations have been reported for ClF_3, BrF_3 and BrF_5 [49]. Still another argon matrix study of BrF_3 [50] has recently appeared, as have argon and nitrogen matrix studies of the vibrational spectrum of ClF_5 [51]. Detailed argon matrix observations of the spectra of B_2Cl_4 and of B_2F_4 [52] have led to the interesting result that both species are isolated in the staggered V_d configuration characteristic of the liquid and gas rather than in the planar V_h configuration obtained in x-ray diffraction studies of the solids. It is believed that the ground-state structure possesses the staggered configuration but that, because of the small rotational barrier, anisotropic van der Waals interactions in the crystalline material result in the stabilisation of the planar configuration. Such interactions would, of course, be minimised in the matrix studies.

(c) *Monomers of strongly hydrogen-bonding species* — Quite recently, the matrix isolation technique has been applied to studies of the infrared spectra of isolated molecules of species which form highly hydrogen-bonded crystals at room temperature but which vapourise to form monomeric species at slightly elevated temperatures. A sampling device suitable for such studies has been described by King[53]. This apparatus has been used to obtain the spectra of cyanamide and of cyanamide-d_2 isolated in an argon matrix[54]. Several new absorptions obtained upon superheating the cyanamide vapour have been attributed to the isomeric species carbodi-imide. The infrared spectrum of glycine isolated in an argon matrix has also recently been reported[55]. Although this molecule exists in the zwitterion form in both the crystalline solid and aqueous solution, the infrared spectrum of the matrix-isolated material is characteristic of the molecular rather than of the zwitterion form.

(d) *High-temperature species* — There have been a great many studies of the infrared spectra of the matrix-isolated species characteristic of high-temperature vapours. Most of the molecules reported previously have been oxides and halides, produced either by the vapourisation of a solid compound containing the elements of interest or by the reaction of oxygen or halogen at a hot metal surface. Infrared and optical spectroscopic data obtained before 1969 have been summarised by Weltner[56]. A somewhat more recent review has been presented by Hastie, Hauge and Margrave[57]. The emphasis of this latter review is on the stabilities and structures of high-temperature species in their ground states. A detailed tabulation of the frequencies obtained in matrix isolation infrared studies of triatomic high-temperature species is included, as is a summary of recent work on such dimeric species as $(LiO)_2$ and $(LiF)_2$ isolated in a matrix environment. The following consideration of the infrared spectra of high-temperature species will be confined to studies published since these two reviews were completed.

In an attempt to shed further light on the structure of $(LiF)_2$, for which matrix isolation data have suggested the stabilisation of both a cyclic and a linear form, Snelson and co-workers[58] have studied the infrared spectra of the vapours over pure NaF and over NaF + LiF mixtures trapped in a neon matrix. The observations were consistent with the stabilisation of planar cyclic forms of both Na_2F_2 and $LiNaF_2$. A detailed normal coordinate analysis of the $LiNaF_2$ data has permitted the estimation of the infrared-inactive frequencies of planar cyclic Li_2F_2 and Na_2F_2.

The infrared observation of matrix-isolated BaO has recently been reported by Abramowitz and Acquista[59]. Studies of the infrared spectra of a number of heavy metal halides, including $SrCl_2$, $BaCl_2$, $EuCl_2$, EuF_2, $PbCl_2$ and UCl_2, isolated in neon, argon, krypton and nitrogen matrices have also been reported[60]. Since both stretching fundamentals have been observed for all of these species, as well as the bending fundamental for several of them, all of them must possess a bent ground-state structure. A two-stage Knudsen cell designed in the laboratory of Günthard[61] has permitted the study of the composition of superheated vapours over $FeCl_2$ and $FeCl_3$. A detailed assignment of the absorptions which were observed between 20 and 680 cm^{-1} for $FeCl_2$, Fe_2Cl_4, $FeCl_3$ and Fe_2Cl_6 isolated in argon and nitrogen matrices has been offered by these workers. The infrared spectra of matrix-isolated ScF_3, YF_3 and LaF_3 and of a number of the other lanthanum series trifluorides have been obtained in neon, argon and nitrogen matrices[62]. All of the vibrational fundamentals of these species have been identified and all of the molecules except ScF_3 have been found to possess a pyramidal ground-state structure. The infrared spectra of a large number of rare earth oxide molecules isolated in argon and neon matrices at 4 K have recently been reported[26], as have been the fundamental absorption of the species UO and one stretching absorption of the species UO_2 in an argon matrix[63].

Considerable research activity has been focused on the infrared spectra of the aluminium family suboxides. The first matrix isolation studies on Al_2O [64] established that the molecule is bent. However, the bending fundamental absorption, believed to lie below the limit of observation (250 cm^{-1}) was not identified. Subsequent observations of the infrared spectra of normal and ^{18}O-substituted Ga_2O, In_2O and Tl_2O isolated in a nitrogen matrix[65] indicated that these molecules also possess a bent ground-state structure. Snelson[66] searched unsuccessfully for the bending absorption of Al_2O in the 190–250 cm^{-1} spectral region. On the basis of a normal coordinate analysis of the data for the stretching frequencies of the normal and ^{18}O-substituted species, a bending frequency of 120 ± 30 cm^{-1} was suggested. A recent study in the laboratory of Franzen[67] of the infrared spectra of $Tl_2^{16}O$ and of $Tl_2^{18}O$ isolated in argon, krypton and nitrogen matrices led to the assignment of the two stretching fundamentals of $Tl_2^{16}O$ at 571 and 642 cm^{-1} and to the suggestion that the molecule is cyclic, with a strong Tl—Tl bond. Support for such a structure was presented from earlier mass spectrometric observations. In a parallel study of the infrared spectra of matrix-isolated thallous halides[68], these workers presented evidence for the stablisation of linear dimeric X—Tl—Tl—X structures (X = F, Cl), in addition to the monomers, also supporting the occurrence of strong Tl—Tl bonds. Concurrent work in the laboratory of Carlson[69] on the infrared spectra of all of the aluminium family suboxides and of the mixed species InGaO isolated in neon, argon, krypton and nitrogen matrices has led to the assignment of the bending fundamental in the 380–510 cm^{-1} spectral range. Such a high frequency for this fundamental would require that it be contributed by the stretching vibration of a strong metal–metal bond in a cyclic molecule. Unfortunately, there remains a discrepancy between the position of the absorption assigned to v_1 of Tl_2O in this work, 510 cm^{-1}, and the value found by Franzen and co-workers, 571 cm^{-1}. Further detailed studies of the ^{18}O-enriched species would be extremely valuable in providing a definite confirmation of the assignment of the bending (or metal–metal stretching) fundamentals of these species and in resolving the discrepancy between the two assignments of v_1 of Tl_2O.

A series of publications has appeared presenting the infrared spectra of matrix-isolated SiO, GeO and SnO and of their simple polymers. The first report presented the assignment of absorptions due to Si_2O_2, Ge_2O_2 and Sn_2O_2 in a matrix environment, supported by extensive ^{18}O-substitution data[70]. This was followed by detailed assignments of the vibrational fundamentals of $(SiO)_n$ $(n = 1-3)$[71], of $(GeO)_n$ $(n = 1-4)$[72] and of $(SnO)_n$ $(n = 1-4)$[73] appearing above 200 cm^{-1} in argon and nitrogen matrix studies. An independent study has yielded an assignment of the fundamentals of SiO, Si_2O_2 and Si_3O_3 down to 33 cm^{-1}, together with tentative data on a pentameric species[74].

Unlike their carbon and silicon counterparts, the heavier Group IV dihalides are stable in the crystalline state. Andrews and Frederick[75] have isolated $GeCl_2$, $SnCl_2$ and $PbCl_2$ vapours in an argon matrix in concentration sufficient to permit observation of both stretching fundamentals. Chlorine isotopic splittings were resolved.

1.1.2 Electronic spectra

1.1.2.1 Atoms

The literature on the spectra of atoms isolated in rare-gas matrices has very recently been summarised both by Gruen and Carstens[76] and by Mann and Broida[77]. As these workers have noted, all of the previous studies had been focused on an understanding of the relatively simple spectra of atoms with ground S states and most of the transitions studied have been of the P–S type. In contrast to observations for the low-lying electronic transitions of most simple molecules isolated in a rare-gas matrix environment, which are typically shifted to the red by at most a few hundred cm^{-1}, the electronic transitions of these atoms isolated in rare-gas matrices are generally shifted to the blue, sometimes by as much as 3000 cm^{-1}. The absorptions of matrix-isolated atoms are broad and often show multiplet structure which has been attributed to the existence of different types of trapping sites and to the removal of orbital degeneracy in the matrix environment. An illustration of the difficulty in determining more precisely the types of site which might be responsible for such splittings has been provided by a pair of notes by Freed-hoff and Duley[78] and by McCarty[79] supporting the assignment of three components for the $Hg(^3P_1 - {}^1S_0)$ transition in rare-gas matrices to atoms trapped in substitutional and substitutional-one vacancy sites and to atoms trapped in substitutional and weakly bound nearest-neighbour sites, respectively.

Gruen and Carstens[76] have conducted a detailed study of the absorption spectrum of titanium atoms isolated in argon and xenon matrices. Correlation with the gas-phase data for titanium atoms has dictated an assignment in which all of the argon matrix absorptions are shifted to the blue, as are most of the xenon matrix absorptions. Two emission band systems were also observed on near-ultraviolet excitation of argon matrix samples, but not of xenon matrix samples. These two groups of bands were tentatively attributed to atomic transitions shifted to somewhat higher energies in the argon matrix environment. Mann and Broida[77] have studied the spectra between 2200 and 4000 Å of Cr, Mn, Fe, Co, Cu, Ni, Sn and Pd atoms in an argon matrix, as well as of Fe and Cu atoms in krypton and xenon matrices. The blue shifts and splittings typical of atoms with simpler electronic structures have also been observed in their studies. They noted that, with few exceptions, the blue shifts were greater for the lighter rare-gas matrix atoms, as was also noted in the titanium atom study. A significantly smaller blue shift was noted for atoms with an odd number of 3d electrons. In general, the blue shift increased with increasing energy of the transition. Further studies of the transitions of matrix-isolated atoms having relatively complicated electronic structures should be of considerable interest.

1.1.2.2 Molecules

Electronic specta have been reported for a great many molecules isolated in a wide variety of matrix materials. The discussion in this Section is necessarily limited to a general description of the characteristics of the electronic spectra of small molecules effectively isolated in inert matrices such as neon or argon, with relatively small interactions between the molecule of interest and the matrix material. A few recent results illustrating specific principles will be considered. Further examples will be provided in the subsequent discussion of the spectra of reactive species.

(a) *Characteristics of matrix isolation spectra* — In contrast to the usual observation for atoms isolated in rare-gas matrices, the band origins of electronic transitions lying below c. 50 000 cm^{-1} are typically shifted to the red by as much as a few hundred cm^{-1} upon isolation of the molecule in a rare-gas environment. Exceptions to this generalisation include C_2^- [24], for which the observed electronic transition is somewhat blue-shifted in a neon matrix and considerably more so in an argon matrix. As previously noted, spectra observed in a neon matrix often approximate most closely to those characteristic of the gas phase. However, observations in a neon matrix are not devoid of site perturbations and generalisations about these are difficult to make.

Vibrational separations characteristic of excited electronic states, like those of the ground state, are virtually unchanged by isolation of the molecule in a rare-gas matrix. In matrix isolation studies, the electronic absorption bands are frequently quite broad, with half-widths of more than 100 cm^{-1}. However, there are several important exceptions to this generalisation. For example, it has been found possible to resolve the ^{13}C isotopic splittings of the 5206 Å band system of C_2^- isolated in an argon matrix[80], as well as emission band structure assigned to rotational levels of the ground electronic state of OH in a neon matrix[81] and absorptions assigned to rotation of NH[12, 82] and of NH$_2$[11, 83] in an argon matrix. Nevertheless, the breadth typical of most such bands generally precludes the detection of heavy isotopic shifts other than those characteristic of deuterium substitution in the molecule. The loss of most of the heavy isotopic data severely complicates the definitive assignment of electronic transitions to previously unidentified species. Typically, the electronic spectra of reactive molecules are identified either by their correspondence with the spectrum previously assigned in gas-phase studies or by demonstration that the band system of interest appears, disappears, or changes in intensity under the same conditions which give such a change in infrared absorptions which have been assigned to a reactive molecule through extensive isotopic substitution experiments.

Relatively little is known about the behaviour of electronic transitions of matrix-isolated molecules which occur at energies above $c.$ 50 000 cm^{-1}. Schnepp and Dressler[201] have studied the Schumann–Runge band system of relatively concentrated solid solutions of oxygen in argon and nitrogen. A small red shift was found for the origin of the transition in both matrices. For low vibrational quantum numbers, the vibrational structure was similar to that of oxygen in the gas phase, but the more highly excited upper-state vibrational levels were found to converge more slowly than those of the gas-phase molecule. This deviation could be accounted for by means of a simple potential function including the oxygen–argon or the oxygen–nitrogen pair interactions. Roncin, Damany and Romand[202] reported vibrational separations for the fourth positive group of CO, between about 1550 and 1250 Å, which were in good agreement with the gas-phase values. A red shift of only 65 cm^{-1} was reported for the band system observed in a neon matrix. Increasingly larger red shifts were observed as the mass of the rare-gas matrix atoms was increased. Similar results were reported for the $B^2\Pi$, $B'2\Delta$ and $G^2\Sigma^-$ ← $X^2\Pi$ transitions of NO isolated in rare-gas matrices. However, the irregular vibrational spacings observed in the $B^2\Pi$ and $B'^2\Delta$ states of the gas-phase molecule did not appear in the matrix studies. These irregularities have been attributed to perturbations by Rydberg states of the same symmetry. It was suggested that, because of the large size of the Rydberg orbitals, transitions to the valence orbitals, less affected by the environment, would be more probable in matrix experiments.

Such failures to detect Rydberg transitions of matrix-isolated molecules led to some question whether Rydberg transitions might be completely suppressed by the matrix. However, Katz and co-workers[84] have presented considerable evidence that at least the first Rydberg levels of molecules are detectable in rare-gas matrices. Using the Wannier model for impurity states in rare-gas solids, they have found that the lowest Rydberg level should be shifted to the blue in the matrix and that the vibrational structure should be similar to that observed for the Rydberg transition of the free molecule. The blue shift which they observed for the absorption assigned to the first level of this lowest-energy Rydberg series is greatest for an argon matrix, intermediate for a krypton matrix and smallest for a xenon matrix, in accordance with the predictions of the theory. The broadening is both predicted and observed to be greatest for the argon matrix. These workers also tentatively identified the $n = 2$ Wannier impurity states for benzene isolated in krypton and xenon matrices. On the basis of such an assignment, they have estimated that the first ionisation potential for benzene in argon, krypton and xenon matrices is lower by 6000, 9200 and 13 000 cm^{-1}, respectively, than the value found for benzene in the gas phase. If this interpretation is correct, it should have interesting implications for the photoionisation of other molecules isolated in rare-gas matrices.

(b) *Spectra of high-temperature species* — Gas-phase observations of the electronic spectra of high-temperature molecules have presented particular difficulty, since the temperatures required for such observations lead to extensive thermal excitation of the ground-state rotational and vibrational levels and even of low-lying electronic levels. Under these conditions, transitions which may possess extensive vibrational structure often appear as continuous absorption regions, and there may be great uncertainty as to the ground electronic state of the molecule. The matrix isolation technique is of obvious utility for circumventing this difficulty, since the molecules are very effectively deactivated to their ground states. Results which have been obtained in optical spectroscopic studies of matrix-isolated high-temperature molecules of astrophysical interest have been surveyed by Weltner[85]. Optical data on a number of metal oxide molecules have also been surveyed in the later review by Weltner[56]. To these data may be added those obtained in the recent study of the optical spectra of a number of rare earth oxides[26]. The absorption and laser-excited fluorescence specta of CuO isolated in rare-gas matrices have been reported by Shirk and Bass[86]. Evidence was obtained for two fluorescence band systems, for each of which a vibrational separation was observed. The resulting energy levels were consistent with those previously derived from the analysis of the very complicated gas-phase spectrum. Pairs of bands associated with each of the transitions from the excited state to the ground-state vibrational levels were observed, with a separation of c. 200 cm^{-1} in two different matrix environments. On the basis of this similar band separation, it was suggested that the band pairs might be associated with transitions to the two components of the ground $^2\Pi$ state of CuO. However, it is quite possible that molecules were trapped in two different types of site in each of the matrices studied and that, fortuitously, the site splittings were nearly equal for both of them. Further studies would be of considerable interest.

An experimental programme in the laboratory of Meyer has been concerned with the study of the enhancement of phosphorescence due to intersystem crossing in a matrix and with the determination of phosphorescence
lifetimes in a wide variety of matrix materials over an extended temperature
range. Among the relatively simple molecules for which results have been
reported are SO_2 and S_2O [87, 88]. These studies have led to the identification
of the analogue of the Cameron bands of CO for SnO and SnS [89] and for
GeO and GeS[90]. The assignment has received further support from the recent
determination of the phosphorescence decay time of these molecules[91].

An interesting recent study has been focused on the charge-transfer
absorption systems of various alkali halides isolated in neon, argon, krypton
and nitrogen matrices[92]. It was suggested that the vibrational band structure
observed in the spectra of LiI, NaI and NaBr results because of the adiabatic
behaviour of these species in the vicinity of the crossing point of the potential
curves for the covalent and ionic forms, leading to the appearance of a potential minimum in the curve for the predominantly covalent upper state.

Gruen and co-workers have conducted extensive studies of the electronic
spectra of various first-series transition metal dihalides, both in the gas phase
and isolated in rare-gas matrices, and have achieved considerable success in
interpreting these spectra in terms of ligand field theory. Their results are
considered in detail in a very recent review by Gruen[93]. Spectra for matrix-
isolated UCl_4 and UBr_4 are also considered. Although the agreement between theory and the observed spectra is impressive, significant deviations
do occur. Jacox and Milligan[25] have noted that more low-lying electronic
transitions of $NiCl_2$ appear than can be accounted for in terms of present
theory and have suggested that the participation of the 3d nickel electrons in
the bonding of this molecule could account for the extra band systems. It is
apparent that much further work is needed.

1.2 PRODUCTION AND SPECTRA OF FREE RADICALS AND MOLECULAR IONS

The results already described certainly suffice to establish the matrix isolation technique as a very useful tool for both the chemist and the spectroscopist. However, there remain to be considered two important areas of
investigation — the spectroscopic study of free radicals and reaction intermediates and the formation and study of molecular ions — in which the matrix
isolation technique has permitted the acquisition of a large body of data in-
accessible by other means. Studies of the vibrational and electronic spectra
of these species are important not only for their own sake, but also as a means
of testing the implications of molecular orbital theory. Both free radicals
and molecular ions provide suitable tests for the predicted behaviour of
electron-deficient species, and studies of molecular anions provide still
further tests for the behaviour of species with a completed set of bonding
orbitals and excess electrons in outer orbitals which often possess appreciable
antibonding character.

The data obtained in matrix isolation studies of free radicals are, in general, complementary to those obtained in studies of the high-resolution electronic spectra of free radicals produced in the gas phase by flash photolysis. A comparison of the results obtained by the two techniques has been given by Milligan and Jacox[94]. Electronic spectra have been observed for a number of free radicals both in the gas phase and in matrix isolation experiments, helping to establish the most satisfactory conditions for the stabilisation of the species of interest. An example is provided by the studies of the NCN radical in a matrix environment. It was expected that, as commonly occurs for azides, cyanogen azide, N_3CN, should photolyse with the elimination of N_2, leaving the species NCN trapped in the matrix. A 3290 Å band system of NCN was known from the gas-phase flash photolysis studies of Herzberg and Travis[95]. However, on photolysis of N_3CN isolated in an argon matrix with the full light of a medium-pressure mercury arc, it was found[96] that this 3290 Å band system was present after the first few minutes of photolysis, but disappeared upon prolonged irradiation of the sample. On the other hand, this band system continued to grow in intensity when light of wavelength shorter than 2800 Å was excluded from the sample. Infrared absorptions at 423 and at 1475 cm^{-1} behaved similarly, and studies on ^{13}C- and ^{15}N-enriched samples confirmed their assignment to NCN. In addition, a new electronic band system with similar photolytic behaviour appeared between 3000 and 2400 Å. It was suggested that this band system was also contributed by NCN. The recent gas-phase flash photolysis studies of Kroto and co-workers[97] have provided still further support for the assignment of this second electronic band system to NCN.

Heretofore very little spectroscopic data have been reported for small molecular ions. A few charged species, including CO_2^+, CS_2^+, N_2O^+ and HCl^+, have yielded to gas-phase spectroscopic investigation. Until recently, no gas-phase spectroscopic data were available for negatively charged species. Of course, spectra had been reported for a wide variety of small molecular ions in pure ionic crystals, in dilute solid solution in such ionic materials as the alkali halides and in polar solvents such as water and acetonitrile. However, a significant perturbation in the energy levels of such ionic species might be expected to result from their interaction with other ions and with polar molecules. Recently, photoelectron spectroscopy has permitted the determination of approximate values for totally symmetric vibrational fundamentals of a considerable number of positively charged species in their ground and low-lying excited states. However, quite large uncertainties are inherent in the determination of relatively small energy differences, and interpretation of progressions involving more than two vibrational fundamentals is often difficult.

In the course of our studies of the *in situ* production of free radicals and molecular ions in an inert matrix environment, we have found the cage effect to be an extremely important factor in determining which processes may occur. Electrons and atoms are, in general, able to diffuse through a rare-gas matrix even at 4 K, whereas even simple diatomic molecules cannot diffuse under isolation conditions satisfactory for the trapping of reactive species. Two diatomic or polyatomic molecular fragments can be trapped in adjacent sites only if there is an appreciable activation energy for their recombination. It is useful in the following discussion to subdivide the presentation of the results of recent studies according to the process by which it has been found possible to stabilise the species of interest.

1.2.1 Free radicals

1.2.1.1 Trapped products of gas-phase processes

Among the early applications of the matrix isolation technique was that to the study of the electronic spectra of the products of gas discharge processes trapped in rare-gas matrices at 4 K. A number of prominent electronic band systems which had previously been assigned to such species as NH[9,12], OH[9,10], NH_2[8,11] and HNO[9] were identified. Infrared absorptions of HNO and of DNO were also identified by Harvey and Brown[98] in experiments in which a discharge was passed through a mixture of argon and hydrogen, a small pressure of NO was introduced downstream from the discharge and the reaction products were frozen onto the cryostat sample window.

Free radicals have also been observed among the matrix-isolated products of high-temperature vaporisation. The first such species to be reported was C_3, presented in sufficient concentration in the matrix-isolated products of the vaporisation of graphite for the observation not only of its electronic absorption spectrum but also of its ground-state antisymmetric stretching fundamental[22,23].

It has been found possible to isolate free radicals produced in gas-phase chemical reactions, as well. Noteworthy among such studies is the stabilisation of SiF_2, produced by the high-temperature reaction of SiF_4 with silicon[33,99].

Studies of reactive species produced by pyrolysis would, of course, be of considerable interest. Because of the complexity of the chemical processes which can occur in such systems, the technique has rather severe limitations for the primary identification of free-radical species. The importance of extensive isotopic substitution studies in the assignment of free-radical absorptions which appear in such experiments cannot be overemphasised. Snelson[100] has studied the spectra of the products of the pyrolysis of C_2F_4 and of CF_3I isolated in a neon matrix and has reported the appearance of the infrared absorptions previously assigned to CF_2 and to CF_3[101]. Similar studies of the pyrolysis of CH_3I, CD_3I and $Hg(CH_3)_2$ have led to the isolation of CH_3 and of CD_3[102]. Again, the primary identification of the product radicals depended upon the appearance of absorptions previously assigned to CH_3 and to CD_3[103]. Studies of the spectra of the partially deuterium-substituted species would be highly desirable in confirming the assignment of two newly identified fundamentals of CH_3 offered in the pyrolysis study. Kaldor and Porter[104] have presented infrared spectroscopic evidence for the stabilisation of the previously unidentified BH_3 free radical among the matrix-isolated products of the pyrolysis of BH_3CO. The pyrolysis conditions were adjusted to provide the maximum yield of BH_3, as determined in supplementary mass spectrometric observations, and the assignment was supported by studies of the boron isotopic splitting and by the identification of absorptions assigned to BD_3. As in the pyrolysis studies of CH_3, further studies of the mixed deuterium isotopic species would be extremely valuable in supporting the identification, and would yield further information regarding the vibrational potential function, as well. Still another recent pyrolysis study is that of Maltsev and co-workers[105], who obtained extremely well-resolved chlorine isotopic splittings in the spectra of CCl_2 and of CCl_3 stabilised on trapping the products of the pyrolysis of $C_6H_5HgCCl_3$ and of $Hg(CCl_3)_2$ in an argon matrix. The spectrum of CCl_2 had previously been reported in matrix isolation studies of the reaction of carbon atoms with Cl_2 [106], of the abstraction of chlorine atoms from CCl_4 by lithium atoms[107] and of the vacuum-ultraviolet photolysis of CH_2Cl_2 [108]. Two absorptions which appeared in studies of the abstraction of chlorine atoms from CCl_4 by lithium atoms were assigned by Andrews[109, 110] as the two stretching fundamentals of CCl_3. The identification of the degenerate stretching fundamental has been confirmed in other studies using both the lithium-atom abstraction of chlorine and the vacuum-ultraviolet photolysis of $HCCl_3$ as sources of CCl_3 [111]. However, despite excellent yields of CCl_3, the absorption previously attributed to v_1 of CCl_3 did not appear either in these studies or in those of Maltsev and co-workers.

1.2.1.2 *Photoprocesses yielding radical + molecule*

Because of the limitations imposed by the cage effect, in rigid matrices there is no net photodecomposition into two diatomic or polyatomic fragments unless the activation energy for their recombination is appreciable. An important class of compounds for which the decomposition products are often detectable is the simple azides, which photolyse with the elimination of molecular nitrogen. The vibrational and electronic absorption spectra of NH appear upon ultraviolet photolysis of HN_3 [112, 113], the vibrational absorptions of the various halonitrenes upon photolysis of the corresponding halogen azide[114, 115] and the vibrational and electronic absorption spectra of NCN upon photolysis of cyanogen azide, N_3CN, with radiation of wavelength longer than 2800 Å [96] or with the 2288 Å cadmium resonance line[116]. Elimination of molecular nitrogen also occurs upon photolysis of matrix-isolated CH_3N_3, but the nitrene fragment undergoes rearrangement to $H_2C=NH$ [117]. The situation is more complicated for the photolysis of diazomethane and related compounds in a matrix environment. Although in the gas phase the photolysis of diazomethane yields $CH_2 + N_2$, in a matrix environment recombination of these two species occurs[118]. On the other hand, CF_2N_2 provides an excellent source of CF_2 in argon and nitrogen matrix experiments[119].

The elimination of molecular hydrogen also provides a route to the stabilisation of a number of reactive species. Although gas-phase photolysis studies[120, 121] have established that a large fraction of the methane undergoes primary photolysis to $CH_2 + H_2$, in argon and nitrogen matrix studies of the vacuum-ultraviolet photolysis of methane[103] a prominent absorption due to CH_3 appeared, but no absorptions due to CH_2. Photoproduction of CH_2 was, however, evidenced by its reaction with a nitrogen matrix to form CH_2N_2, several infrared absorptions of which were identified. It is inferred that in the argon matrix experiments CH_2 can recombine with H_2 trapped in an adjacent site. This conclusion is supported by the detection of CH_4 in studies[122] of the reaction of carbon atoms with molecular hydrogen trapped in an argon matrix at 4 K.

Compounds of formula H_2MX_2, with M a Group IV element and X a halogen, frequently photolyse to give a high yield of MX_2 but only weak absorptions attributable to HMX_2. Because of the strength of the H—H bond, photodetachment of molecular hydrogen might be expected to have its onset at a lower energy than detachment of a hydrogen atom. On the other hand, the extent to which the reaction of MX_2 with H_2 to re-form the parent molecule occurs is not known, and MX_2 may, alternatively, result from the successive photodetachment of two hydrogen atoms. Examples of such processes are provided by the stabilisation of $SiCl_2$[123], SiF_2[124] and $GeCl_2$[125] upon vacuum-ultraviolet photolysis of the corresponding H_2MX_2 or D_2MX_2 molecule in a matrix.

Further consideration of the species of formula CX_2 and HCX formed in such systems is of some interest. Although CH_2 has long been known to possess a ground triplet state, CF_2 possesses a ground singlet state. The very prominent band system of singlet CF_2 which appears in the gas phase between 2700 and 2300 Å has also been observed in a matrix environment, with upper-state vibrational spacings which agree within experimental error with those of the gas phase [101, 119]. In the matrix experiments, an infrared absorption appears very close to the lower-state bending fundamental frequency derived from the gas-phase observations. Recently, it has been found possible to excite a fluorescence band system of matrix-isolated CF_2 using the short wavelength tail of the 2537 Å emission of a medium-pressure mercury arc lamp[126]. The fluorescence band spacings also agree within experimental error with the lower-state vibrational spacings observed in the gas-phase studies. It is concluded that the lower state of the CF_2 transition is indeed the ground state of the molecule. An electronic band system between 5600 and 4400 Å has been assigned to matrix-isolated CCl_2 produced either by the reaction of photolytically produced carbon atoms with chlorine[106] or by the vacuum-ultraviolet photolysis of CH_2Cl_2 [108]. As for CF_2, an extended progression in the upper-state bending fundamental appears in the absorption studies. Shirk[127] has observed the laser-excited fluorescence from this transition of CCl_2 in a matrix environment and has deduced the frequency of the ground-state bending fundamental from the data. The mixed halogen species ClCF has been studied.[128] as well, and an electronic transition has been observed near 3900 Å, intermediate between the transitions of CF_2 and CCl_2. A progression in the upper-state bending vibration of ClCF appears in absorption, and a fluorescence spectrum involving an extended progression in the ground-state bending vibration has been detected. The similarities of the three transitions suggest that the two chlorine-containing species, like CF_2, may possess ground singlet states. Absorption band systems have also been observed in matrix isolation studies of the related species HCF [129] and HCCl [130]. The positions and band spacings of these systems are closely similar to those assigned to singlet HCF[131] and HCCl[132]. The position of the ground-state bending fundamental of HCF observed in the matrix studies is also in satisfactory agreement with the value deduced for this vibrational fundamental in the gas-phase studies. Thus, it seems likely that all of the CX_2 and HCX species possess singlet ground states.

1.2.1.3 Photoprocesses yielding radical + atom

The limitations imposed by the cage effect are largely surmounted for photo-processes involving atom detachment, since at least the lighter atoms can diffuse through rare-gas matrices even at 4 K. The extent of this diffusion is, as might be expected, dependent upon the mass and size of the atom. Milligan and Jacox[21] have studied the vacuum-ultraviolet photolysis of a series of compounds of formula XCN isolated in an argon matrix. When X is hydrogen, photolysed samples show very prominent absorptions due to the violet system of CN. There is a marked decrease in the yield of CN in studies of the vacuum-ultraviolet photolysis of FCN, a still further decrease when ClCN is studied and very little evidence for CN in experiments on BrCN. On the other hand, evidence for extensive photolysis of all of these species is provided by the appearance of prominent infrared absorptions assigned to the isomeric XNC species.

A large number of free radicals have been stabilised by the photodetach-ment of a hydrogen atom from the parent molecule. The first such species to be stabilised was NH_2[82], produced by the vacuum-ultraviolet photolysis of ammonia in sufficient yield for direct infrared detection of two of the vibrational fundamentals. Evidence for the rotation of NH_2 in an argon matrix has previously been noted.

Of especial interest has been the stabilisation of CH_3 upon vacuum-ultraviolet photolysis of methane in a rare-gas or a nitrogen matrix[103]. As previously noted, the back reaction of CH_2 with H_2 leads to the predominance of the hydrogen-atom detachment process in the matrix experiments. Secondary photolysis of CH_3 also occurs, and the familiar electronic tran-sitions of CH have been observed in the argon matrix experiments. The appearance of the out-of-plane deformation fundamental of CH_3 in the infrared is of further interest because of the occurrence of a significant 'negative anharmonicity' as deuterium atoms are substituted in the molecule. This observation requires that the molecule be planar. The previous gas-phase studies[133] were consistent with a planar or a near-planar structure for the molecule. Riveros[134] has obtained a very satisfactory fit of the matrix data to a mixed quadratic–quartic potential.

The electron spin resonance observations of Fessenden and Schuler[135] indicate that H_2CF deviates from planarity by no more than 5 degrees. Although H_2CF has also been detected in studies[129] of the infrared spectrum of the products of the vacuum-ultraviolet photolysis of matrix-isolated CH_3F, unfortunately the totally symmetric deformation fundamental of the pyramidal structure, or the out-of-plane deformation fundamental of the planar structure, has not been identified. On the other hand, a prominent absorption due to the corresponding fundamental of H_2CCl has been observed and shows a large 'negative anharmonicity'[108, 136], which can only occur for a non-totally symmetric vibration. Therefore, H_2CCl, like CH_3, must be planar in its ground state.

The vacuum-ultraviolet photolysis of $HCCl_3$ is of considerable interest, since CCl_3 is the predominant product on photolysis of $HCCl_3$ using 1216 Å radiation, whereas $HCCl_2$ rather than CCl_3 is produced when 1067 Å radiation is used for photolysis[137]. This observation is consistent with the vacuum-ultraviolet absorption studies of Zobel and Duncan[138], who noted an apparent progression in a highly anharmonic upper-state C—H stretching fundamental with origin near 1330 Å, another transition with absorption maximum near 1170 Å and still another region of continuous absorption beyond 1110 Å in the spectrum of $HCCl_3$.

A number of other species of formula MX_3, including CBr_3 [111], $SiCl_3$ [139], SiF_3 [140] and $GeCl_3$ [125] have also been produced in sufficient concentration for direct infrared study upon vacuum-ultraviolet photolysis of the corresponding HMX_3 species in a matrix environment.

The vacuum-ultraviolet photolysis of matrix-isolated SiH_4, like that of matrix-isolated CH_4, leads to the stabilisation of significant concentrations of secondary photolysis products[141]. Despite this similarity, there are several important points of contrast in the properties of the free-radical products of the two systems. All four vibrational fundamentals of SiH_3 appear in the infrared, requiring that the molecule possess a pyramidal rather than a planar structure. SiH_2 is stabilised in sufficient concentration for assignment of its three vibrational fundamentals. The bending absorption appears very close to the vibrational separation characteristic of the lower state of the singlet SiH_2 electronic transition[142], suggesting that the lower singlet state of this transition is the ground state of the molecule. However, the electronic absorption system of SiH_2 has not been observed in the matrix experiments, leaving open the possibility that a ground triplet state of SiH_2 might also have a similar bending fundamental frequency. Sufficient SiH was stabilised for direct observation of its vibrational fundamental. Even more extensive 'stripping' of hydrogen was also observed in these studies; appreciable concentrations of Si_2H_6 and of Si_2 appeared in photolysed samples. Indeed, the concentration of Si_2 was sufficiently great to permit the tentative assignment of a previously unreported electronic transition of this species. There are, of course, a number of reaction sequences which would lead ultimately to the stabilisation of Si_2 in these experiments. One of the more probable processes involves the reaction of SiH_2 with SiH_4 to produce Si_2H_6, which then undergoes extensive photodecomposition, leading finally to the stabilisation of Si_2. The feasibility of such extensive photolysis was established by the observation of prominent electronic absorptions of Si_2 upon vacuum-ultraviolet photolysis of matrix-isolated Si_2H_6. Still another example of such a photolytic 'stripping' process is provided by the recent studies of the vacuum-ultraviolet photolysis of H_3BNH_3 in argon and neon matrices[143], leading to the appearance of the previously identified[144] BN triplet system.

Recently, Meyer and his co-workers[145, 146] have provided support for a detailed spectroscopic analysis of the composition of sulphur liquid and vapour at various temperatures by studies of the electronic spectra of species of formula S_n produced on photolysis of S_nCl_2 precursors isolated in a krypton matrix at 20 K. In these experiments, as in many of the hydrogen-detachment photolysis studies, it is possible for either molecular detachment or two successive atomic detachments to account for the products. However, for

molecules with several intervening sulphur atoms photolysis by the detachment of chlorine atoms, rather than by the elimination of Cl_2 from the parent
molecule, might be predicted to play a significant role.

1.2.1.4 Atom–molecule reactions

A large number of free-radical species have been produced in a matrix
environment by the reaction of an atom with a molecule present in small
concentration in an inert matrix or with a molecule of the matrix material.
These reactions fall into two general classes, depending upon whether an
atomic beam, generally of an alkali metal, is co-deposited with the sample
or whether the reactant atom is produced by photolysis of a suitable precursor in the matrix environment.

(a) *Halogen-atom abstraction by alkali metals* – A number of free-radical
species have been stabilised in significant concentration by the interaction of
alkali metal atoms with a halogen-substituted methane isolated in an argon
matrix. Typically, lithium atoms are the most effective in such halogen-atom
abstractions, but sodium and potassium atoms often react to some extent.
Indeed, Andrews has noted the stabilisation of appreciable concentrations
not only of CCl_3 [109, 110], but also of CCl_2 [107] upon reaction of lithium atoms
with CCl_4 in an argon matrix. Similar results were obtained in a companion
study of the interaction of lithium atoms with CBr_4 [147], leading to the stabilisation not only of CBr_3 but also of CBr_2. The spectra reported in these studies
are quite complicated and other incompletely characterised products are also
present. In another series of experiments on these two systems[111], very high
yields of CCl_3 and of CBr_3 were obtained, but very little CCl_2 or CBr_2 and
no other detectable products, except for the lithium halide. Other free
radicals which have been stabilised as a result of such halogen-abstraction
reactions include $HCBr_2$ [148], $HCCl_2$ [149], HCF_2 [150], H_2CCl [136] and PCl_2
and PBr_2 [151]. In several of the halomethane studies, support for the free-
radical identification was obtained from supplementary observation of the
product spectrum when lithium was allowed to react with a mixed halogen
species. Although the heavier halogen is usually abstracted more readily,
products of the abstraction of the lighter halogen are also observed. Thus,
both H_2CF and H_2CCl have been identified among the products of the
reaction of lithium atoms with H_2CFCl, and both H_2CBr and H_2CCl in
the reaction of lithium atoms with H_2CClBr [136].

(b) *Atom produced by photolysis* – A number of free radicals have been
formed by the reaction of photolytically produced H atoms with matrix-
isolated molecules. Indeed, the first simple polyatomic free radical to be
stabilised in a low-temperature environment in sufficient concentration for
direct infrared spectroscopic observation, HCO, was first obtained by
photolysis of HI or HBr in a CO matrix[27]. With the advent of vacuum-
ultraviolet photolysis lamps suitable for use in matrix isolation studies, a
very large number of molecules was added to the list of suitable H-atom
sources. Indeed, it became possible to use the appearance of HCO upon
photolysis of a hydrogen-containing molecule in a CO matrix as a test for
the occurrence of photolysis by H-atom detachment.

For many years, it had been presumed that a very complicated emission band system which appears between c. 2500 and 4100 Å in the flames characteristic of the combustion of a wide variety of hydrocarbons was contributed by HCO[152-154]. However, the complexity of these hydrocarbon flame bands was so great that they defied all attempts at a detailed assignment. Recently, Dixon[155] has conducted a rotational analysis of several of the bands observed in high resolution in the gas phase and has obtained lower-state moments of inertia which correspond very well with those previously determined[156] for HCO in its ground state. At about the same time, Milligan and Jacox[157] obtained a prominent absorption spectrum between 2600 and 2100 Å upon photolysis of a number of H-atom precursors in a CO matrix or in an argon matrix to which a small concentration of CO had been added. Deuterium enrichment studies confirmed the presence of a single hydrogen atom in the molecule. It was possible to analyse the spectrum in terms of the excitation of two upper-state vibrations. With a knowledge of these two upper-state vibrational separations and of the frequencies of the ground-state vibrational fundamentals of HCO and of DCO [27, 158], it became possible to propose a detailed vibrational assignment of the hydrocarbon flame bands. In the matrix experiments there also appeared at shorter wavelengths a number of unclassified bands, which could be fitted into a separate array. The longest wavelength unclassified band corresponds closely to the origin postulated by Dixon for still another electronic transition of HCO.

The first simple polyatomic free radical to be stabilised in a rare-gas matrix was HO_2, for which all three vibrational fundamentals were assigned by Milligan and Jacox[159]. As was found to be true for HCO, the appearance of this radical has proved to be a very good test for the occurrence of H-atom photodetachment. Very high yields of HO_2 have since been obtained in experiments in which vacuum-ultraviolet photolysis is used to produce H atoms. Despite such high yields, as yet no discrete electronic transition of HO_2 has been identified. As yet unpublished normal-coordinate calculations using data for eight isotopic species of HO_2, performed in this laboratory, are consistent with an HOO valence angle of 105 ± 5 degrees, in good agreement with the value of 110.7 degrees obtained in the INDO calculations of Gordon and Pople[160].

The 2537 Å photolysis of NCN in a matrix environment has been found to provide a good source of carbon atoms, with the added advantage that ^{13}C-enriched NCN is readily obtained, permitting detailed isotopic studies of the products. Carbon atom addition to various simple molecules has provided a route to the stabilisation of CNN[161], CCO[162], CCl_2[106], CF_2[101], HCCl[130], HCF [129] and NCO [163] in a matrix environment in sufficient concentration for direct infrared study.

There have also been a number of studies of the reactions of oxygen atoms with various matrix-isolated molecules. Of especial interest is the stabilisation of CO_3 upon vacuum-ultraviolet photolysis of solid CO_2 [164] or ultraviolet photolysis of O_3 in the presence of CO_2 [164, 165]. The number of absorptions observed for CO_3 was too great to permit a structure with threefold symmetry. The possibility that this decrease in symmetry results from the presence of one or more hydrogen atoms in the molecule has been excluded by studies of the vacuum-ultraviolet photolysis of D_2O in a CO_2 matrix[166], in which the product absorption pattern was identical to that of the other studies. The detailed isotopic data of Moll, Clutter and Thompson[164] have been fitted to a valence-force potential, assuming a planar $O_2C=O$ structure with O—C—O valence angles of 80 and 65 degrees, corresponding to non-bonded and bonded O atoms, respectively. A reasonable set of potential constants has been obtained for the 65 degrees structure[166], but not for the 80 degree structure.

A number of halogen atom reactions have also been observed in a matrix environment. All three vibrational fundamentals and two electronic band systems of FCO appear upon reaction of F atoms with CO [167], and the three vibrational fundamentals of ClCO have been identified in a study of the reaction of Cl atoms with CO [168]. The reaction of F atoms with CF_2 to produce CF_3 has led to the stabilisation of a sufficient concentration of CF_3 in an argon matrix for identification of all four vibrational fundamentals[101].

The reaction of OH with CO has long been known to be the principal route by which CO_2 is formed in combustion processes. This reaction requires a small activation energy, $c.$ 1 kcal mol^{-1}. In view of the reactivity of F atoms and of Cl atoms with a CO matrix, it was of considerable interest to study the reaction of the isoelectronic species OH with a CO matrix, in the hope of trapping a reaction intermediate. Upon vacuum-ultraviolet photolysis of H_2O in a CO matrix, prominent absorptions due to CO_2, HCO, H_2CO and HCOOH appeared. In addition, a number of absorptions due to previously unidentified reactive molecules were observed[169]. Among these were absorptions in the spectral regions characteristic of O—H and C=O stretching absorptions, suggesting an H—O—C=O structure. Since most of the absorptions appeared in pairs, the stabilisation of *cis* and *trans* stereoisomers of such a reaction intermediate was also suggested. Extensive isotopic substitution studies provided further support for such an assignment, and it was possible to obtain sets of valence-force potential constants which were consistent with a proposed assignment of virtually all of the absorptions to the two stereoisomeric H—O—C=O reaction intermediates.

1.2.2 Molecular ions

For a number of years a band system with origin near 5200 Å which was first observed in studies of the spectra of the trapped products of a discharge through various hydrocarbons[170] and later in studies of the spectra of the matrix-isolated vapours from graphite[22, 23, 171-174] was assigned to the Swan transition of triplet C_2. Subsequent to the initial assignment, Ballik and Ramsay[175] demonstrated that the lower state of the Swan transition is not the ground state of the C_2 molecule, but that a singlet state lies some 600 cm^{-1} lower. Furthermore, the separation between the absorption bands in the matrix studies was $c.$ 200 cm^{-1} greater, and that between the emission bands about 150 cm^{-1} greater than the vibrational separations characteristic of the Swan system in the gas phase. Both the trapping of molecules in excited states and such large deviations from the gas-phase vibrational separations are unprecedented in matrix isolation studies. In an attempt to clarify the assignment, the vibrational and electronic spectra resulting on vacuum-ultraviolet photolysis of acetylene were studied[80]. Both the Phillips and the Mulliken systems of singlet C_2 were prominent in the photolysed samples, demonstrating that the lowest singlet state remains the ground state in rare-gas matrices. However, the 5200 Å band system was also present, and ^{13}C-substitution studies demonstrated the presence of two carbon atoms in the molecule which contributes it.

Subsequently, Herzberg and Lagerqvist[176] observed a new band system with origin near 5400 Å in the spectrum of the transient products of a flash discharge through gaseous methane. Isotopic studies required that this new band system be contributed by a species of formula C_2. Both the upper- and the lower-state vibrational spacings were closely similar to those of the 5200 Å band system of the matrix experiments. Since it was not possible to find a pair of low-lying electronic states of C_2 which could account for the observed bands, it was suggested that the band system could suitably be contributed by C_2^-. The electron affinity of C_2 is known to be appreciable, and supplementary mass-spectrometric observations demonstrated the presence of a significant concentration of C_2^- in the system. Although a spin splitting was not detected, the high J lines were significantly broadened.

The correspondence between the vibrational spacings of this new gas-phase transition and those of the matrix experiments suggested a series of matrix experiments in which a small concentration of an alkali metal was added to the Ar/C_2H_2 sample[24]. Upon vacuum-ultraviolet photolysis of the resulting deposit, the intensities of the singlet C_2 absorptions were markedly diminished, and the intensity of the 5200 Å band system was dramatically enhanced. Although in the absence of an alkali metal the 5200 Å band system could be completely destroyed by exposing the sample to 2000–3000 Å radiation, in the presence of an alkali metal its intensity was unchanged even upon prolonged exposure of the sample to such radiation. The persistence of this band system in the samples containing an alkali metal is consistent with the occurrence of a steady state involving the photodetachment of electrons from C_2^- and the capture by other C_2 molecules of electrons produced by photo-ionisation of the alkali metal. Recently, Frosch[177] has identified both the fluorescence emission from the Swan system of triplet C_2 and the C_2^- emission bands in Ar/C_2H_2 samples which had been exposed to x-rays prior to the fluorescence study. All of these data support the conclusion that the 5200 Å band system is indeed contributed by C_2^- and that charged species may be trapped in rare-gas matrices in sufficient concentration for direct spectroscopic detection. Furthermore, the demonstration that alkali metal atoms are a suitable source of photoelectrons for matrix isolation experiments has provided a technique useful for the stabilisation of still other negatively charged species.

1.2.2.1 Photoionisation

The demonstration that photoionisation can occur in rare-gas matrices and that detectable concentrations of ionic species may be stabilised in this environment of course suggests that it may be possible to study the spectra of molecular cations formed on photoionisation of suitable precursors. Most small molecules possess first ionisation potentials in excess of 11.8 eV, the cut-off of lithium fluoride photolysis lamp windows. However, the ionisation potential of CCl_3 is only c. 8.8 eV, and that of $HCCl_2$ is c. 9.5 eV. In studies of the 1216 Å photolysis of $HCCl_3$ in an argon matrix, in which the major photolysis product is CCl_3, an absorption appeared at 1037 cm^{-1} which isotopic data indicated to be contributed by still another product of formula CCl_3. This absorption has been assigned to CCl_3^+, formed by the photo-ionisation of the CCl_3 product[137]. When 1067 Å photolysis radiation was used, very little CCl_3 was formed, but $HCCl_2$ absorptions appeared, and the 1037 cm^{-1} absorption was replaced by an absorption at 1038 cm^{-1} which had chlorine isotopic splittings characteristic of a molecule containing two chlorine atoms. Detailed isotopic studies have led to the assignment of this absorption and of a weaker absorption at 1290 cm^{-1} to the two fundamentals of b_1 symmetry of planar $HCCl_2^+$.

NO_2 also possesses an exceptionally low ionisation potential and in studies of the vacuum-ultraviolet photolysis of NO_2 in an argon matrix a product absorption at 1244 cm^{-1} has been demonstrated to be contributed by NO_2^- [178]. Presumably, the photoionisation of NO_2 provides a source of electrons, which are captured by other NO_2 molecules. However, no absorptions of NO_2^+ have been detected.

1.2.2.2 Electron capture

Examples of associative electron capture, leading to the stabilisation of negatively charged species in the matrix, are provided by the already mentioned studies of the spectra of matrix-isolated C_2^- and NO_2^-. Electron bombardment of NO_2 isolated in an argon matrix also leads to the appearance of the NO_2^- absorption[178].

In a number of other systems, dissociative electron capture has been observed. The electrons formed on photoionisation of either CCl_3 or $HCCl_2$ have been found to be captured by $HCCl_3$ molecules isolated in the argon matrix, resulting in the stabilisation of $HCCl_2^-$ [137]. In the gas phase, electron capture by a number of chlorinated hydrocarbons has been postulated to lead to the production of the neutral radical and of Cl^-. However, in the matrix it might be expected that Cl^- formed in such a reaction would polarise the $HCCl_2$ fragment, leading to an enhancement in the probability of their recombination. On the other hand, it might be anticipated that the electron-deficient species $HCCl_2$ would have an appreciable electron affinity, so that dissociative electron capture to produce $HCCl_2^- + Cl$ atoms would occur at an energy quite close to that of the more familiar process in which Cl^- is produced. The Cl atoms would be relatively free to migrate from the site of their production, leaving the observed $HCCl_2^-$ trapped in the matrix.

Gas-phase studies have provided abundant evidence that N_2O also undergoes dissociative electron capture, producing $N_2 + O^-$, upon interaction with very low-energy electrons. In studies of the photoionisation of an alkali metal in an argon matrix to which a small concentration of N_2O had been added[179], a dramatic decrease in the N_2O absorptions was noted when the sample was exposed to mercury arc radiation. A peak at 1205 cm^{-1}, demonstrated to be contributed by a species which possesses two oxygen atoms, grew during the first few minutes of exposure of the sample to radiation of wavelength shorter than c. 3000 Å. On prolonged irradiation of the sample this 1205 cm^{-1} peak disappeared and absorptions of dimeric NO became prominent.

1.2.2.3   Ion–molecule reactions

The cage effect would, of course, be expected to place severe limitations on the occurrence of reactions between molecular ions and other molecules in a matrix environment; these would be possible only between an ion and a molecule trapped in adjacent sites in the matrix. However, charged atoms would be expected to undergo at least limited migration through the matrix. In the studies of the interaction of photoelectrons with matrix-isolated N_2O [179], the 1205 cm^{-1} absorption has been assigned to the ONO anti-symmetric stretching fundamental of $O_2N{=}N^-$, produced either by the reaction of N_2O^- with an adjacent N_2O molecule or by the reaction of O^- with N_2O. Although both processes may occur in this system, the appearance of the 1205 cm^{-1} absorption even at Ar : N_2O mole ratios of 2000 suggests that the O^- reaction does play a significant role. When a small concentration of NO was also present in the sample, exposure to radiation effective in photoionising the alkali metal led to the appearance of the NO_2^- absorption, suggesting that the reaction of O^- with NO also occurred.

In unpublished studies in this laboratory of the photoproduction of electrons in Ar/N_2O samples to which a small concentration of O_2 had been added, a prominent absorption appeared at 802 cm^{-1}. This new absorption was neither shifted nor changed in contour in studies using a $^{15}N^{15}NO$ sample, indicating that nitrogen atoms do not participate in the vibration. Detailed ^{18}O-substitution studies have demonstrated that three oxygen atoms are present in the product molecule. Analysis of the isotopic data suggests that the absorption is contributed by the antisymmetric stretching fundamental of O_3^-, with a valence angle of 110 ± 5 degrees.

1.2.2.4   Charge-transfer interaction

In addition to serving as a source of photoelectrons, alkali metal atoms may serve as electron donors in charge-transfer interactions with molecules isolated in a matrix environment. It appears necessary to invoke such charge-transfer interactions to explain the results of experiments by Kasai[180], who studied the electron spin resonance spectra of a number of metals trapped at 4 K in an argon matrix to which a small concentration of HI had been added to provide an electron acceptor. Although the ionisation potential of cadmium corresponds to 1380 Å radiation, the hydrogen atom doublet and the signals due to the various isotopic species of Cd^+ were observed after exposure of the sample to radiation in the 2300–4200 Å spectral range. In studies of Cr atoms with an ionisation potential corresponding to 1830 Å radiation under similar conditions, the signal due to neutral Cr atoms diminished in intensity and the signal due to Cr^+ grew, as did the H-atom doublet. Similar results were also obtained using Mn atoms, with an ionisation potential corresponding to 1670 Å radiation. In each of these systems, the electron spin resonance signals were characteristic of the free atom or ion; there was no evidence for perturbations due to interactions in the matrix. Kasai also studied the photoexcitation of Na atoms in Ar/HI samples, using radiation of wavelength longer than 5000 Å. In the absence of Na atoms, this radiation was ineffective in photolysing the HI present in the sample. In the presence of Na atoms, irradiation of the sample led to the complete disappearance of the initially present Na-atom spectrum and to the appearance of prominent H-atom peaks. The technique has since been extended to permit electron spin resonance observation of still other singly charged metal cations, as well as of molecular anions formed by associative electron capture and of free radicals formed by dissociative electron capture[181].

The extent of charge-transfer interaction varies greatly, depending on the donor and acceptor. In the alkali metal series, caesium, with the lowest ionisation potential, is the most efficient electron donor, and lithium, with the highest ionisation potential, is the least efficient donor. Although the electron affinity of NO_2 has not been definitely established, it is known to be exceptionally high. Thus, strong charge-transfer interaction between alkali metal atoms and NO_2 would be expected to occur. The absorption due to NO_2^- isolated in the argon lattice appeared on co-deposition of an Ar/NO_2 sample with an atomic beam of any of the alkali metals, even lithium[178, 179]. The isolated NO_2^- absorption in the lithium atom experiments was relatively weak, but grew in intensity upon exposure of the sample to radiation effective in photoionising lithium atoms. Absorptions also appeared at somewhat lower frequencies, with positions and relative intensities which were dependent on the alkali metal and on the temperature of the sample window. These absorptions have been assigned to $M^+ \cdots NO_2^-$ and to $M_2^+ \cdots NO_2^-$ ion pairs.

A number of other molecules also undergo sufficiently strong charge-transfer interaction with alkali metal atoms for the appearance of absorptions due to molecular anions in the initial sample deposit. Among these molecules is SO_2, which forms a collision complex in which the participation of the alkali metal cation is sufficiently great to cause an alkali metal dependence in the positions of all three SO_2^- vibrational fundamentals[182]. When caesium atoms are used, the alkali metal dependence is sufficiently small to permit the calculation of a set of force constants for SO_2^- which fit all of the isotopic data within experimental error without taking into account the participation of the caesium cation in the vibrations. An absorption at $1352 \, cm^{-1}$, formed on co-deposition of an Ar/NO sample with an atomic beam of lithium atoms, was initially assigned to covalent LiON by Andrews and Pimentel[183]. However, recent studies have shown that this absorption must be contributed by a different species from the other two absorptions which were assigned to LiON fundamentals[179]. An absorption appears between 1350 and $1375 \, cm^{-1}$ upon co-deposition of an Ar/NO sample with any of the alkali metals, suggesting its assignment to the NO^- stretching fundamental of an $M^+ \cdots NO^-$ complex. Spence and Schulz[184] have recently determined a vibrational spacing of c. $1355 \, cm^{-1}$ for NO^- in the gas phase. Matrix isolation studies by Andrews of the interaction between lithium[185], sodium[186] and potassium and rubidium[187] atoms and O_2 have led to the identification of a weak absorption near 1100 cm^{-1} with the O—O stretching vibration of various $M^+\cdots O_2^-$ species. The position of this absorption is in reasonable agreement with the first vibrational spacing of c. 1065 cm^{-1} reported[188] for ground-state O_2^- in the gas phase and with the 1090 cm^{-1} value obtained in Raman studies of O_2^- trapped in alkali halide crystals[189].

Upon co-deposition of Ar/HCl or Ar/HBr samples with an atomic beam of an alkali metal, Milligan and Jacox[190, 191] noted the appearance of broad low-frequency absorptions. Upon subsequent exposure of the sample to mercury arc radiation, these absorptions diminished in intensity and new absorptions grew with frequencies and contours identical to those reported by Pimentel and co-workers[192, 193] for species which they identified as linear, centrosymmetric ClHCl and BrHBr radicals. These absorptions also appeared upon vacuum-ultraviolet photolysis of relatively concentrated (c. 1%) samples of HCl or HBr in an argon matrix[190, 191]. The appearance of these products under conditions which strongly favour the stabilisation of charged species in the matrix, the close agreement between the assigned values for v_1 and those for v_1 of the XHX$^-$ species in crystalline materials, and the proximity of the intermediate complex absorptions in the caesium atom experiments to the absorptions of free 'XHX' formed on subsequent irradiation of the sample, all support the alternate assignment of the absorptions to linear, centrosymmetric ClHCl$^-$ and BrHBr$^-$ [190, 191]. The strongest charge-transfer interaction should, of course, occur for caesium atoms, the most efficient electron donors, and the $Cs^+\cdots XHX^-$ complexes, like $Cs^+\cdots SO_2^-$, would be expected to approximate most closely the behaviour of the free anions.

An interesting series of experiments in which charge-transfer interaction may also play a significant role is that of the interaction between alkali metal atoms and the various methyl halides in an argon matrix environment. The gas-phase reaction between alkali metal atoms and alkyl halides to produce the alkali halide and the alkyl radical has been known for many years and there have been an impressive number of molecular beam investigations of these reactions. Within the past two years, detailed molecular beam studies of the interaction between lithium[194], sodium[195], potassium[196] and caesium[197] atoms and methyl iodide have appeared. Characteristically, there is considerable back scattering of the products, suggesting a short-lived collision complex. In the matrix isolation studies by Andrews and Pimentel[198] of the reaction between lithium atoms and CH_3Br and CH_3I, there appeared an absorption near 730 cm^{-1} which behaved appropriately in isotopic substitution experiments for assignment to the v_2 fundamental of planar CH_3. However, a subsequent extension of the study to the reactions of sodium and potassium atoms with the matrix-isolated methyl halides[199] showed that the absorption was quite strongly dependent on the alkali metal atom chosen for the experiment. The large alkali metal shift and the small shift in going from the iodide to the bromide suggested an H_3C—M—X structure, with three-centre bonding. On the other hand, the molecular beam studies have yielded results consistent with an H_3C—$X^-\cdots M^+$ type interaction. Further studies would be of considerable interest.

Finally, in a number of systems the infrared spectrum of the initial deposit of the molecule of interest in an argon matrix is unchanged by the presence of a small concentration of an alkali metal in the sample, yet the spectrum which appears upon mercury arc irradiation of the sample is characteristic of one or more negatively charged species produced upon photoionisation of the alkali metal. Generally, the spectrum observed for the negatively charged species in such systems is insensitive to the alkali metal chosen for the experiment and might readily be explained by postulating photoionisation of the alkali metal, diffusion of the electron through the argon matrix and subsequent electron capture by a molecule in a site removed from the influence of the cation. This interpretation encounters some difficulty, since sodium atoms frequently participate in such processes, yet their ionisation potential corresponds to a wavelength of 2412 Å, considerably in excess of the radiation output characteristic of the mercury arc lamp. Indeed, in studies of the interaction between N_2O and sodium atoms in an argon matrix[179] the product absorption at 1205 cm^{-1}, assigned to $N_2O_2^-$, appeared even on mercury arc irradiation of the sample through a filter which excluded radiation of wavelength shorter than 3000 Å from the sample. Such results suggest the occurrence of weak charge-transfer interaction between the sodium atom and the N_2O molecule.

1.3 DIRECTIONS OF FUTURE RESEARCH

With the recent results on the spectra of charged species in a rare-gas matrix environment, the potential of the matrix isolation technique for the spectroscopic study of both neutral and charged reaction intermediates envisioned by Lewis and Lipkin[2] has at last been realised. Nevertheless, much work remains to be done. The technique will continue to provide a powerful tool for the detailed assignment of the spectra of relatively complicated stable molecules, as well as for spectroscopic studies of highly hydrogen-bonded species and of high-temperature molecules. The application of the technique to analytical chemistry can be expected to be considerably extended. Laser Raman studies will also begin to contribute to our knowledge of the spectra of these molecules. The guidelines for the conditions favouring the production of matrix-isolated free radicals and molecular ions will be subjected to considerable further testing and extension. There is likely to be an increasing emphasis on the trapping of transient species produced in such gas-phase processes as pyrolysis, as well as on the extension of photolysis techniques into the wavelength region beyond the 1050 Å lithium fluoride cut-off. The results of further experiments yielding spectroscopic data for electron-deficient species and for species in which electrons are added to outer antibonding orbitals are to be eagerly awaited, since they provide fertile ground for the testing and extension of molecular orbital theory.

References

1. Lewis, G. N., Lipkin, D. and Magel, T. T. (1941). *J. Amer. Chem. Soc.,* **63,** 3005
2. Lewis, G. N. and Lipkin, D. (1942). *J. Amer. Chem. Soc.,* **64,** 2801
3. Whittle, E., Dows, D. A. and Pimentel, G. C. (1954). *J. Chem. Phys.,* **22,** 1943

4. Becker, E. D. and Pimentel, G. C. (1956). *J. Chem. Phys.*, **25**, 224
5. Van Thiel, M., Becker, E. D. and Pimentel, G. C. (1957). *J. Chem. Phys.*, **27**, 95
6. Van Thiel, M., Becker, E. D. and Pimentel, G. C. (1957). *J. Chem. Phys.*, **27**, 486
7. Milligan, D. E. and Jacox, M. E. (1972). *Advan. High Temp. Chem.*, **4**, 1
8. Robinson, G. W. and McCarty, M., Jr. (1958). *J. Chem. Phys.*, **28**, 349
9. Robinson, G. W. and McCarty, M., Jr. (1958). *J. Chem. Phys.*, **28**, 350
10. Robinson, G. W. and McCarty, M., Jr. (1958). *Can. J. Phys.*, **36**, 1590
11. Robinson, G. W. and McCarty, M., Jr. (1959). *J. Chem. Phys.*, **30**, 999
12. McCarty, M., Jr. and Robinson, G. W. (1959). *J. Amer. Chem. Soc.*, **81**, 4472
13. Pimentel, G. C. (1960). *Formation and Trapping of Free Radicals,* ed. by Bass, A. M. and Broida, H. P., 109. (New York: Academic Press)
14. Linevsky, M. J. (1961). *J. Chem. Phys.*, **34**, 587
15. Katz, B., Ron, A. and Schnepp, O. (1967). *J. Chem. Phys.*, **46**, 1926
16. Barnes, A. J., Davis, J. B., Hallam, H. E., Scrimshaw, G. F., Hayward, G. C. and Milward, R. C. (1969). *Chem. Commun.*, 1089
17. Von Holle, W. G. and Robinson, D. W. (1970). *J. Chem. Phys.*, **53**, 3768
18. Mason, M. G., Von Holle, W. G. and Robinson, D. W. (1971). *J. Chem. Phys.* **54**, 3491
19. Bowers, M. T. and Flygare, W. H. (1966). *J. Chem. Phys.*, **44**, 1389
20. Barnes, A. J., Hallam, H. E. and Scrimshaw, G. F. (1969). *Trans. Faraday Soc.*, **65**, 3172
21. Milligan, D. E. and Jacox, M. E. (1967). *J. Chem. Phys.*, **47**, 278
22. Weltner, W., Jr., Walsh, P. N. and Angell, C. L. (1964). *J. Chem. Phys.*, **40**, 1299
23. Weltner, W., Jr. and McLeod, D., Jr. (1964). *J. Chem. Phys.*, **40**, 1305
24. Milligan, D. E. and Jacox, M. E. (1969). *J. Chem. Phys.*, **51**, 1952
25. Jacox, M. E. and Milligan, D. E. (1969). *J. Chem. Phys.*, **51**, 4143
26. De Kock, R. L. and Weltner, W., Jr. (1971). *J. Phys. Chem.*, **75**, 514
27. Ewing, G. E., Thompson, W. E. and Pimentel, G. C. (1960). *J. Chem. Phys.*, **32**, 927
28. Pacansky, J. and Calder, V. (1970). *J. Chem. Phys.*, **53**, 4519
29. Berney, C. V., Redington, R. L. and Lin, K. C. (1970). *J. Chem. Phys.*, **53**, 1713
30. Milligan, D. E., Jacox, M. E. and McKinley, J. D. (1965). *J. Chem. Phys.*, **42**, 902
31. Allavena, M., Rysnik, R., White, D., Calder, G. V. and Mann, D. E. (1969). *J. Chem. Phys.*, **50**, 3399
32. Hastie, J. W., Hauge, R. H. and Margrave, J. L. (1969). *J. Inorg. Nucl. Chem.*, **31**, 281
33. Hastie, J. W., Hauge, R. H. and Margrave, J. L. (1969). *J. Amer. Chem. Soc.*, **91**, 2536
34. Shirk, J. S. and Claassen, H. H. (1971). *J. Chem. Phys.*, **54**, 3237
35. Tursi, A. J. and Nixon, E. R. (1970). *J. Chem. Phys.*, **52**, 1521
36. Tursi, A. J. and Nixon, E. R. (1970). *J. Chem. Phys.*, **53**, 518
37. Barnes, A. J., Hallam, H. E. and Scrimshaw, G. F. (1969). *Trans. Faraday Soc.*, **65**, 3150
38. Barnes, A. J. and Hallam, H. E. (1970). *Trans. Faraday Soc.*, **66**, 1920
39. Barnes, A. J. and Hallam, H. E. (1970). *Trans. Faraday Soc.*, **66**, 1932
40. Redington, R. L. and Lin, K. C. (1971). *J. Chem. Phys.*, **54**, 4111
41. Taddei, G., Castellucci, E. and Verderame, F. D. (1970). *J. Chem. Phys.*, **53**, 2407
42. Fateley, W. G., Bent, H. A. and Crawford, B., Jr. (1959). *J. Chem. Phys.*, **31**, 204
43. Guillory, W. A. and Hunter, C. E. (1969). *J. Chem. Phys.*, **50**, 3516
44. Lannon, J. A., Verderame, F. D. and Anderson, R. W., Jr. (1971). *J. Chem. Phys.*, **54**, 2212
45. Baldeschwieler, J. D. and Pimentel, G. C. (1960). *J. Chem. Phys.*, **33**, 1008
46. Hall, R. T. and Pimentel, G. C. (1963). *J. Chem. Phys.*, **38**, 1889
47. Guillory, W. A. and Hunter, C. E. (1971). *J. Chem. Phys.*, **54**, 599
48. Aljibury, A. L. K. and Redington, R. L. (1970). *J. Chem. Phys.*, **52**, 453
49. Frey, R. A., Redington, R. L. and Aljibury, A. L. K. (1971). *J. Chem. Phys.*, **54**, 344
50. Christe, K. O., Curtis, E. C. and Pilipovich, D. (1971). *Spectrochim. Acta*, **27A**, 931
51. Christe, K. O. (1971). *Spectrochim. Acta*, **27A**, 631
52. Nimon, L. A., Seshadri, K. S., Taylor, R. C. and White, D. (1970). *J. Chem. Phys.*, **53**, 2416
53. King, S. T. (1970). *J. Phys. Chem.*, **74**, 2133
54. King, S. T. and Strope, J. H. (1971). *J. Chem. Phys.*, **54**, 1289
55. Grenie, Y., Lassegues, J-C. and Garrigou-Lagrange, C. (1970). *J. Chem. Phys.*, **53**, 2980
56. Weltner, W., Jr. (1969). *Advan. High Temp. Chem.*, **2**, 85
57. Hastie, J. W., Hauge, R. H. and Margrave, J. L. (1970). *Ann. Rev. Phys. Chem.*, **21**, 85
58. Cyvin, S. J., Cyvin, B. N. and Snelson, A. (1970). *J. Phys. Chem.*, **74**, 4338

59. Abramowitz, S. and Acquista, N. (1971). *J. Res. Nat. Bur. Std. (U.S.)*, **A75**, 23
60. Hastie, J. W., Hauge, R. H. and Margrave, J. L. (1971). *High Temp. Sci.*, **3**, 56
61. Frey, R. A., Werder, R. D. and Günthard, H. H. (1970). *J. Mol. Spectrosc.*, **35**, 260
62. Hauge, R. H., Hastie, J. W. and Margrave, J. L. (1971). *J. Less-Common Metals*, **23**, 359
63. Abramowitz, S., Acquista, N. and Thompson, K. R. (1971). *J. Phys. Chem.*, **75**, 2283
64. Linevsky, M. J., White, D. and Mann, D. E. (1964). *J. Chem. Phys.*, **41**, 542
65. Hinchcliffe, A. J. and Ogden, J. S. (1969). *Chem. Commun.*, 1053
66. Snelson, A. (1970). *J. Phys. Chem.*, **74**, 2574
67. Brom, J. M., Jr., Devore, T. and Franzen, H. F. (1971). *J. Chem. Phys.*, **54**, 2742
68. Brom, J. M., Jr. and Franzen, H. F. (1971). *J. Chem. Phys.*, **54**, 2874
69. Makowiecki, D. M., Lynch, D. A. and Carlson, K. D. (1971). *J. Phys. Chem.*, **75**, 1963
70. Anderson, J. S., Ogden, J. S. and Ricks, M. J. (1968). *Chem. Commun.*, 1585
71. Anderson, J. S. and Ogden, J. S. (1969). *J. Chem. Phys.*, **51**, 4189
72. Ogden, J. S. and Ricks, M. J. (1970). *J. Chem. Phys.*, **52**, 352
73. Ogden, J. S. and Ricks, M. J. (1970). *J. Chem. Phys.*, **53**, 896
74. Hastie, J. W., Hauge, R. H. and Margrave, J. L. (1969). *Inorg. Chim. Acta*, **3**, 601
75. Andrews, L. and Frederick, D. L. (1970). *J. Amer. Chem. Soc.*, **92**, 775
76. Gruen, D. M. and Carstens, D. H. W. (1971). *J. Chem. Phys.*, **54**, 5206
77. Mann, D. M. and Broida, H. P. (1971). *J. Chem. Phys.*, **55**, 84
78. Freedhoff, H. S. and Duley, W. W. (1971). *J. Chem. Phys.*, **54**, 3244
79. McCarty, M. M., Jr. (1971). *J. Chem. Phys.*, **54**, 3245
80. Milligan, D. E., Jacox, M. E. and Abouaf-Marguin, L. (1967). *J. Chem. Phys.*, **46**, 4562
81. Tinti, D. S. (1968). *J. Chem. Phys.*, **48**, 1459
82. Milligan, D. E. and Jacox, M. E. (1965). *J. Chem. Phys.*, **43**, 4487
83. Robinson, G. W. (1962). *Advan. Chem. Ser.*, **36**, 10
84. Katz, B., Brith, M., Sharf, B. and Jortner, J. (1969). *J. Chem. Phys.*, **50**, 5195
85. Weltner, W., Jr. (1967). *Science*, **155**, 155
86. Shirk, J. S. and Bass, A. M. (1970). *J. Chem. Phys.*, **52**, 1894
87. Meyer, B., Phillips, L. F. and Smith, J. J. (1968). *Proc. Nat. Acad. Sci. U.S.*, **61**, 7
88. Phillips, L. F., Smith, J. J. and Meyer, B. (1969). *J. Mol. Spectrosc.*, **29**, 230
89. Smith, J. J. and Meyer, B. (1968). *J. Mol. Spectrosc.*, **27**, 304
90. Meyer, B., Jones, Y., Smith, J. J. and Spitzer, K. (1971). *J. Mol. Spectrosc.*, **37**, 100
91. Meyer, B., Smith, J. J. and Spitzer, K. (1970). *J. Chem. Phys.*, **53**, 3616
92. Oppenheimer, M. and Berry, R. S. (1971). *J. Chem. Phys.*, **54**, 5058
93. Gruen, D. M. (1971). *Progr. Inorg. Chem.*, **14**, 119
94. Milligan, D. E. and Jacox, M. E. (1970). *Physical Chemistry—An Advanced Treatise*, Vol. 4, *Molecular Properties*, 193. (New York: Academic Press)
95. Herzberg, G. and Travis, D. N. (1964). *Can. J. Phys.*, **42**, 1658
96. Milligan, D. E., Jacox, M. E. and Bass, A. M. (1965). *J. Chem. Phys.*, **43**, 3149
97. Kroto, H. W., Morgan, T. F. and Sheena, H. H. (1970). *Trans. Faraday Soc.*, **66**, 2237
98. Harvey, K. B. and Brown, H. W. (1959). *J. Chim. Phys.*, **56**, 745
99. Bassler, J. M., Timms, P. L. and Margrave, J. L. (1966). *Inorg. Chem.*, **5**, 729
100. Snelson, A. (1970). *High Temp. Sci.*, **2**, 70
101. Milligan, D. E. and Jacox, M. E. (1968). *J. Chem. Phys.*, **48**, 2265
102. Snelson, A. (1970). *J. Phys. Chem.*, **74**, 537
103. Milligan, D. E. and Jacox, M. E. (1967). *J. Chem. Phys.*, **47**, 5146
104. Kaldor, A. and Porter, R. F. (1971). *J. Amer. Chem. Soc.*, **93**, 2140
105. Maltsev, A. K., Mikaelian, R. G., Nefedov, O. M., Hauge, R. H. and Margrave, J. L. (1971), in the press *Proc. Nat. Acad. Sci. U.S.*, **68**, 3238
106. Milligan, D. E. and Jacox, M. E. (1967). *J. Chem. Phys.*, **47**, 703
107. Andrews, L. (1968). *J. Chem. Phys.*, **48**, 979
108. Jacox, M. E. and Milligan, D. E. (1970). *J. Chem. Phys.*, **53**, 2688
109. Andrews, L. (1967). *J. Phys. Chem.*, **71**, 2761
110. Andrews, L. (1968). *J. Chem. Phys.*, **48**, 972
111. Rogers, E. E., Abramowitz, S., Jacox, M. E. and Milligan, D. E. (1970). *J. Chem. Phys.*, **52**, 2198
112. Milligan, D. E. and Jacox, M. E. (1964). *J. Chem. Phys.*, **41**, 2838
113. Rosengren, K. and Pimentel, G. C. (1965). *J. Chem. Phys.*, **43**, 507
114. Milligan, D. E. (1961). *J. Chem. Phys.*, **35**, 372
115. Milligan, D. E. and Jacox, M. E. (1964). *J. Chem. Phys.*, **40**, 2461.

116. Milligan, D. E. and Jacox, M. E. (1966). *J. Chem. Phys.*, **45**, 1387
117. Milligan, D. E. (1961). *J. Chem. Phys.*, **35**, 1491
118. Moore, C. B. and Pimentel, G. C. (1964). *J. Chem. Phys.*, **41**, 3504
119. Milligan, D. E., Mann, D. E., Jacox, M. E. and Mitsch, R. A. (1964). *J. Chem. Phys.*, **41**, 1199
120. Mahan, B. H. and Mandel, R. (1962). *J. Chem. Phys.*, **37**, 207
121. Ausloos, P., Gorden, R., Jr. and Lias, S. G. (1964). *J. Chem. Phys.*, **40**, 1854
122. Moll, N. G. and Thompson, W. E. (1966). *J. Chem. Phys.*, **44**, 2684
123. Milligan, D. E. and Jacox, M. E. (1968). *J. Chem. Phys.*, **49**, 1938
124. Milligan, D. E. and Jacox, M. E. (1968). *J. Chem. Phys.*, **49**, 4269
125. Guillory, W. A. and Smith, C. E. (1970). *J. Chem. Phys.*, **53**, 1661
126. Milligan, D. E., Jacox, M. E. and Smith, C. E., unpublished data
127. Shirk, J. S. (1971). *J. Chem. Phys.*, **55**, 3608
128. Smith, C. E., Milligan, D. E. and Jacox, M. E. (1971). *J. Chem. Phys.*, **54**, 2780
129. Jacox, M. E. and Milligan, D. E. (1969). *J. Chem. Phys.*, **50**, 3252
130. Jacox, M. E. and Milligan, D. E. (1967). *J. Chem. Phys.*, **47**, 1626
131. Merer, A. J. and Travis, D. N. (1966). *Can. J. Phys.*, **44**, 1541
132. Merer, A. J. and Travis, D. N. (1966). *Can. J. Phys.*, **44**, 525
133. Herzberg, G. (1961). *Proc. Roy. Soc. (London)*, **A262**, 291
134. Riveros, J. M. (1969). *J. Chem. Phys.*, **51**, 1269
135. Fessenden, R. W. and Schuler, R. H. (1965). *J. Chem. Phys.*, **43**, 2704
136. Andrews, L. and Smith, D. W. (1970). *J. Chem. Phys.*, **53**, 2956
137. Jacox, M. E. and Milligan, D. E. (1971). *J. Chem. Phys.*, **54**, 3935
138. Zobel, C. R. and Duncan, A. B. F. (1955). *J. Amer. Chem. Soc.*, **77**, 2611
139. Jacox, M. E. and Milligan, D. E. (1968). *J. Chem. Phys.*, **49**, 3130
140. Milligan, D. E., Jacox, M. E. and Guillory, W. A. (1968). *J. Chem. Phys.*, **49**, 5330
141. Milligan, D. E. and Jacox, M. E. (1970). *J. Chem. Phys.*, **52**, 2594
142. Dubois, I. (1968). *Can. J. Phys.*, **46**, 2485
143. Mosher, O. A. and Frosch, R. P. (1970). *J. Chem. Phys.*, **52**, 5781
144. Douglas, A. E. and Herzberg, G. (1940). *Can. J. Res.*, **18A**, 179
145. Meyer, B., Stroyer-Hansen, T., Jensen, D. and Oommen, T. V. (1971). *J. Amer. Chem. Soc.;* **93**, 1034
146. Meyer, B., Oommen, T. V. and Jensen, D. (1971). *J. Phys. Chem.*, **75**, 912
147. Andrews, L. and Carver, T. G. (1968). *J. Chem. Phys.*, **49**, 896
148. Carver, T. G. and Andrews, L. (1969). *J. Chem. Phys.*, **50**, 4223
149. Carver, T. G. and Andrews, L. (1969). *J. Chem. Phys.*, **50**, 4235
150. Carver, T. G. and Andrews, L. (1969). *J. Chem. Phys.*, **50**, 5100
151. Andrews, L. and Frederick, D. L. (1969). *J. Phys. Chem.*, **73**, 2774
152. Vaidya, W. M. (1934). *Proc. Roy. Soc. (London)*, **A147**, 513
153. Vaidya, W. M. (1951). *Proc. Phys. Soc. (London)*, **A64**, 428
154. Vaidya, W. M. (1964). *Proc. Roy. Soc. (London)*, **A279**, 572
155. Dixon, R. N. (1969). *Trans. Faraday Soc.*, **65**, 3141
156. Johns, J. W. C., Priddle, S. H. and Ramsay, D. A. (1963). *Discuss. Faraday Soc.*, **35**, 90
157. Milligan, D. E. and Jacox, M. E. (1969). *J. Chem. Phys.*, **51**, 277
158. Milligan, D. E. and Jacox, M. E. (1964). *J. Chem. Phys.*, **41**, 3032
159. Milligan, D. E. and Jacox, M. E. (1963). *J. Chem. Phys.*, **38**, 2627
160. Gordon, M. S. and Pople, J. A. (1968). *J. Chem. Phys.*, **49**, 4643
161. Milligan, D. E. and Jacox, M. E. (1966). *J. Chem. Phys.*, **44**, 2850
162. Jacox, M. E., Milligan, D. E., Moll, N. G. and Thompson, W. E. (1965). *J. Chem. Phys.*, **43**, 3734
163. Milligan, D. E. and Jacox, M. E. (1967). *J. Chem. Phys.*, **47**, 5157
164. Moll, N. G., Clutter, D. R. and Thompson, W. E. (1966). *J. Chem. Phys.*, **45**, 4469
165. Weissberger, E., Breckenridge, W. H. and Taube, H. (1967). *J. Chem. Phys.*, **47**, 1764
166. Jacox, M. E. and Milligan, D. E. (1971). *J. Chem. Phys.*, **54**, 919
167. Milligan, D. E., Jacox, M. E., Bass, A. M., Comeford, J. J. and Mann, D. E. (1965). *J. Chem. Phys.*, **42**, 3187
168. Jacox, M. E. and Milligan, D. E. (1965). *J. Chem. Phys.*, **43**, 866
169. Milligan, D. E. and Jacox, M. E. (1971). *J. Chem. Phys.*, **54**, 927
170. McCarty, M., Jr. and Robinson, G. W. (1959). *J. Chim. Phys.*, **56**, 723
171. Weltner, W., Jr. and McLeod, D., Jr. (1966). *J. Chem. Phys.*, **45**, 3096

172. Barger, R. L. and Broida, H. P. (1965). *J. Chem. Phys.*, **43,** 2364
173. Barger, R. L. and Broida, H. P. (1965). *J. Chem. Phys.*, **43,** 2371
174. Brabson, G. D. (1965). *Ph.D. Thesis,* University of California, Berkeley
175. Ballik, E. A. and Ramsay, D. A. (1963). *Astrophys. J.,* **137,** 61
176. Herzberg, G. and Lagerqvist, A. (1968). *Can. J. Phys.,* **46,** 2363
177. Frosch, R. P. (1971). *J. Chem. Phys.,* **54,** 2660
178. Milligan, D. E., Jacox, M. E. and Guillory, W. A. (1970). *J. Chem. Phys.,* **52,** 3864
179. Milligan, D. E. and Jacox, M. E. (1971). *J. Chem. Phys.,* **55,** 3404
180. Kasai, P. H. (1968). *Phys. Rev. Lett.,* **21,** 67
181. Kasai, P. H. and McLeod, D., Jr. (1971). *Accounts Chem. Res.,* **4,** 329
182. Milligan, D. E. and Jacox, M. E. (1971). *J. Chem. Phys.,* **55,** 1003
183. Andrews, W. L. S. and Pimentel, G. C. (1966). *J. Chem. Phys.,* **44,** 2361
184. Spence, D. and Schulz, G. J. (1971). *Phys. Rev. A,* **3,** 1968
185. Andrews, L. (1969). *J. Chem. Phys.,* **50,** 4288
186. Andrews, L. (1969). *J. Phys. Chem.,* **73,** 3922
187. Andrews, L. (1971). *J. Chem. Phys.,* **54,** 4935
188. Boness, M. J. W. and Schulz, G. J. (1970). *Phys. Rev. A,* **2,** 2182
189. Holzer, W., Murphy, W. F., Bernstein, H. J. and Rolfe, J. (1968). *J. Mol. Spectrosc.,* **26,** 543
190. Milligan, D. E. and Jacox, M. E. (1970). *J. Chem. Phys.,* **53,** 2034
191. Milligan, D. E. and Jacox, M. E. (1971). *J. Chem. Phys.,* **55,** 2550
192. Noble, P. N. and Pimentel, G. C. (1968). *J. Chem. Phys.,* **49,** 3165
193. Bondybey, V., Pimentel, G. C. and Noble, P. N. (1971). *J. Chem. Phys.,* **55,** 540
194. Parrish, D. D. and Herm, R. R. (1971). *J. Chem. Phys.,* **54,** 2518
195. Birely, J. H., Entemann, E. A., Herm, R. R. and Wilson, K. R. (1969). *J. Chem. Phys.,* **51,** 5461
196. Kwei, G. H., Norris, J. A. and Herschbach, D. R. (1970). *J. Chem. Phys.,* **52,** 1317
197. Parrish, D. D. and Herm, R. R. (1970). *J. Chem. Phys.,* **53,** 2431
198. Andrews, L. and Pimentel, G. C. (1967). *J. Chem. Phys.,* **47,** 3637
199. Tan, L. Y. and Pimentel, G. C. (1968). *J. Chem. Phys.,* **48,** 5202
200. Norman, I. and Porter, G. (1955). *Proc. Roy. Soc. (London),* **A230,** 399
201. Schnepp, O. and Dressler, K. (1965). *J. Chem. Phys.,* **42,** 2482
202. Roncin, J-Y., Damany, N. and Romand, J. (1967). *J. Mol. Spectrosc.,* **22,** 154

2

Pressure-induced Absorption Spectra of Hydrogen

H. L. WELSH
University of Toronto, Ontario

2.1   INTRODUCTION

Homonuclear diatomic molecules, such as hydrogen, have a centre of symmetry in the ground electronic state and are therefore inactive in rotational or vibrational dipole absorption. In 1949[1], in an attempt to observe an infrared vibrational absorption of oxygen dimers*, a more general type of spectrum, due to intermolecular forces operative during collisions, was identified. Such pressure- or collision-induced spectra arise from the dipole moments produced by the distortion of the electron distributions of molecules mutually interacting in binary, ternary and higher-order collisions. The induced dipole is modulated by the vibration and rotation of the collision partners and, because of its strong dependence on the intermolecular separation, also by their relative translational motion. Thus, in pure gases and in mixtures of gases a variety of pressure-induced spectra are known, ranging from pure translational and rotational spectra in the far infrared to fundamental and overtone rotation–vibrational spectra in the near infrared. Pressure-induced electronic spectra are also clearly possible; it is probable that the well-known diffuse near-infrared and visible bands of oxygen should be thus described†, and a pressure-induced band of nitrogen in the vacuum-ultraviolet region has been identified[4]. Normally infrared-inactive vibrations and rotations of polyatomic molecules with a high degree of symmetry have also been detected in pressure-induced absorption.

Most studies of pressure-induced spectra have been confined to the homonuclear diatomic molecules and, among these, the case of hydrogen has assumed special significance. Pressure-induced spectra are in general very diffuse, basically because the lifetime of the induced dipole is very short. However, the moment of inertia I of the H_2 molecule is relatively small and the rotational constant ($B = h/8\pi^2 cI \approx 60\ cm^{-1}$) sufficiently large to give some separation of the various rotational components of a band; the structure of the bands of hydrogen is therefore more clearly delineated than for other molecules. The spectra of hydrogen have for this reason been largely instrumental in clarifying the processes involved in collision-induced absorption and checking the results of theoretical studies. These processes are now so well understood that the H_2 molecule can, through its induced spectrum, be used in a definite sense as a probe of its environment in a gas, liquid or solid. It seems likely that such studies may be of considerable importance in the investigation of the molecular dynamics of dense media.

Several reviews of the earlier work have been published[5–8]. In the present account the main emphasis will therefore be placed on more recent work, especially such subjects as the collisional interference effect, the diffusion narrowing of quadrupole-induced lines, the structure of the H_2 overtone bands and the spectra of H_2–X van der Waals complexes. Because of space limitations the discussion of induced spectra in the liquid and solid states will

---

*It is interesting to note that more than 20 years elapsed before definite spectroscopic evidence of $(O_2)_2$ complexes was obtained[2].

†The quadratic variation of the intensity of these bands with pressure, established by Janssen[3] in 1885, may have been the first observation of a pressure-induced spectrum.

have to be omitted. Within these limits it is believed that the bibliography given at the end of the chapter is reasonably complete.

2.2 THE GENERAL PROPERTIES OF PRESSURE-INDUCED INFRARED ABSORPTION

The main properties of pressure-induced absorption have been established experimentally and will be summarised briefly in the following paragraphs.

2.2.1 Variation of the integrated intensities with the gas density

Over a limited range of low gas pressures, only binary, i.e. two-body, collisions are important; under these conditions, the integrated intensity of a pressure-induced absorption band varies quadratically with the gas density. More generally, in a mixture of an absorbing gas at the partial density ρ_a and a foreign gas at the partial density ρ_f, the integrated absorption coefficient of a given transition or set of transitions can be expanded in a power series in the densities in the form

$$\int A \, dv = a_1 \rho_a^2 + a_1 \rho_a^3 + \ldots \\ + b_1 \rho_a \rho_f + b_2 \rho_a \rho_f^2 + b_2' \rho_a^2 \rho_f + \ldots \tag{2.1}$$

where $A \equiv A(v) = (1/l) \ln (I_0/I)$ is the absorption coefficient at the frequency v^*, l is the path length and $a_1, a_2, \ldots$ etc. are temperature-dependent constants. The first set of terms in equation (2.1) arises from collisions of molecules of type a, i.e. from the base density of the absorbing gas, whereas the second set constitutes the *enhancement* of the absorption by collisions between a and f molecules. In the lower pressure region, where binary collisions predominate, only the quadratic terms in the density expansion are important. Thus, for the fundamental band of hydrogen (1.7–2.7 µm) at room temperature and 100 atm pressure the term $a_2 \rho_{H_2}^3$ contributes not more than 3% to the total absorption[9]. For comparison with theory it is often useful to use the absorption coefficient *per unit wavelength*, $\tilde{A} \equiv (1/v)A$, which has the nature of a transition probability.

Figure 2.1 shows enhancements of the H_2 fundamental in H_2–He, H_2–Ar and H_2–Xe mixtures at 298 K and total densities in the range 62–117 amagat†, i.e. in the region of predominantly binary collisions[10]. For comparison purposes, the profiles can therefore be approximately normalised by plotting $\tilde{A}/\rho_{H_2}\rho_f$(f = He, Ar, Xe) as ordinate. Both the shape of the band and the integrated absorption intensity evidently depend in a marked fashion on the perturbing gas; in particular, the areas under the curves (b_1 in equation (2.1)) are in the proportion 1 : 3.3 : 5.3 for helium, argon and xenon as perturbing gases.

---

*Frequency will usually be denoted below as a wavenumber (cm^{-1}): on occasion, however, v (s^{-1}) will be used.

†The relative density in amagat units is the ratio of the density in question to the density of the gas at N.T.P.

2.2.2   The rotational selection rules

The rotational selection rules for transitions of a homonuclear diatomic molecule induced by a static electric field were already derived by Condon[11] in 1932* as $\Delta J = 0, \pm 2$, where J is the rotational angular momentum quantum number. It might be expected that these rules, the same as for the Raman effect, hold also for transitions induced by the intermolecular forces. The triangles on the frequency axis in Figure 2.1 (and in following diagrams) are the positions of the $Q_1(J)$, $S_1(J)$ and $O_1(J)$ transitions† calculated from

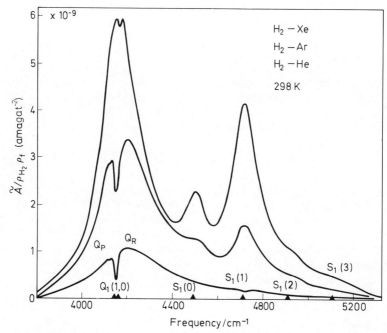

Figure 2.1   Absorption profiles of the enhancement of the H_2 fundamental by helium, argon and xenon at 298 K. The partial densities in amagat were: $\rho_{H_2} = 10.3$, $\rho_{He} = 105$; $\rho_{H_2} = 4.5$, $\rho_{Ar} = 113$; $\rho_{H_2} = 11.1$, $\rho_{Xe} = 51$

the constants of the free molecule[13]. At 300 K the fractional populations of the $J = 0, 1, 2, 3$ states of normal hydrogen are 0.13, 0.66, 0.12 and 0.09, respectively; hence, only the $S_1(0), S_1(1), S_1(2)$ and $S_1(3)$ rotational components are clearly distinguishable as maxima in Figure 2.1. A large fraction of the intensity of the band falls in the Q-branch region and the calculated $Q_1(1)$ frequency corresponds to a minimum in the intensity distribution. This *splitting of the Q branch* to give low- and high-frequency maxima, labelled

---

*The H_2 fundamental induced by a static electric field was first observed in 1953[12].

†$Q_1(J)$, $S_1(J)$ and $O_1(J)$ designate the $J \leftarrow J$, $J+2 \leftarrow J$, and $J-2 \leftarrow J$ transitions, respectively, where J is the initial rotational quantum number; the subscript indicates the change Δv in the vibrational quantum number, i.e. 0 for the pure rotational band, 1 for the fundamental band, etc.

Q_P and Q_R in Figure 2.1, is a characteristic density-dependent effect of the H_2 fundamental.

2.2.3   Translational broadening and temperature effects

A striking feature of the spectra in Figure 2.1 is the great breadth of the individual transitions; this is a consequence of the short duration of the collision and the Heisenberg uncertainty principle. The broad induced transition is in effect a continuum of summation and difference tones, $v_m \pm v_\kappa$, where v_m is the molecular frequency and hcv_κ is the continuum of relative kinetic energies of the colliding pair. The intensities in the low- and high-frequency wings at frequencies displaced by $\mp \Delta v$ from v_m are therefore related by a Boltzmann relation of the form[14]

$$I(v_m - \Delta v)/I(v_m + \Delta v) = \exp(-\Delta v h c / kT) \qquad (2.2)$$

which imparts a characteristic asymmetry to the profile of each transition.

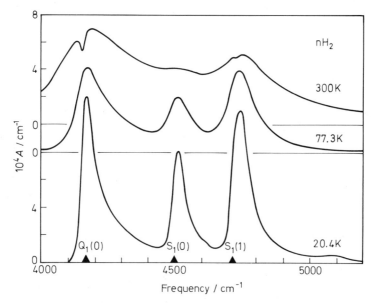

Figure 2.2 Profiles of the H_2 fundamental for pure hydrogen at a constant density of 13.6 amagat at various temperatures
(From Watanabe, A. and Welsh, H. L.[15], by courtesy of the National Research Council of Canada)

The participation of the relative kinetic energy in the absorption process also leads to a marked temperature variation of the intensity distribution in the fundamental band, as shown in Figure 2.2[15]. Since the collision duration increases as the temperature is lowered, the half-width of a given transition decreases and the asymmetry due to the Boltzmann relation becomes more pronounced. Lowering the temperature also reduces the Q-branch splitting and the relative intensity of the Q_P component; thus, the Q_P maximum is no longer evident at the lower temperatures in Figure 2.2.

2.2.4   Pressure effects on the band profile

Over large ranges of pressure there are two marked pressure effects on the intensity distribution in the profile of the fundamental band; these are illustrated in Figure 2.3 for a H_2–Ar mixture[16].

The more striking of the two effects is the splitting of the Q branch, already mentioned, which becomes very large at high densities. As the splitting increases a relatively narrow component located at the $Q_1(1)$ frequency is uncovered[14]; this component has a shape which resembles that of the $S_1(0)$

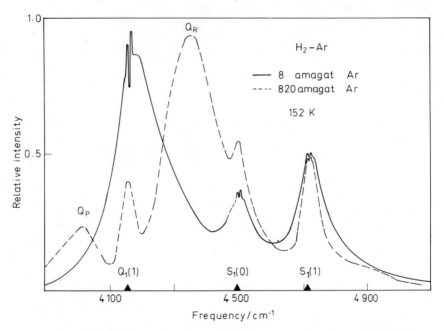

Figure 2.3   Enhancement of the H_2 fundamental in an H_2–Ar mixture at low and high densities at 152 K; the profiles are normalised to give S_1 (1) the same peak intensity. The density splitting of the Q branch and the density narrowing of the S lines is apparent
(From De Remigis, J., et al.[16], by courtesy of the National Research Council of Canada)

and $S_1(1)$ components. Thus, the experimental evidence is that there are two distinct mechanisms involved in the induction of the transitions of the fundamental band, one producing the very broad split Q branch and the other producing the narrower $Q_1(1)$, $S_1(0)$ and $S_1(1)$ components. The second pressure effect is the narrowing of the $S_1(1)$ line at very high densities[16]; narrowing of the $Q_1(1)$ and $S_1(0)$ lines also occurs, but this is not so apparent in Figure 2.3.

As will be shown below, the Q branch splitting, which is explained as an intercollisional interference effect[17], and the pressure narrowing of the S lines, which can be interpreted as a diffusion effect[18], may be of importance in the study of molecular motions in dense fluids.

2.2.5   Double transitions

When both molecules in a binary collision possess internal degrees of freedom, each of them can perform a molecular transition with the absorption of a single photon. The absorption of hydrogen in the overtone region (1.05 – 1.33 μm) furnished the first clear example of such a double transition[19]: the band consists of the superposition of a pure overtone band, in which the $v = 2 \leftarrow v = 0$ vibrational transition occurs in one molecule of a pair, and a band in which both molecules perform the $1 \leftarrow 0$ transition simultaneously. Subsequently, simultaneous vibrational transitions of an H_2 molecule and a different molecule X were observed[20]; the vibration of X may be normally inactive, e.g. nitrogen[21], or infrared-active, e.g. carbon dioxide[20] and hydrogen chloride[22].

Double transitions are also present in the fundamental band but are not readily apparent because of overlapping; they must however be taken into account in a detailed analysis of the band[23]. Thus, the S(1) component for pure hydrogen, for example, consists of three distinct transitions: the single transition $S_1(1)$ at $4713 \, cm^{-1}$ and the double transitions, $Q_1(1) + S_0(1)$ at $4742 \, cm^{-1}$ and $Q_1(0) + S_0(1)$ at $4748 \, cm^{-1}$. Some evidence of this structure may be seen in the 20.4 K profile in Figure 2.2.

2.2.6   Spectra of van der Waals complexes

At any temperature some fraction of the molecules of a gas or gas mixture (except H_2–He and He–He) will exist in bound states in the intermolecular potential well of a pair of molecules. The population of the bound states increases rapidly as the temperature is lowered and under appropriate experimental conditions the existence of such van der Waals complexes becomes *spectroscopically* detectable. The lifetime of an energy eigenstate of a complex is at least as long as the time between collisions; transitions between bound states are therefore much sharper than the translationally broadened free–free transition, and constitute a fine structure on the pressure-induced spectrum. These spectra of weakly bound complexes of simple non-polar molecules are observable only at low temperatures and relatively low pressures, and hence only with very long path lengths; some evidence of the fine structure is present at the peaks of the Q, S(0) and S(1) components of the 8 amagat profile in Figure 2.3. At higher resolution the fine structure due to the complex can be used to study in a direct way the intermolecular potential of such molecular pairs as H_2–H_2 [24] and H_2–Ar [25]. More recently, studies at still higher resolution have shown that these spectra furnish a new method of exploring the anisotropy of the intermolecular potential for H_2–rare gas pairs[26].

2.3 THE BASIC THEORY OF PRESSURE-INDUCED INFRARED ABSORPTION

In this Section we present a brief review of the theory of the induction mechanism and the density and temperature variation of the integrated coefficient of absorption. By 'integrated coefficient' we mean integration over the translational energy transitions, the frequency distribution of which will be considered in a later Section. The basic theory has been developed by Van Kranendonk and co-workers[27-33] and by some other authors[34-37]. The problem reduces in principle to the calculation of the dipole moment induced in given molecular configurations (pairs, triples, etc.), followed by an averaging of the square of the matrix element of the total dipole moment in terms of configurational distribution functions to obtain the integrated absorption coefficient. The discussion here will be limited to the induction mechanism (Section 2.3.1), the binary coefficient (Section 2.3.2) and the ternary coefficient (Section 2.3.3) for the $1 \leftarrow 0$ vibrational transition of a pure gas of homonuclear diatomic molecules. Further discussions of theory will be given in later Sections where they are appropriate.

2.3.1 The induced dipole moment and the rotational selection rules

The electric dipole moment induced by the intermolecular forces in a pair of interacting diatomic molecules, 1 and 2, is a function of the separation $\boldsymbol{R}_{12}$ of the centres of gravity of the two molecules, their orientations $\omega_1 = (\theta_1 \phi_1)$ and $\omega_2 = (\theta_2 \phi_2)$ with respect to a z-axis along $\boldsymbol{R}_{12}$ and their internuclear separations r_1 and r_2

$$\boldsymbol{\mu} = \boldsymbol{\mu}(\boldsymbol{R}_{12}; \omega_1, \omega_2; r_1, r_2) \tag{2.3}$$

This expression can be expanded as a sum of terms, each with a definite angular dependence,

$$\mu_\kappa = 4\pi \sum_{\lambda_1 u_1 \lambda_2 u_2} D_\kappa(\lambda_1 \mu_1 \lambda_2 \mu_2; R_{12}; r_1, r_2) Y_{\lambda_1}^{u_1}(\omega_1) Y_{\lambda_2}^{u_2}(\omega_2) \tag{2.4}$$

where the $Y_\lambda^\mu(\omega_i)$ ($i = 1,2$) are normalised spherical harmonics of the orientations ω_i of the molecule i, and $\kappa = 0, \pm 1$ designates the components of $\boldsymbol{\mu}$ in the system of spherical coordinates $\mp(1\sqrt{2})(x \pm iy)$ and z. For homonuclear diatomic molecules D_κ is zero for λ_1 or λ_2 odd. The D values are of course functions of R_{12}, r_1 and r_2, and are characteristic quantities of the pair of molecules.

To arrive at an explicit form for the D coefficients one must assume a concrete model for the induction process. As for models of the intermolecular potential, a good approximation for the induced moment might be obtained by taking into account only the parts of longest range and shortest range. The long-range part is due to the interaction of one molecule with the electric field produced by the other. For a homonuclear diatomic molecule the multipole of lowest order is the quadrupole moment which produces at large distances a field varying as $1/R_{12}^4$. The induced moment in this case is strongly dependent on the relative orientations of the molecules.

The short-range part of the induced moment is due to overlapping of the electron distributions of the molecules. If the overlap is not too large the dipole moment can be considered as the sum of two terms, an exchange term[36] and a deformation term[27]. For hydrogen, the two parts are approximately equal and the total overlap moment lies mainly along the intermolecular axis z and decreases very nearly exponentially with R_{12}.

This so-called (exp-4) model, involving a short-range isotropic and longer range anisotropic (quadrupolar) induction, has been successful in explaining many of the properties of the induced infrared spectra of hydrogen and other non-polar molecules. It has been argued however that a dispersion term of the form $\mu \sim 1/R_{12}^7$ should not be neglected[38].

Assuming the validity of the (exp-4) model we can now determine the form of the D coefficients in equation (2.4). The $1 \leftarrow 0$ vibrational transition is considered to take place in molecule 1 of the pair and the matrix element of the dipole moment will be expressed as

$$<1|\mu|0> = k_1 M(R_{12}; \omega_1, \omega_2) \tag{2.5}$$

where $M = (\partial\mu/\partial r_1)_0$ is the rate of change of μ with respect to r_1 at the equilibrium internuclear distance, and $k_1 = <1|r-r_e|0> = (h/8\pi^2 m v_0)^{\frac{1}{2}}$ in the harmonic oscillator approximation. The term $D_0(0000)$ in equation (2.4), which is independent of the molecular orientations, can correspond only to overlap induction, and M_0 will have the form:

$$M_0 = 4\pi D_0(0000) Y_0^0(\omega_1) Y_0^0(\omega_2) = D_0(0000)$$
$$= \xi \exp(-R_{12}/\rho) \tag{2.6}$$

where ξ and ρ are parameters which give, respectively, the magnitude and range of the oscillating part of the overlap-induced dipole. The rotational selection rule is evidently $\Delta J = 0$, and the overlap induction thus contributes only to the Q branch of the fundamental band.

Since the compoι. ιts of the field of a quadrupole moment Q at large distances have the forms $(Q/R^4)Y_2^0$ and $(Q/R^4)Y_2^{\pm2}$, it is easy to select and calculate the coefficients in equation (2.4), due to quadrupolar induction, which contribute to M:

$$D_0(2000) = +(3/\sqrt{5})(Q_1'\alpha_2/R_{12}^4); D_{\pm1}(2\pm100) = -(3\sqrt{15})(Q_1'\alpha_2/R_{12}^4); \tag{2.7}$$
$$D_0(0020) = -(3/\sqrt{5})(\alpha_1' Q_2/R_{12}^4) D_{\pm1}(002\pm1) = (3/\sqrt{15})(\alpha_1' Q_2/R_{12}^4) \tag{2.8}$$

where α_i ($i = 1,2$) is the mean polarisability of the molecule i, and the primes indicate differentiation with respect to r_1. The rotational selection rules can be immediately deduced from equations (2.4), (2.7) and (2.8) as $\Delta J_1 = 0, \pm2$ combined with $\Delta J_2 = 0$, or $\Delta J_1 = 0$ combined with $\Delta J_2 = 0, \pm2$; the quadrupolar induction thus gives rise to Q, S, and O transitions in the fundamental band, including such double transitions as $Q_1(1)+S_0(1)$. If the anisotropy, γ, of the polarisability is taken into account there are 10 additional non-zero terms in equation (2.4) of the type

$$D_{\pm1}(2\pm22\mp1) = (\sqrt{2/5})(\gamma' Q_2/R_{12}^4) \tag{2.9}$$

In this case, double rotational transitions such as $\Delta J_1 = 2$ combined with $\Delta J_2 = 2$ can take place.

2.3.2 The binary absorption coefficient

In discussing the calculation of the integrated binary absorption coefficient it is convenient to rewrite equation (2.1) in the form,

$$\int \tilde{A}(v)dv = \tilde{\alpha}_1 n^2 + \tilde{\alpha}_2 n^3 + \ldots \tag{2.10}$$

in which $\tilde{A}(v) = (c/v)A(v)$ with v in s^{-1}; n, the number of molecules per unit volume, is equal to ρn_0, where n_0 is the number density of the gas at N.T.P. Thus, $\tilde{\alpha}_1$ is expressed in the units of s^{-1} cm^{-6} and a_1 in equation (2.1) in units of cm^{-1} amagat^{-2}.

The contribution of the overlap induction to $\tilde{\alpha}_1$ for the fundamental band is obtained by averaging the R_{12}-dependence of $|M_0(R_{12})|^2$ in equation (2.6) over the pair-correlation function $g_0(R)$:

$$\tilde{\alpha}_1 \text{ (overlap)} = (8\pi^3/3h)k_1^2 \int g_0(R)|M_0(R)|^2 dR, \tag{2.11}$$

$$= (\pi/3mv_0)\int_0^\infty g_0(R)\xi^2 e^{-2R/\rho} 4\pi R^2 dR, \tag{2.12}$$

$$= \lambda^2 \mathscr{I}\bar{\gamma} \tag{2.13}$$

In this compact form $\mathscr{I}$ is the temperature-dependent dimensionless integral

$$\mathscr{I} = 4\pi \int_0^\infty e^{-2(x-1)\sigma/\rho} g_0(x)x^2 dx \tag{2.14}$$

in which $x = R/\sigma$ and σ is the diameter of the molecules defined by, say, the Lennard-Jones potential

$$V(R) = 4\varepsilon[(\sigma/R)^{12} - (\sigma/R)^6] \tag{2.15}$$

The other quantities occurring in equation (2.13) are $\lambda = (\xi/e)e^{-\sigma/\rho}$, where λe is the amplitude of the oscillating overlap dipole moment when the molecules are a distance σ apart, and $\bar{\gamma} = (\pi e^2 \sigma^3/3mv_0)$, which has the dimension of an integrated absorption coefficient per (density)^2.

The total quadrupole-induced contribution to the fundamental band has been given by Van Kranendonk[30] in a form similar to equation (2.13),

$$\tilde{\alpha}_1(\text{quad.}) = (\mu_1^2 + \mu_2^2)\mathscr{I}\bar{\gamma} \tag{2.16}$$

The amplitude parameters, μ_1 and μ_2, are derived from equations (2.7) and (2.8) by summing the Q, O and S intensities over all the rotational states J, and can be expressed in the simple form

$$\mu_1 = (1/e\sigma^4)(Q_1'\alpha_2); \mu_2 = (1/e\sigma^4)(\alpha_1'Q_2) \tag{2.17}$$

$\mathscr{I}$ is an integral, analogous to equation (2.14),

$$\mathscr{I} = 12\pi \int_0^\infty x^{-8} g_0(x)x^2 dx \tag{2.18}$$

The total binary coefficient for the fundamental band is obtained by summing equations (2.13) and (2.16),

$$\tilde{\alpha}_1 = \lambda^2 \mathscr{I}\bar{\gamma} + (\mu_1^2 + \mu_2^2)\mathscr{I}\bar{\gamma} \tag{2.19}$$

This 'sum rule' is rigorously true only for purely central intramolecular forces, and the calculation of $\tilde{\alpha}_1$ requires experimental or theoretical values of the molecular constants ξ, ρ, Q, Q', α and α'.

There have been recent theoretical calculations of the matrix elements of the quadrupole moment and the polarisability of hydrogen[39-42] and it is undoubtedly preferable to use these, rather than Q, Q', α and α', in calculations of the quadrupole-induced part of the absorption of hydrogen. The matrix-element form of $\tilde{\alpha}_1$ (quad) for an arbitrary vibrational transition, $(v'_1, v'_2)-(v_1, v_2)$, has been given by Poll[42]:

$$\tilde{\alpha}_1(\text{quad}) = (4\pi^3/3h)\sigma^{-5}\mathscr{I}\Sigma\{P_{J_1}P_{J_2}C(J_12J'_1;00)^2 C(J_20J'_2;00)^2 \times$$
$$<v'_1 J'_1 | Q_1 | v_1 J_1> v'_2 J'_2 | \alpha_2 | v_2 J_2>^2 + \text{cycl}\} \tag{2.20}$$

where the summation is taken over all individual transitions of the band and 'cycl' denotes a term identical to the preceding one except for an interchange of the indices 1 and 2. The Boltzmann factors P_J are defined by

$$P_J = Z^{-1}g_J(2J+1(\exp(-E_J/kT) \tag{2.21}$$

where, for hydrogen, g_J is 1 or 3 for J even or odd, and Z is the rotational partition function; the P_J are thus normalised so that $\Sigma_J P_J = 1$. The quantities $C(J\lambda J';00)$ are Clebsch–Gordan coefficients and are given by

$$C(J0J';00)^2 = \delta_{JJ}; \quad C(J2J-2;00)^2 = \frac{3J(J-1)}{2(2J-1)(2J+1)};$$

$$C(J2J;00)^2 = \frac{J(J+1)}{(2J-1)(2J+3)}; C(J2J+2;00)^2 = \frac{3(J+1)(J+2)}{2(2J+1)(2J+3)} \tag{2.22}$$

The anisotropy of the polarisability can be taken into account by an extension of equation (2.20) [43].

The radial integrals $\mathscr{I}$ and $\mathscr{J}$ in equations (2.14) and (2.18) must be calculated numerically as functions of the temperature. At high temperatures, quantum effects in the translational motion of the molecules can be neglected and the classical expression for $g_0(x)$ can be used:

$$g_0(x) = \exp[-V^*(x)/T^*] \tag{2.23}$$

where $V^* = V/\varepsilon$, with $V(R)$ as in equation (2.15), and $T^* = kT/\varepsilon$. However, for hydrogen at temperatures below $c.$ 100 K quantum effects must be taken into account; these have been calculated recently by Poll and Miller[113].

2.3.3 The ternary absorption coefficient

The ternary absorption coefficient, $\tilde{\alpha}_2$ in equation (2.10), has been discussed in some detail by Van Kranendonk[29, 31]; we shall review here the essential results.

The dipole induced in molecule 1 of a gas of N molecules is, strictly speaking, a function of the internuclear distances r_i, the orientations ω_i and the position vectors R_i of all the molecules in the gas, that is,

$$\mu = \mu(r_1 \dots r_N; \omega_1 \dots \omega_N; R_1 \dots R_N) = \mu(r^N; \omega^N; R^N) \qquad (2.24)$$

However, the induced dipole moment is a finite-range function of the inter-molecular separations, i.e. it decreases more rapidly than R^{-3} with increasing R, and it is possible to expand equation (2.24) in a series of cluster functions. Thus, let $\mu(1 \dots n)$ be the dipole moment induced in the cluster of molecules 1, ..., n when these are present alone in the given volume. Cluster functions $U(1 \dots, n)$ can then be introduced by means of the equations

$$\mu(12) = U(12)$$
$$\mu(123) = U(12) + U(13) + U(23) + U(123)$$

$$\dots\dots\dots\dots\dots\dots\dots\dots\dots\dots\dots\dots\dots\dots\dots\dots\dots\dots \qquad (2.25)$$

$$\mu(1 \dots N) = \sum_{i<j} U(ij) + \sum_{i<j<k} U(ijk) + \dots$$

These equations can be solved successively for the U values in terms of the μ values and, when these are substituted in the last equation of (2.5) this equation reduces to an identity which represents the cluster development of $\mu(1 \dots N)$. Proceeding in this way the coefficients in the density expansion of $\int \tilde{A} dv$ given in equation (2.10) can be found.

The ternary coefficient for the fundamental band consists of the sum of three terms

$$\tilde{\alpha}_2 = \tilde{\alpha}_2^{(1)} + \tilde{\alpha}^{(2)} + \tilde{\alpha}_2^{(3)} \qquad (2.26)$$

We shall comment only on their meanings without giving their explicit forms. The term $\tilde{\alpha}_2^{(1)}$ arises from the density dependence of the pair distribution function ('finite volume' effect) and is normally positive. $\tilde{\alpha}_2^{(2)}$ is the result of an interference effect which appears when there are three molecules present within the range of the induced dipole moment of each other; this *cancellation effect* gives a negative contribution to $\tilde{\alpha}_2$. The third term $\tilde{\alpha}_2^{(3)}$ is due to the non-additive part of the induced moment in clusters of three molecules, and can probably be neglected in most cases.

The ternary coefficient is thus mainly the sum of a positive and negative term, $\tilde{\alpha}_2^{(1)} + \tilde{\alpha}_2^{(2)}$. The calculated value[31] for hydrogen at 300 K is very small and gives a ternary term in the density expansion (equation 2.1) which is less than 1% of the total intensity at a density of 100 amagat; recent experimental data tend to confirm this. Earlier experimental values of $\tilde{\alpha}_2$, obtained from high pressure data[44], are greater and should probably not be compared with an essentially low-density theory.

It is interesting to note that the cancellation effect is present for single transitions but not for double transitions. For a molecule in a centrosymmetric environment of perturbing molecules single transitions should not be present; thus, the phase transition of *ortho*-enriched solid hydrogen from an h.c.p. to an f.c.c. lattice was detected by the disappearance of single transitions from the infrared fundamental band[45].

2.4   THE FAR-INFRARED SPECTRA OF HYDROGEN

2.4.1   The pure rotational spectrum

According to the theory presented above, compressed hydrogen gas should show a pure rotational spectrum of $S_0(J)$ lines in the further infrared. The

$S_0(J)$ frequencies, as measured in the rotational Raman spectrum[13], are 354.4, 587.1, 814.4 and 1034.7 cm^{-1} for $J = 0, 1, 2$ and 3, respectively*.

The S(2) and S(3) transitions were first observed by Ketelaar et al.[47] using KBr optics; in a later paper[48] further data for pure H_2 and mixtures of H_2 with He, Ar, N_2 and CO_2 at 213, 298 and 353 K were given. By using CsBr optics Kiss et al.[49, 50] were able to record the whole rotational spectrum in the range 300–1400 cm^{-1} for a number of perturbing gases with total pressures up to 250 atm at 300 K and, where possible, at 195 and 85 K. The spectra for pure H_2 and the series of rare gases as perturbers are shown in Figure 2.4.

Although the $S_0(J)$ lines are well-defined in the spectra in Figure 2.4 for the heavier perturbing gases and show, more or less, the anticipated intensity

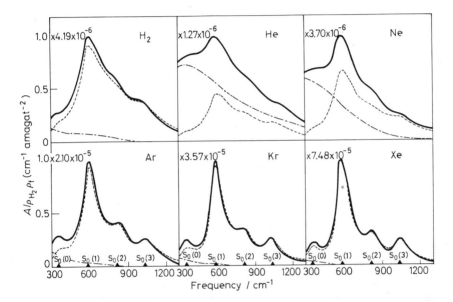

Figure 2.4 Induced spectra in the 300–1300 cm^{-1} region for H_2 and H_2–rare gas mixtures. The solid curves are the experimental profiles; the dashed curves represent the separated pure rotational (—) and pure translational (–·–) parts of the spectrum. It should be noted that the ordinate scales are different for the various cases
(From Kiss, Z. J. and Welsh, H. L.[50], by courtesy of the National Research Council of Canada)

distribution, the spectra for the lighter perturbers, especially He and Ne, are very diffuse and exhibit an unexpected increase in intensity towards lower frequencies. It was therefore concluded that the rotational spectra are super-imposed on an underlying continuum which was ascribed to a pressure-induced pure translational spectrum. To effect a valid comparison of experimental and theoretical intensities in the rotational spectrum it was necessary to separate the pure translational spectrum. The first step in this process was

---

*For the HD molecule the centre of mass does not coincide with the centre of electric charge; the induced rotational spectrum of HD thus shows $R_0(J)$ lines, corresponding to $\Delta J = +1$, in addition to $S_0(J)$ lines[46].

to find a satisfactory empirical analytical expression for the shape of a single rotational line. From data obtained at lower temperatures, where the overlapping of the components is not too great, it was found[50] that the high-frequency part of the line could be represented satisfactorily by the dispersion shape,

$$\tilde{I}(\nu_m + \Delta\nu) = I(\nu_m)/\{(\Delta\nu/\delta)^2 + 1\}, \tag{2.27}$$

where $\tilde{I}(\nu_m)$ is the intensity at the maximum, assumed to occur at the molecular transition frequency ν_m, and δ is the half-width of the dispersion curve at half maximum. The low-frequency wing is represented by the corresponding intensity distribution modified by the Boltzmann relation given in equation (2.2) above. Figure 2.5 shows the separated $S_0(0)$ and $S_0(1)$

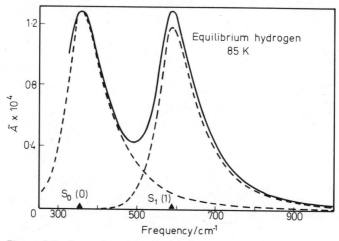

Figure 2.5 Separation of the absorption profile for equilibrium hydrogen (85 K. 118 amagat) into $S_0(0)$ and $S_0(1)$ components with similar Boltzmann-modified dispersion lines shapes
(From Kiss, Z. J. and Welsh, H. L.[50], by courtesy of the National Research Council of Canada)

components for equilibrium hydrogen at 85 K, where the translational spectrum contributes a negligible amount to the intensity. With this empirical line shape and the theoretical relative intensities of the rotational transitions, the separation of the rotational and translational spectra was carried out; the half-width δ, assumed the same for all the lines, was taken as a parameter which could be adjusted to secure the best reproduction of the experimental profile. The separation is shown for the 300 K profiles in Figure 2.4; the substantial translational contribution to the spectra for the lighter perturbing molecules is apparent. The binary coefficients, $\tilde{\alpha}_1(\text{trans})$ and $\tilde{\alpha}_1(\text{rot})$, integrated over the region accessible to observation in these experiments, are given in Table 2.1.

The (exp-4) model suggests that the rotational spectrum arises mainly

Table 2.1   Experimental and calculated values of the binary absorption co-efficients for the H_2 rotational spectrum[50]

T/K	Perturbing gas	Experimental		Calculated	
		$10^{34}\,\tilde{\alpha}_1$ (trans)/ $s^{-1}\,cm^6$	$10^{34}\,\tilde{\alpha}_1$ (rot)/ $s^{-1}\,cm^6$	$10^{34}\,\tilde{\alpha}_1$ (rot)/ $s^{-1}\,cm^6$	δ/cm^{-1}
375	Xe	0.61	10.9	12.5	90
300	n-H_2	0.18	1.19	1.11	155
	N_2	0.45	3.09	3.07	96
	He	0.32	0.12	0.125	145
	Ne	0.44	0.39	0.34	127
	Ar	0.49	3.92	3.42	94
	Kr	0.25	5.97	6.40	83
	Xe	0.20	10.9	12.8	80
195	n-H_2	—	1.19	1.11	124
	N_2	—	3.27	3.31	77
	He	0.091	0.096	0.115	120
	Ne	0.12	0.35	0.330	100
	Ar	—	3.61	3.68	75
85	n-H_2	—	1.25	1.12	84
	e-H_2	—	1.30	1.25	84

from quadrupolar induction. The binary coefficient for an $S_0(J)$ line can be written down from the theory of Section 2.3 in the explicit form[37, 49]

$$\tilde{\alpha}_1^{quad}(J) = (48\pi^2/h)\alpha_{H_2}^2 Q_{H_2}^2 \left\{ \frac{P(J)}{2J+1} - \frac{P(J+2)}{2J+5} \right\} \times$$

$$\frac{(J+1)(J+2)}{(2J+3)} \int_0^\infty e^{-V(R)/kT} dR/R^6, \qquad (2.28)$$

where the additional Boltzmann term $P(J+2)$ is necessary to take account of stimulated emission in the low-frequency region. The same formula is valid for the enhancement by a foreign gas by replacing α_{H_2} by α_f. Equation (2.28) was used to calculate $\tilde{\alpha}_1$ for the rotation band with the results shown in Table 2.1. The original calculations[50] were made with $Q_{H_2} = 0.45ea_0^2$ and $\alpha_{H_2} = 5.7a_0^3$; however in Table 2.1 we have used for Q_{H_2} the calculated value[39], $<03|Q|01> = 0.485ea_0^2$, i.e. the matrix element for the $S_0(1)$ transition, and for α_{H_2} the calculated value[42] of $<01|\alpha|01> = 5.42a_0^3$; the matrix elements show only a small J-dependence in the $v = 0$ state.

The experimental and calculated values of $\tilde{\alpha}_1$(rot) in Table 2.1 differ on the average by only $\pm 7\%$ although the values range over two orders of magnitude. This excellent agreement can be regarded as confirming not only the theory and the essentially quadrupolar nature of the induction, but also the combining rules used to obtain the Lennard-Jones constants for mixed species, i.e.

$$\sigma_{AB} = (\tfrac{1}{2})(\sigma_A + \sigma_B); \; \varepsilon_{AB} = \sqrt{(\varepsilon_A \varepsilon_B)} \qquad (2.29)$$

which were used to calculate the integral in equation (2.28) for the gas mixtures. It might be noted that the tendency of $\tilde{\alpha}_1(\text{rot})$ to increase with decreasing temperature arises not only from the temperature dependence of the pair distribution function but also from the Boltzmann term in equation (2.28).

Refinements in the calculation of $\tilde{\alpha}_1(\text{rot})$ were introduced by Van Kranendonk and Kiss[32] in that they took into account the anisotropy of the overlap interaction and of the polarisability, γ_{H_2}, as well as quantum effects in the pair distribution function. The angle-dependent overlap moment and γ_{H_2} were found to contribute only about 1% and 2%, respectively, to the value of $\tilde{\alpha}_1(\text{rot})$ for pure hydrogen; an interference effect between the overlap- and quadrupole-induced moments is more important and reduces the total intensity by c. 8%.

At higher gas densities, i.e. 200–400 amagat, the double rotational transitions, $S_0(0)+S_0(1)$ and $S_0(1)+S_0(1)$, could be observed[50] near 941 and 1174 cm^{-1}, respectively. The measured value of $\tilde{\alpha}_1[S_0(1)+S_0(1)]$ was 3.2×10^{-37} s^{-1} cm^6, as compared with the value 2.1×10^{-37} s^{-1} cm^6, calculated with $\gamma_{H_2} = 1.6\, a_0^3$. The double transition has about 1/150 of the intensity of the $S_1(1)$ transition.

Finally, we draw attention to the values of the half-width δ, obtained from the decomposition of the experimental profile into separate components; these values are given in the last column of Table 2.1. From the graph in Figure 2.6(a) it is seen that for a given molecular pair, δ varies as the square

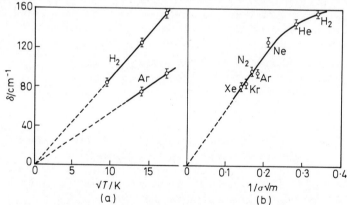

Figure 2.6 Variation of the dispersion half-width δ for the $S_0(J)$ lines of hydrogen with (a) temperature T, and (b) the reduced mass μ and the Lennard-Jones diameter σ of the collision pair
(From Kiss, Z. J. and Welsh, H. L.[50], by courtesy of the National Research Council of Canada)

root of the temperature, and is consequently proportional to the average relative velocity of the molecules of the pair. The variation of δ for different collision pairs at the same temperature is illustrated in Figure 2.6(b), in which δ is plotted again $1/\sigma\sqrt{m}$ where m is the reduced mass and σ the Lennard-Jones 'diameter' of the pair. Since the factor $1/\sqrt{m}$ is proportional to the average relative velocity and σ is a measure of the distance of closest approach, it might be argued that $1/\sigma\sqrt{m}$ is inversely proportional to the time spent in the region of interaction. The predominantly linear relationship in Figure 2.6(b) indicates that δ is inversely proportional to the duration of the collision, in accordance with the Heisenberg uncertainty principle.

The dispersion line form (equation (2.27)) becomes unphysical for large values of δ since the higher moments of the intensity distribution are infinite; it is not surprising therefore that it gives too high an intensity in the high-frequency tail of the line as was already observed in the first investigation[50]. In the application of this line form to the translational spectrum Bosomworth and Gush[51] found it necessary to graft an exponential tail onto the dispersion form at $\Delta v = v_m + 1.5\delta$. In a careful study of the high-frequency wing of the rotational spectrum Mactaggart and Hunt[52] found the exponential tail too extreme and proposed instead the truncation of the dispersion shape by a power-law tail of the form $(\Delta v)^{-d}$.

2.4.2 The pure translational spectrum

The existence of a pure translational spectrum induced by intermolecular forces, which was postulated to explain the continuum underlying the rotational spectrum of hydrogen, was confirmed by the observation of translational spectra in binary mixtures of the rare gases[53]. The initial observations, made with CsBr optics and thus confined to frequencies greater than ~ 300 cm^{-1}, showed a continuum with the intensity increasing towards lower frequencies and varying as the product of the partial densities of the two gases. It is clear that a translational spectrum should be observable for any pure gas of non-spherically symmetrical molecules or for the mixture of any pair of gases. Even a pure gas of spherically symmetrical molecules should show a translational spectrum with intensity varying as the third power of the density, but such spectra have apparently not yet been observed.

Translational spectra in the far infrared (i.e. below 300 cm^{-1}) have been studied for various gases and gas mixtures[51, 54–57]. They show in general a broad maximum in the 50–150 cm^{-1} region, the position varying with the masses of the molecules and the temperature, with the intensity falling off towards zero frequency and more slowly towards higher frequencies. In mixtures of the rare gases the far-infrared spectrum is of course a *pure translational* spectrum. For the heavier diatomic molecules, unlike hydrogen, the translational and rotational bands are not, in general, separated from one another. In principle, in the presence of anisotropic intermolecular inter-action, the translational and rotational motions of the molecules are coupled and no rigorous separation of the far-infrared band into separate parts can be made even if, as for hydrogen, the parts are to a large extent separated in frequency. However, for hydrogen the anisotropy of the intermolecular forces is evidently not large.

The far-infrared spectrum of hydrogen at 300 K is shown in Figure 2.7, which combines the results of Kiss *et al.*[49] for the 340–1400 cm^{-1} region with those of Bosomworth and Gush[51] for the 20–440 cm^{-1} region, the latter obtained by Fourier transform spectroscopy. The translational spectrum, separated out by the method to be described, is rather weak as compared with the rotational spectrum. The shape and intensity of the translational spectrum is of course strongly dependent on the stimulated emission at lower frequencies where the energies are comparable with $kT (= 208 \text{ cm}^{-1}$ at 300 K). If $G(v)$ is the probability of absorption (or stimulated emission) between states separated by the energy hcv, $G(v)$ and the absorption co-efficient $A(v)$ are related as follows:

$$A(v) = v\tilde{A}(v) = vG(v)[1 - \exp(-hcv/kT)] \qquad (2.30)$$

or

$$G(v) = (1/v)A(v)[1 - exp(-hcv/kT)]^{-1} \qquad (2.31)$$

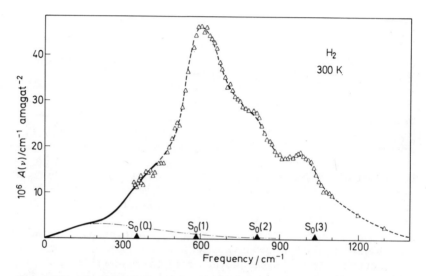

Figure 2.7 The far-infrared spectrum of hydrogen at 300 K showing the separated pure translational component (–·–·). The experimental curve combines the results of Kiss, Z. J. et al.[49] (––––) and Bosomworth, D. R. and Gush, H. P.[51] (———)
(Reproduced by courtesy of the National Research Council of Canada)

The graph of $G(v)$ v. v up to 700 cm^{-1} is given in Figure 2.8. The separation of the rotational components in Figure 2.8 was effected by the method described in Section 2.4.1, but with the dispersion line form modified with an exponential tail. Attempts to find an empirical analytical representation of $G(v)$ for the translational component were not very successful even for rare-gas mixtures where the translational spectrum is unencumbered by a rotational component.

The translational component of hydrogen at 77 K, shown in Figure 2.9, is weaker but less overlapped by the rotational spectrum than at room temperature. An interesting effect, evident in Figure 2.9, is the displacement of the peak of the $S_0(0)$ line by 17 cm^{-1} towards higher frequencies from the Raman value of 354 cm^{-1}; the reason for this is discussed in Section 2.5.1.2.

The integrated binary absorption coefficient for the far-infrared spectrum of monatomic gas mixtures, diatomic gases and diatomic–monatomic gas

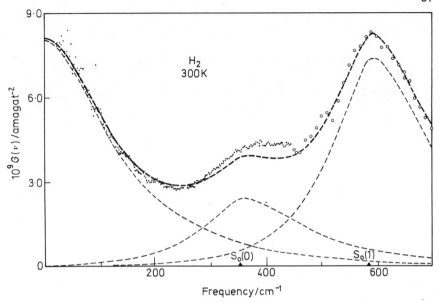

Figure 2.8 The line shapes in the far-infrared spectrum of hydrogen at 300 K using the data of Figure 2.7. The lightly dashed curves are the components of the fitted profile shown as the heavily dashed curve

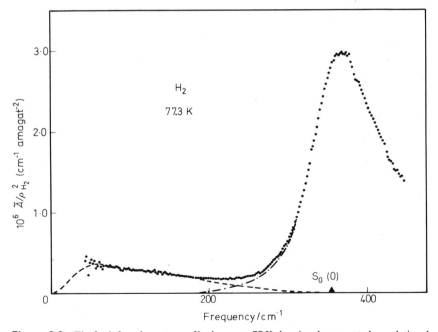

Figure 2.9 The far-infrared spectrum of hydrogen at 77 K showing the separated translational component (---)
(From Bosomworth, D. R. and Gush, H. P.[51], by courtesy of the National Research Council of Canada)

mixtures was calculated by Poll and Van Kranendonk[33]. The integrated absorption coefficient per unit path length is given by

$$\int A(v)\mathrm{d}v = (8\pi^3/3hc)V^{-1} \sum_{i<f} (P_i - P_f) |\mu_{if}|^2 v_{if} \qquad (2.32)$$

where $hv_{if} = E_f - E_i$ and V is the volume of the absorbing gas. The condition $i<f$ means that the sum is to run over all states for which $E_i < E_f$. Since the translational spectrum extends to zero frequency it is convenient to calculate $A(v)$ rather than $\tilde{A}(v)$. By using a trace technique to carry out the summation in equation (2.32) over the complete set of states and with the assumption of isotropic intermolecular forces, the binary coefficient for the pure translational band, α_1(trans), can be evaluated for different types of gases and the (exp-4) model in an expression of the general type (equation (2.19)). For hydrogen at 300 K it was found that to good approximation α_1(trans) $= (3.3 + 0.7) \times 10^{-33}$ s^{-1} cm^5, where the first term represents the quadrupolar induction and the second the interference effect between the quadrupolar and overlap moments; the overlap induction in itself gives a negligible contribution. The calculated value, 4.0×10^{-33}, is somewhat smaller than the value 5.6×10^{-33} s^{-1} cm^5 measured from the spectrum in Figure 2.6, probably because of the error involved in separating out the translational component. For the spectrum at 77 K (Figure 2.7) the translational component is better defined and the calculated value, α_1(trans) $= 2.7 \times 10^{-33}$, agrees very well with the experimental value, 2.5×10^{-33} s^{-1} cm^5.

The pressure-induced absorption of hydrogen in the far infrared plays an important part in determining the energy balance and the structure of the atmospheres of the major planets[58]; it is therefore desirable to know the binary absorption coefficient as a function of the frequency. By direct integration of the Schrödinger equation the theory outlined above was extended by Trafton[59] to a numerical computation of the profile of the translational absorption of hydrogen; the results show good agreement with the experimental data in Figure 2.7 in the region in which these are reasonably definite, i.e. 80–180 cm^{-1}. The thermal opacity provided by the pressure-induced absorption of hydrogen was then used in constructing model atmospheres for the major planets[60]. The pressure-induced opacity due to molecular hydrogen and H_2–He mixtures in late-type stars was calculated by Linsky[61]; for temperatures less than about 2500 K the translational, rotational and vibrational transitions of hydrogen provide the main source of opacity in stellar photospheres between 1 and 10 μm. Emissivity contributions by pressure-induced transitions of hydrogen in a plasma have been discussed by Olfe[62]. A revision and extension of the basic theory, suitable for high temperatures, has been given recently by Patch[63]; calculations of the binary absorption were made for the region 100–40 000 cm^{-1} with temperatures up to 7000 K.

2.4.3 The band shape in overlap-induced absorption

The translational absorption of mixtures of the rare gases has interesting possibilities for studying, on the one hand, rare-gas interatomic potentials through the temperature variation of the spectra[64] and, on the other, the shape of overlap-induced band profiles. There have been several theoretical treatments of the latter problem[65-72], but we confine the discussion here to only two of these.

Levine and Birnbaum[66] use a classical approach, neglect the intermolecular potential and represent the induced moment as a modified Gaussian function of R_{12},

$$\mu(R_{12}) = \mu_0 \gamma R_{12} \exp(-\gamma^2 R_{12}^2) \tag{2.33}$$

where γ^{-1} characterises the range of the dipole moment and μ_0 its strength. The factor γR_{12} 'shuts off' the induction for $R_{12} \lesssim \gamma^{-1}$. The binary absorption as calculated from these rather unphysical assumptions is

$$A(v) = \frac{\mu_0 \pi^3 n_1 n_2}{12 c \gamma^2} \left(\frac{2}{\pi m k T}\right)^{\frac{1}{2}} x^4 K_2(x) \tag{2.34}$$

where $x = (2\pi c v/\gamma)(m/kT)^{\frac{1}{2}}$ and $K_2(x)$ is a modified Bessel function of the second kind. For $v \to 0$, $A(v) \propto v^2$, as required, and for large v the initial-state-averaged transition probability, $G(v)$ in equation (2.31), decreases exponentially as was observed experimentally[51].

In the treatment of Sears[69] the expression for $A(v)$ in equation (2.32) is written

$$A(v) = \kappa(v/V) \sum_{i<f} (P_i - P_f) |M_{if}|^2 \{\delta(v - v_{if}) - \delta(v + v_{if}) \tag{2.35}$$

where $\kappa = 16\pi^2/3hc$ and the term $\delta(v + v_{if})$ vanishes identically for $i < f$ and is therefore redundant. However, the inclusion of negative frequencies allows a symmetry to be introduced into the formulation. By introducing the Fourier integral representation for the δ-functions in equation (2.35), we can express $A(v)$ as[73]

$$A(v) = 2\kappa n_1 n_2 \left[\frac{1 - \exp(-hv/kT)}{1 + \exp(-hv/kT)}\right] W(v) \tag{2.36}$$

where $W(v)$ is a reduced line-shape function. $W(v)$ is the Fourier transform of the autocorrelation function of the dipole moment

$$C(t) = (\tfrac{1}{2}) < \mu(0) \cdot \mu(t) + \mu(t) \cdot \mu(0) > \tag{2.37}$$

where the square brackets indicate an ensemble average and $\mu(t)$ is the operator μ in the Heisenberg picture. If the translational motion is treated classically, the ensemble average can be replaced by a time average

$$C(t) = < \mu(0) \cdot \mu(t)>, \tag{2.38}$$

and $W(v)$ can be expressed as

$$W(v) = \int_{-\infty}^{+\infty} e^{2\pi i v t} C(t) dt \tag{2.39}$$

The problem thus reduces to the calculation of the intracollisional dipole correlation function $C(t)$. Sears assumes the exponential variation of the induced dipole moment and expands $C(t)$, and hence $W(v)$, as a power series in ρ/σ.

The parameters of both of the line forms discussed above can be adjusted to fit quite well the experimental data for rare-gas mixtures. Because of the different simplifying assumptions used, the parameters obtained from these and other theories can differ widely[69]. However, as will be shown later, the analytical expressions obtained for the line shapes have in themselves an important use in analysing more complex induced spectra.

2.5   THE NEAR-INFRARED SPECTRA OF HYDROGEN

2.5.1   The fundamental rotation–vibrational band

2.5.1.1   General characteristics

The fundamental band of hydrogen for the pure gas and in mixtures with foreign gases has been investigated extensively[9, 14, 15, 22, 44, 74–84]; the corresponding band of deuterium at 2.6–3.7 μm has also been studied[15, 85, 86]. The experimental conditions have been varied over wide ranges of pressure and temperature, e.g. up to 5000 atm at room temperature[44], and from 420 K[78, 79] to 18 K[15]; to obtain useful data under these various conditions absorption path lengths from 4 mm to 13 m were necessary.

The main characteristics of the band and its variation with experimental conditions are now well understood in terms of the basic theory and some recent extensions. The narrower components [$Q_1(1)$, $S_1(0)$ and $S_1(1)$ in Figure 2.1] arise mainly from quadrupolar induction; the absence of a quadrupole-induced $Q_1(0)$ component is of course due to the spherical symmetry of the H_2 molecule in the $J = 0$ state. The increasing intensity of the $S_1(J)$ lines in going from the H_2–He to the H_2–Xe mixture in Figure 2.1 is chiefly a consequence of the form of the quadrupole-induced dipole (equation (2.16)) and the large increase in the atomic polarisability in going from helium ($\alpha = 1.4a_0^3$) to xenon ($\alpha = 27.4a_0^3$). The overlap induction produces a broad Q branch which has $Q_1(0)$, $Q_1(1)$, etc. components. Although these are shifted somewhat from another in frequency because of rotation–vibrational interaction, the individual components are not distinguishable except, to a certain extent, at higher temperatures[79]. For normal hydrogen at room temperature the $Q_1(1)$ overlap component is predominant and the minimum of the split Q branch coincides with the $Q_1(1)$ frequency.

Table 2.2 gives experimental values from various sources for the binary absorption coefficient of the H_2 fundamental for different perturbing molecules and temperatures. Although $\tilde{\alpha}_1(\text{quad})$ in equation (2.20) can be calculated rather accurately, $\tilde{\alpha}_1(\text{overlap})$ in equation (2.13) can be only roughly estimated for the pure gas and is not known for H_2–rare gas mixtures; it is therefore not possible to make a very meaningful comparison of experimental and theoretical values of $\tilde{\alpha}_1(\text{total})$. We note, however, that the value of $\lambda = \xi e^{-\sigma/\rho}$ for pure hydrogen calculated from simple wave functions is 8.5×10^{-3} with $\rho = 0.145\,\sigma$, whereas the value of λ obtained by fitting the experimental value of $\tilde{\alpha}_1$ at 300 K is 9.0×10^{-3}.

The temperature variation of $\tilde{\alpha}_1$ for pure hydrogen, as shown by the experimental data in Table 2.2, consists of a decrease with decreasing temperature down to about 80 K, followed by a rise at low temperatures. The initial decrease arises from the greater distance of closest approach of the molecules as the temperature is lowered; the rate of decrease is more rapid for the overlap-induced than for the quadrupole-induced intensity, i.e. $\mathscr{I}$ in equation (2.19) decreases more rapidly than $\mathscr{J}$. The increase in $\tilde{\alpha}_1$ at the lowest temperatures is due to the effect of the van der Waals minimum and the existence of bound states on the pair distribution function. Lower temperature values of $\tilde{\alpha}_1$ calculated by Poll[87] using a quantum-mechanical pair distribution function reproduce very well the trend of the experimental data. The value of $\tilde{\alpha}_1$ for deuterium at 300 K is $\sim 60\%$ of that of hydrogen; the difference appears to be due mainly to the smaller contribution of the overlap induction for deuterium[85].

The profile of the fundamental band in pure hydrogen is, of course, the superposition of various single and double transitions. A first attempt to

Table 2.2 **Experimental values of the binary absorption coefficient for the fundamental band of normal hydrogen**

Gas	T/K	$10^{35}\,\tilde{\alpha}_1/s^{-1}\,cm^6$	Gas	T/K	$10^{35}\,\tilde{\alpha}_1/s^{-1}\,cm^6$
H_2–He	300	1.03*	H_2	300	2.24§
H_2–Ne	300	2.37†	H_2	195	1.89§
H_2–Ar	300	3.86*	H_2	78	1.32§,||
H_2–Kr	300	7.56†	H_2	40	1.38||
H_2–Xe	300	11.34||	H_2	24	1.51||
H_2–N_2	300	5.11*	H_2	18	1.67||
H_2–O_2	300	6.12‡	D_2	300	1.40¶

*Hare and Welsh[44]; †Reddy and Lee[81]; ‡Varghese and Reddy[82]; §Hunt and Welsh[23]; \|\|Watanabe and Welsh[80]; ¶Reddy and Cho[85].

separate the profile into individual components, using the Boltzmann-modified line form for the overlap- as well as the quadrupole-induced components, was fairly satisfactory, in spite of the fact that the splitting of the Q branch had to be ignored[23, 88]. The procedure was more successful in fitting the band profile at lower temperatures and pressures where the splitting of the Q branch is not so marked; an example of the decomposition of the profile is shown in Figure 2.10[80]. The eleven components of the band at 20.4 K include double $Q_Q(1)$ and $Q_Q(0)$ transitions in which the second molecule, which must be in the $J = 1$ state, performs an orientational transition. The dispersion half-widths, δ_q and δ_o of the quadrupolar and overlap components, respectively, were taken as parameters in a computational curve-fitting procedure, δ_o was found to vary from 205 to 128 cm^{-1} in going from 77 to 18 K, and δ_q from 62 to 36 cm^{-1}; the latter varies rather accurately as $\sqrt{T}$. The values of $1/\delta_o$ and $1/\delta_q$ at a given temperature are indicative of the relative ranges of the overlap and quadrupole interactions, respectively.

2.5.1.2 *The intercollisional interference effect*

An explanation of the splitting of the Q branch as an intercollisional interference effect has been given recently by Van Kranendonk[17]. An earlier explanation[14, 89, 90] of the shape of the overlap-induced Q branch as a continuous rotational spectrum due to the turning of the dipole moment during the collision has been shown to be a negligible effect[33].

According to Van Kranendonk[17] the radiative processes occurring in successive binary collisions are coherent under certain conditions; the dipole moment correlation function, $C(t)$ in equation (2.37), must therefore include an *intercollisional* as well as an intracollisional part. If the induced dipole shows the same dependence on R as the intermolecular force, i.e.

$$\mu(R) = \alpha f(R) \qquad (2.40)$$

where α is independent of R, the correlation in successive collisions is a negative one and leads to destructive interference. The simple physical

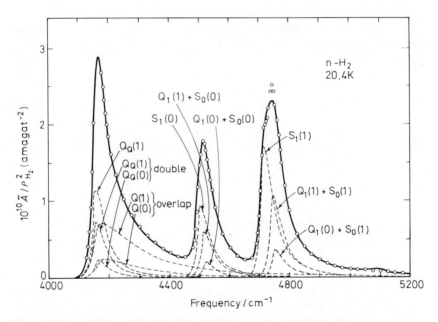

Figure 2.10 Analysis of the fundamental band of normal hydrogen into its 11 components at 20.4 K. The solid curve is the experimental profile, the dashed curves the individual components, and the dots the summation of these.
(From Watanabe, A. and Welsh, H. L.[80], by courtesy of the National Research Council of Canada)

picture is that, if the induced dipole is along the intermolecular line, the dipole induced in the second collision is on the average oppositely directed to that in the first collision and interference can take place.

If the correlation is taken over only the immediately successive collision and if ternary and higher-order collisions are neglected, the Fourier transform of the intercollisional correlation function has the form

$$D(\omega) = 1 - \gamma(1 + \omega^2 \tau_c^2)^{-1} \qquad (2.41)$$

where τ_c is the mean time between collisions and γ is equal to 1 if equation

(2.40) is exactly fulfilled; thus, $D(\omega)$ varies from $\omega^2\tau_c^2$ at small ω to 1 at large ω. The reduced line form $W(\omega)$ is given by

$$W(\omega) = D(\omega)W^0(\omega) \qquad (2.42)$$

where $W^0(\omega)$ is the intracollisional reduced line-shape function.

It is clear from (2.40) that intercollisional interference can be expected for isotropic overlap induction since, for this case, the form of $\mu(R_{12})$ is exponential and is thus similar to that of the intermolecular force in the range of R_{12} for which overlap induction is significant. The intercollisional interference effect should therefore be present in the low-frequency region of the translational spectra of mixtures of rare gases; in other words, the function $G(\nu)$ in equation (2.31) should tend to zero at $\nu = 0$. This effect has been observed by Marteau et al.[56, 57] by extending the translational spectrum down to 16 cm^{-1}; for Xe–Ar at a total pressure of 1800 atm, for example, $G(\nu)$ begins to decrease rapidly for frequencies less than $\sim 40 \text{ cm}^{-1}$. An attempt has been made recently by Muc et al.[91] to observe the effect at 1 cm^{-1} by absorption in a microwave cavity. Intercollisional interference should not, of course, be present in the translational spectrum of pure hydrogen which, as has been shown, is mainly quadrupole induced; here the $G(\nu)$ curve should approach a finite value for $\nu \to 0$, as indicated in Figure 2.8.

In the fundamental band of hydrogen the intercollisional interference effect produces the splitting of the overlap-induced part of the Q branch. Thus from equations (2.41) and (2.42) we see that, in the neighbourhood of the $Q(J)$ frequency, $W(\nu - \nu_m)$ shows a dip which has the form of an inverted dispersion curve of half-width δ_c where

$$\delta_c(\text{cm}^{-1}) = 1/2\pi c\tau_c \qquad (2.43)$$

From the kinetic theory formula

$$\tau_c = (2n_2\sigma_{12}^2)^{-1}\sqrt{(m/2\pi kT)} \qquad (2.44)$$

where σ_{12} is the mutual 'hard-sphere' diameter, n_2 the number density of the perturbing molecules and m the reduced mass of the H_2-perturber pair, it is evident that δ_c varies linearly with the density n_2 and inversely as $\sqrt{T}$. The interference dip should persist to the lowest densities and should show a marked temperature variation. If $\gamma = 1$ in equation (2.41), there should be zero absorption at $\nu = \nu_m$.

These conclusions are confirmed in a striking fashion by the spectrum of the fundamental band of a para-H_2–He mixture shown in Figure 2.11; this spectrum was obtained[92] at 26.8 K with a path length of 110 m. The interference dip is very sharp and deep; the value of δ_c is $\sim 3 \text{ cm}^{-1}$. There is of course no $Q_1(0)$ quadrupolar component in the spectrum of Figure 2.11 to obscure the dip; the fact that the dip does not go to zero intensity is probably due partly to the limited spectral resolution ($\sim 0.2 \text{ cm}^{-1}$) but mostly to a γ value in equation (2.41) slightly less than unity.

The $S_1(0)$ component in Figure 2.11 is relatively much stronger than the $S_1(J)$ components in the H_2–He profile in Figure 2.1 because of the differential temperature dependence of the overlap and quadrupolar induction. It will be noted that the maxima of the $Q_1(0)$ and $S_1(0)$ components in Figure 2.11(a) occur at frequencies higher than the molecular frequencies. However, when the intensity distributions in these components are 'symmetrised' [cf. equation (2.36)] by multiplying them by the factor $\{1+\exp[-hc(v-v_m)/kT]\}$, the reduced line-shape functions $W(v)$ are symmetrical about the molecular frequencies v_m. In earlier investigations at high pressures[14,44] the width of the splitting Δv_{PR} of the Q branch showed a linear variation with the density, which gave by extrapolation a non-zero value of the splitting at zero density. The linear variation of Δv_{PR} reflects the density variation of δ_c; however, the non-zero value of Δv_{PR} at zero density is evidently accounted for by the essential asymmetry of the observed line shape which becomes clearly

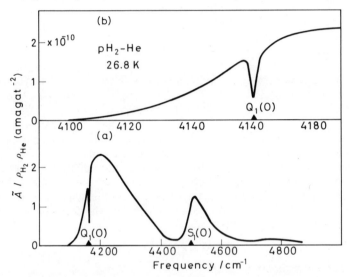

Figure 2.11 (a) The H_2 fundamental for a H_2–He mixture ($\rho_{H_2} = 0.7$, $\rho_{He} = 47.4$ amagat) at 26.8 K in a path length of 110 m. (b) The Q_1 (0) region on an expanded frequency scale
(Reproduced by courtesy of A. R. W. McKellar[92])

apparent only at low densities. The shift of the $S_0(0)$ component in Figure 2.9 can presumably be explained in the same way.

In an extension of the theory of intercollisional interference, Lewis and Van Kranendonk[93] have taken into account the correlations between *all* collisions in the collision sequence of a molecule. In this case δ_c is shown to be related to τ_c by the equation

$$\delta_c = (1-\tilde{\Delta})/2\pi c\tau_c \qquad (2.45)$$

where $\tilde{\Delta}$ is the mean persistence-of-velocity ratio for the types of molecules involved. For heavy perturbers $\tilde{\Delta}$ is negligibly small, but for H_2–He mixture $\tilde{\Delta}$ is 0.24.

2.5.1.3   *Profile analysis of the fundamental band*

These new developments in the theory of line shapes have given renewed impetus to attempts to analyse the profile of the fundamental band under a variety of experimental conditions. Transforming the results of the preceding discussions into a form suitable for application to experimental data on the fundamental band we write the absorption coefficient $\tilde{A}(v)$ in the form

$$\tilde{A}(v) = \sum_{n,m} \frac{\alpha_{nm} W_n(\Delta v_m)}{1 + \exp[-hc(\Delta v_m)/kT]} \tag{2.46}$$

where $\Delta v_m = v - v_m$, m numbers a particular molecular transition, n gives the type of induction (0 for overlap, 1 for quadrupolar induction), and the denominator converts the 'symmetrised' line $W_n(\Delta v_m)$ into the observed Boltzmann-modified line shape.

For the overlap-induced contributions to the Q branch we use equation (2.42), $W_0^0(\Delta v_m) = D(\Delta v_m) W^0(\Delta v_m)$, and write equation (2.41) in the form

$$D(\Delta v_m) = 1 - \gamma\{1 + (\Delta v_m/\delta_c)^2\}^{-1} \tag{2.47}$$

In the experiments described below, the intracollisional line form $W^0(\Delta v_m)$ in the region of low densities, where $D(\Delta v_m)$ has only a small effect on the band profile, was found to be reproduced somewhat better by the line form of Levine and Birnbaum[66] (equation (2.34)) than by that of Sears[69]. $W_0^0(\Delta v_m)$ was therefore taken to be

$$W_0^0(v) = \left(\frac{\Delta v_m}{\delta_d/2}\right)^2 K_2\left(\frac{\Delta v_m}{\delta_d/2}\right) \tag{2.48}$$

where δ_d, the intracollisional half-width, can be shown to be approximately equal to the half-width at half-height of $W_0^0(v)$. For quadrupole-induced components, $W_1(\Delta v_m)$ was assumed to be the usual dispersion form with δ_q as the half-width parameter. The relative intensity coefficients α_{nm} in equation (2.46) were obtained from the basic theory in Section 2.3.2.2.

A computer program was written to sum equation (2.46) over all the transitions of the band for a given set of adjustable parameters and to adjust the intensity of the calculated profile to give the best least-squares fit to the experimental profile[84]. In addition to δ_c, δ_d and δ_q, the relative intensity of the overlap-induced to the quadrupole-induced transitions was treated as a parameter; minor adjustments in the molecular frequencies were also made.

The H_2 fundamental induced by the rare gases was selected as a suitably simple example for an initial study. The band for H_2–He, H_2–Ar, H_2–Kr and H_2–Xe mixtures was recorded for densities ranging from 3 to 1200 amagat and temperatures from 20 to 298 K, the density and temperature ranges varying for the different perturbing gases; path lengths from 0.013 to 15 m and pressures up to 2000 atm were used[95]. For the highest densities some alteration in the line shape, equations (2.47) and (2.48), had to be introduced; however, the main object of the study was to obtain values of the half-widths δ_c, δ_d and δ_q without particular regard to the physical significance of the line form.

Typical examples of the computerised analysis of the band profile are shown in Figure 2.12 for a low- and a high-density nH_2–Kr mixture at 212 K; the density effects already noted in Section 2.2.4 are evident. The final results of the analysis for H_2–Ar mixtures over a wide range of densities at 160 K are shown in Figure 2.13 in which the values obtained for δ_d, δ_c and δ_q are plotted against the argon density. The value of δ_d is essentially density-independent even at high densities; this is an indication of the very short range of the overlap induction.

The intercollisional half-width δ_c increases linearly with the density in the lower range but more rapidly at the higher densities. From the experimental values of δ_c the collision times τ_c could be calculated from equation (2.43) and

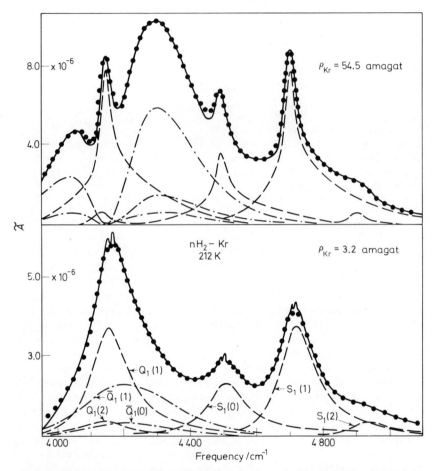

Figure 2.12 Analysis of the H_2 fundamental band for nH_2–Kr mixtures at 212 K: (——) the experimental profile, (—·—·) the overlap-induced transitions $[\overline{Q}_1 (0), \overline{Q}_1 (1)]$, (— — —) the quadrupole-induced transitions, (top) the summation of the separate components[10]

compared with kinetic theory values given by equation (2.44). It is however more interesting to use the relation

$$D_{12} = (3kT/4m)\tau_c = (3kT/8\pi mc)(1/\delta_c) \qquad (2.49)$$

where D_{12} is the mutual diffusion coefficient of hydrogen in the rare gas, to calculate D_{12} from the experimental values of δ_c. The values of D_{12} thus obtained are plotted against the rare-gas density in Figure 2.14, and are compared with kinetic theory values of D_{12}. The latter were calculated from equations (2.49) and (2.44) with τ_c corrected for high densities by the method given by Chapman and Cowling[94]. The agreement of the calculated and observed values of D_{12} is satisfactory except at the highest densities. The discrepancy may be in the calculated kinetic theory values or, on the other hand, the intercollisional interference theory may begin to fail when successive collisions begin to overlap as is the case for the higher densities. It is however remarkable that the theory remains at least approximately valid at such high

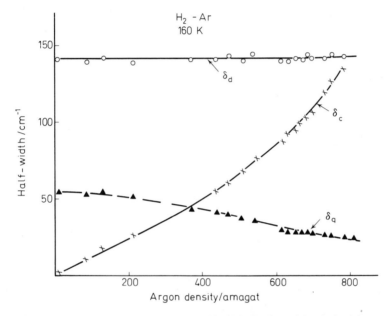

Figure 2.13 Density variation of the half-widths δ_d, δ_c, and δ_q obtained from the analysis of the H_2 fundamental band for H_2–Ar mixtures at 160 K
(From Mactaggart, J. W. and Welsh, H. L.[84], by courtesy of the National Research Council of Canada)

densities; this is again a manifestation of the short range of the overlap induction.

Two sets of experimental points for H_2–He are given in Figure 2.14, one using the simple theory and the other incorporating the correction given in (2.45) which takes account of correlations in the whole collision sequence. The validity of the correction is evident.

2.5.1.4 Diffusion narrowing of quadrupole-induced lines

Figure 2.13 shows that the half-width δ_q of the quadrupole-induced transitions of dilute H_2–Ar mixtures remains constant with increasing density up to ~ 300 amagat and then decreases. This behaviour, first observed by De Remigis *et al.*[16] for the H_2–Ar system, has also been demonstrated for H_2–Kr and H_2–Xe mixtures[95]. The nature of this characteristic variation of δ_q is clearer in the log–log plot in Figure 2.15, from which it is seen that for a certain range of high densities δ_q is approximately inversely proportional to the density.

At low densities, where $\tau_d \ll \tau_c$, the induction takes place in isolated binary collisions and the constant value of δ_q is characteristic of the range of the

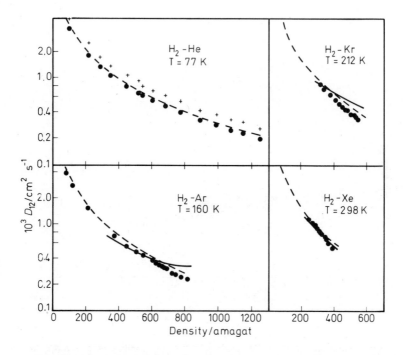

Figure 2.14 Semi-logarithmic plot of the diffusion constants D_{12} of hydrogen in H_2–rare mixtures obtained from analyses of the H_2 fundamental band: ● obtained from experimental values of δ_c [+ without the correction of equation (2.45) for H_2–He], (——) obtained from the experimental density variation of δ_q (Section 2.5.1.4), (– – –) calculated from the density-corrected kinetic theory formula (From Mactaggart, J. W. and Welsh, H. L.[84], by courtesy of the National Research Council of Canada)

quadrupolar induction. The narrowing of the transitions at higher densities has been interpreted by Zaidi and Van Kranendonk[18] in terms of the mutual diffusion constant D_{12} of hydrogen molecules in the rare gas. The physical picture of the narrowing is that the duration of a given H_2–rare gas collision is effectively lengthened by the proximity of other rare-gas molecules. A simple model of the diffusional narrowing gives the relation

$$\delta_q = \xi D_{12}/\pi c \overline{R}^2 \qquad (2.50)$$

where $\overline{R}$ is the effective range of the quadrupole induction and ξ is a 'correction factor' to allow for the simplicity of the model. The value of $\overline{R}$ can be estimated from the half-width δ_q^0 at low densities,

$$\overline{R} = \overline{v}/2\pi\delta_q^0 \tag{2.51}$$

where $\overline{v}$ is the average relative velocity of the colliding pair. At densities which are not too high D_{12} varies inversely as the density; thus, the behaviour

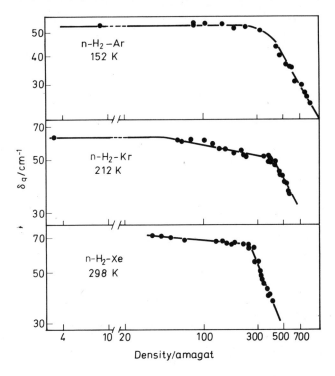

Figure 2.15 Log–log plot of the variation of δ_q with density for values of δ_q obtained from analyses of the H_2 fundamental for H_2-rare gas mixtures
(From Mactaggart, J. W. *et al.*[95], by courtesy of the National Research Council of Canada)

of δ_q after the onset of the pressure narrowing is explained. If $\overline{R}$ is calculated from δ_q^0 according to equation (2.51) and ξ is assumed to be unity, we can calculate D_{12} from equation (2.50) using the experimental data of Figure 2.15. The variation of D_{12} for the density region in which the theory applies is plotted in Figure 2.14 above; the agreement with the values obtained from the intercollisional interference effect is surprisingly good, although there is some tendency for the two sets of data to deviate at the highest densities.

The fact that, through the medium of pressure-induced absorption and with rather simple theories, one can study molecular motions in relatively high density gases, encourages the hope that the method may be applicable to gases at still higher densities and to liquids. At such densities the kinetic theory approach may no longer be appropriate, and models such as those considered, for example, by Desai and Yip[96] may have to be used; however, the consideration of these is outside the scope of the present Review.

2.5.2　The overtone bands

The interpretation of the pressure-induced spectrum of hydrogen in the first

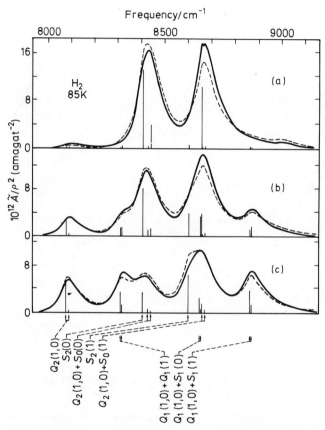

Figure 2.16 The pressure-induced spectrum of hydrogen in the first overtone region: (a) *para* concentration $C_{para} \approx 0.95$, 85 K; (b) $C_{para} \approx$ 0.57, 85 K; (c) $C_{para} = 0.25$, 86 K. The solid curves are experimental, the dashed curves are fitted theoretical profiles
(From McKellar, A. R. W. and Welsh, H. L.[43], by courtesy of the Royal Society)

overtone region as the superposition of a pure overtone and a double vibrational transition[19] was confirmed by the observation of Herzberg[98] of the induced spectrum in the region of the second overtone, 0.77–0.84 μm, as the superposition of the single $3 \leftarrow 0$ and the $(2 \leftarrow 0)+(1 \leftarrow 0)$ double transition. The identification by Herzberg[97, 98] of the $S_3(0)$ component of the laboratory spectrum with a diffuse feature observed by Kuiper[99] in spectra of Uranus and Neptune constituted the first positive identification of hydrogen in planetary atmospheres. The first overtone region has also been investigated at very high pressures (2300–4300 atm) at 300 K and the conclusion was reached that the absorption is predominantly quadrupole-induced since no Q-branch splitting was observed as in the fundamental[44]. The enhancement

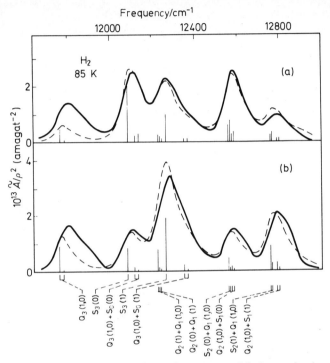

Figure 2.17 The pressure-induced spectrum of hydrogen in the second overtone region: (a) *para*-concentration $C_{para} \approx 0.57$, 85 K; (b) $C_{para} = 0.25$, 85 K. The solid curves are experimental; the dashed curves are theoretical profiles fitted except in the Q_3 region
(From McKellar, A. R. W. and Welsh, H. L.[43], by courtesy of the Royal Society)

of the single $2 \leftarrow 0$ transition of hydrogen by foreign gases has also been studied[100].

Because of the increasing interest of the pressure-induced spectra of hydrogen in planetary atmosphere studies[58, 101, 102] there have been some new laboratory investigations of the first and second overtones. Using a spectrum obtained at 24 K in a path length of 13.6 m Watanabe *et al.*[103] showed that quadrupole induction alone accounts for the observed profile with 14 single and double transitions of different frequencies contributing to the intensity profile. In another investigation McKellar and Welsh[43], using a path length of 137 m, obtained improved spectra over the temperature range, 85–116 K, more applicable to planetary atmosphere studies; three examples for hydrogen with different *ortho/para* ratios are shown in Figure 2.16. The relative intensities of the various transitions shown by the stick spectra were calculated from equation (2.20) using theoretical matrix elements of the H_2 quadrupole moment and polarisability[39, 42]. Each transition was then represented by a Boltzmann-modified dispersion line shape with the half-width δ varied to obtain the best overall fit. The general agreement of the experimental and calculated profiles is satisfactory apart from a small discrepancy in the relative intensities of the single and double transitions. Watanabe[83] has suggested an explanation of the discrepancy in terms of different density dependences for single and double transitions.

The absorption in the second overtone region obtained under the same experimental conditions is illustrated in the two absorption profiles in Figure 2.17; 19 single and double transitions contribute significantly to the intensity. The calculated profile is in reasonable agreement with the observed

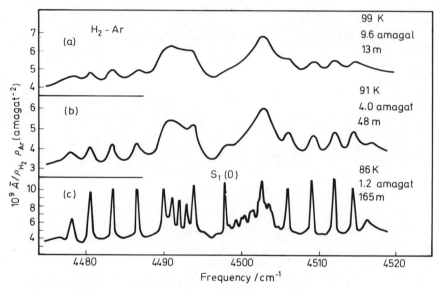

Figure 2.18 The spectrum due to H_2–Ar complexes accompanying the $S_1(0)$ transition in *para*-H_2–Ar mixtures under three different experimental conditions
(From McKellar, A. R. W. and Welsh, H. L.[26], by courtesy of the American Institute of Physics)

spectrum except in the neighbourhood of the $Q_3(0)$ and $Q_3(1)$ transitions. The extra intensity observed experimentally is undoubtedly due to a small amount of overlap induction present in the second overtone but not in the first overtone; the reason for this is not clear.

The $S_3(0)$ component in the spectra of Figure 2.17 is the one which appears in spectra of the major planets[98], the rest of the region being obscured by allowed absorption bands of methane. It will be noted from the calculated spectrum that the $S_3(0)$ component is superimposed on two weaker transitions, $Q_3(1, 0) + S_0(0)$; this probably accounts for the shift of the maximum in the planetary spectrum from the calculated $S_3(0)$ frequency.

The theory of quadrupolar induction not only accounts for the intensity distribution in the overtone bands but also for the absolute intensity of the bands. Thus, for the first overtone region the calculated integrated intensity is 5.23×10^{-9} cm^{-1} amagat^{-2} as compared with the observed value of $\sim 4.9 \times 10^{-9}$. For the second overtone the agreement is not so close but the experimental data are less accurate.

A weak diffuse feature at ~ 6400 Å in spectra of Uranus and Neptune has been ascribed by Spinrad[104] to the $S_4(0)$ feature of the induced third overtone band. This band has not been observed in the laboratory but profiles for various temperatures have been computed[43]. Poll[42] has recently used the estimated strength[105] of the planetary band to compute a value for the H_2 abundance in the atmosphere of Uranus.

2.6 SPECTRA OF H_2-X VAN DER WAALS COMPLEXES; ANISOTROPY OF THE INTERMOLECULAR FORCES

The fine structure accompanying the various components of the H_2 fundamental band (Figure 2.3) is interpreted as arising from transitions between bound states of binary van der Waals complexes; the analysis of these spectra thus furnishes a new quantitative approach to the study of the intermolecular forces between non-polar molecules. The analysis of the $(H_2)_2$ spectrum by Watanabe and Welsh[24] was followed by studies of the H_2-Ar, H_2-N$_2$ and H_2-CO systems by Kudian et al.[25, 106]. Detailed calculations of the energy levels of complexes, based on an isotropic Lennard-Jones potential, were made for $(H_2)_2$ by Gordon and Cashion[107], and for H_2-Ar and the general case by Cashion[108, 109]. A more extensive experimental study of the H_2-Ar, H_2-Kr and H_2-Xe complexes was carried out by Kudian and Welsh[110] and a higher resolution investigation of H_2- and D_2-rare gas systems has recently been completed by McKellar and Welsh[26]. High-resolution studies of $(H_2)_2$, $(D_2)_2$, H_2-D$_2$ [111] and H_2-Ne [112] have also been made.

The experimental condition for observing well-resolved spectra of complexes are rather critical: the temperature of the gas should not greatly exceed ε/k for the complex and the pressure should be kept as low as possible; thus a very long absorption path length at a low temperature is necessary. The effect of the gas density is illustrated by the spectra in Figure 2.18, in which the $S_1(0)$ structures for a para-H_2-Ar mixture at three different densities are shown. The width of the lines decreases approximately proportionally with the density, and an analysis of the half-widths shows that the lifetime is of the order of the collision time, τ_c, as might be expected since ε/k for H_2-Ar is ~ 67 K and the gas temperature was ~ 90 K.

Complete spectra of H_2-Xe complexes are shown in Figure 2.19; the $Q_1(0)$ and $S_1(0)$ spectra were obtained with 98% para-H_2 and the $Q_1(1)$ and $S_1(1)$ spectra with 93% ortho-H_2 as the hydrogen component of the mixture. The overall structure of the bands can be understood from a non-rigid rotator model, the individual lines arising from transitions between the rotational energy states, $F(l) = Bl(l+1) - Dl^2(l+1)^2$, where l is the rotational angular momentum and B and D the rotational constants of the complex. The bands accompanying the $Q_1(1)$, $S_1(0)$ and $S_1(1)$ transitions of the H_2 molecule are quadrupole-induced and the selection rule is $\Delta l = \pm 1, \pm 3$; the band therefore has R(l), P($l+1$) and T(l), N($l+3$) branches, where l is the rotational quantum number in the initial state. For the $Q_1(0)$ band, which is overlap-induced, the selection rule is $\Delta J = \pm 1$ and only P and R branches are observed. The assignments of the transitions are indicated in Figure 2.19. The arrows in Figure 2.19 mark the positions of the H_2 transitions; for the $Q_1(1)$, $S_1(0)$ and $S_1(1)$ transitions these coincide with sharp lines which are due to *quadrupole absorption* of the free H_2 molecule and are not part of the spectrum of the complex.

Although transitions involving l or l' values up to 11 can be identified in the spectra, the higher values correspond to pseudo-bound or virtual states which

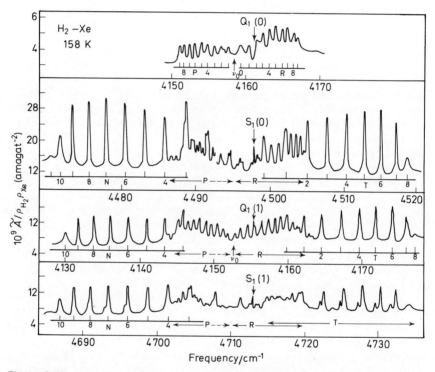

Figure 2.19 Spectra of H_2–Xe complexes accompanying H_2 transitions in the H_2 fundamental band. The gas densities were $\rho_{H_2} = 0.70$, 0.43, 0.42 and 0.42, $\rho_{Xe} = 0.56$, 0.34, 0.52, and 0.52 amagat, for the $Q_1(0)$, $S_1(0)$, $Q_1(1)$, and $S_1(1)$ spectra, respectively (From McKellar, A. R. W. and Welsh, H. L.[26], by courtesy of the American Institute of Physics)

have a shorter lifetime than the bound states, as indicated by the greater breadth of the lines involving transitions to or from these states. These metastable rotational states lie above the asymptote of the intermolecular potential curve and are bound in the effective potential which is the sum of the intermolecular potential energy and the rotational kinetic energy. An inspection of the spectra indicates that the maximum bound-state l value is 8 or 9; calculation with the Lennard-Jones potential gives $l_{max} = 8$.

The splitting of many of the lines in Figure 2.19, which is especially marked in the $S_1(1)$ spectrum, shows that the simple rotator model, which assumes that there is no coupling between the rotational motions of the H_2 molecule and the complex, is inadequate. It is necessary, on the contrary, to assume that the J and l vectors are coupled to give in the limit of weak coupling a total angular momentum j,

$$j = J + l \qquad (2.52)$$

The energy level for a given l is thus split into sub-levels with different j values, the magnitude of the splitting depending on the anisotropy of the intermolecular forces. By applying the selection rules, $\Delta j = 0, \pm 1$ and $\Delta l = \pm 1, \pm 3$ for the quadrupole-induced lines, it is easy to evaluate the multiplicity of the various transitions[110]. Thus, it is found that the T and N lines of the $Q_1(1)$ and $S_1(0)$ bands remain single, in agreement with the observed spectra; the T and N lines of the $S_1(1)$ band and the P and R lines of the $Q_1(1)$, $S_1(0)$ and $S_1(1)$ bands should show varying degrees of multiplicity which is also in qualitative accord with the experimental profiles. The evaluation of the magnitude of the splitting would require a model for the anisotropy of the intermolecular potential. It is evident that the wealth of detail observed in these spectra can be used to check the validity of such models and to provide information on the anisotropy which is difficult to obtain by other methods. The calculations should include the transitions involving the pseudo-bound states of the complex, since these are particularly sensitive to the nature of the intermolecular potential and to its depth.

Since no anisotropic effects can be present for the $Q_1(0)$ spectrum of the complex, this spectrum can be analysed in terms of a simple non-rigid rotator model and an isotropic intermolecular potential. An analysis of the $Q_1(0)$ spectra using Cashion's calculations[109] for a Lennard–Jones 12–6 potential gave values of the constants ε and σ for the H_2–rare gas potential in fair agreement with those obtained from the combining rules. The effective B value obtained from the analysis could be used to calculate R_0, the mean intermolecular distance in the zero-point vibrational state of the complex; for H_2–Ne, H_2–Ar, H_2–Kr and H_2–Xe complexes the values of R_0 were 3.99, 3.94, 4.07 and 4.25 Å, respectively. Finally, it might be noted that no evidence of bound H_2–He complexes has been found, as might be expected from the small depth of the intermolecular potential well and the small masses of the component molecules.

References

1. Crawford, M. F., Welsh, H. L. and Locke, J. L. (1949). *Phys. Rev.*, **75**, 1067
2. Long, C. A. and Ewing, G. E. (1971). *Chem. Phys. Lett.*, **9**, 225
3. Janssen, J. (1885). *Compt. Rend.*, **101**, 642
4. Lutz, B. L. (1969). *J. Chem. Phys.*, **51**, 706
5. Ketelaar, J. A. A. (1959). *Record of Chemical Progress*, **20**, 1
6. Filimonov, V. N. (1959). *Usp. Fiz. Nauk*, **69**, 565 [(1960). *Sov. Phys.-Usp.*, **2**, 565]
7. Colpa, J. P. (1965). *Physics of High Pressures and the Condensed Phase*, 490. (Amsterdam: North-Holland Publishing Company)
8. Tonkov, M. V. (1970). *Spektroskopiya Vzaimodeistvuyushchikh Molekul (Spectroscopy of Interacting Molecules)*, 5 (Leningrad: Leningrad University Press)
9. Hunt, J. L. (1959). *Ph.D. Thesis*. (University of Toronto)
10. Mactaggart, J. W. (1971). *Ph.D. Thesis*. (University of Toronto)
11. Condon, E. U. (1932). *Phys. Rev.*, **41**, 759
12. Crawford, M. F. and Dagg, I. R. (1953). *Phys. Rev.*, **91**, 1569
13. Stoicheff, B. P. (1957). *Can. J. Phys.*, **35**, 730

14. Chisholm, D. A. and Welsh, H. L. (1954). *Can. J. Phys.*, **32,** 291
15. Watanabe, A. and Welsh, H. L. (1965). *Can. J. Phys.*, **43,** 818
16. De Remigis, J., Mactaggart, J. W. and Welsh, H. L. (1971). *Can. J. Phys.*, **49,** 381
17. Van Kranendonk, J. (1968). *Can. J. Phys.*, **46,** 1173
18. Zaidi, H. R. and Van Kranendonk, J. (1971). *Can. J. Phys.*, **49,** 385
19. Welsh, H. L., Crawford, M. F., MacDonald, J. C. F. and Chisholm, D. A. (1951). *Can. J. Phys.*, **83,** 1264
20. Fahrenfort, J. and Ketelaar, J. A. A. (1954). *J. Chem. Phys.*, **22,** 1631
21. Vodar, B. (1959). *Spectrochim. Acta,* **14,** 213
22. Coulon, R., Robin, J. and Vodar, B. (1955). *Compt. Rend.,* **240,** 956
23. Hunt, J. L. and Welsh, H. L. (1964). *Can. J. Phys.*, **42,** 873
24. Watanabe, A. and Welsh, H. L. (1964). *Phys. Rev. Lett.,* **13,** 810
25. Kudian, A., Welsh, H. L. and Watanabe, A. (1965). *J. Chem. Phys.*, **43,** 3397
26. McKellar, A. R. W. and Welsh, H. L. (1971). *J. Chem. Phys.*, **55,** 595
27. Van Kranendonk, J. and Bird, R. B. (1951). *Physica,* **17,** 953; 968
28. Van Kranendonk, J. (1952). *Doctoral dissertation,* University of Amsterdam (unpublished)
29. Van Kranendonk, J. (1957). *Physica,* **23,** 825
30. Van Kranendonk, J. (1958). *Physica,* **24,** 347
31. Van Kranendonk, J. (1959). *Physica,* **25,** 337
32. Van Kranendonk, J. and Kiss, Z. J. (1959). *Can. J. Phys.*, **37,** 1137
33. Poll, J. D. and Van Kranendonk, J. (1961). *Can. J. Phys.*, **39,** 189
34. Mizushima, M. (1949). *Phys. Rev.,* **76,** 1268
35. Mizushima, M. (1950). *Phys. Rev.,* **77,** 149; 150
36. Britton, F. R. and Crawford, M. F. (1958). *Can. J. Phys.*, **36,** 761
37. Colpa, J. P. and Ketelaar, J. A. A. (1958). *Mol. Phys.*, **1,** 343
38. Levine, H. B. (1968). *Phys. Rev. Lett.,* **21,** 1512
39. Karl, G. and Poll, J. D. (1967). *J. Chem. Phys.*, **46,** 2944
40. Birnbaum, A. and Poll, J. D. (1969). *J. Atmos. Sci.,* **26,** 943
41. Dalgarno, A., Allison, A. C. and Browne, J. C. (1969). *J. Atmos. Sci.,* **26,** 946
42. Poll, J. D. (1970). *Proc. I.A.U. Symposium No. 40 on planetary atmospheres, Marfa, Texas, October 1969* (Dordrecht: Reidel)
43. McKellar, A. R. W. and Welsh, H. L. (1971). *Proc. Roy. Soc. London,* **A322,** 421
44. Hare, W. F. J. and Welsh, H. L. (1958). *Can. J. Phys.*, **36,** 88
45. Clouter, M. and Gush, H. P. (1965). *Phys. Rev. Lett.,* **15,** 200
46. Trefler, M., Cappel, A. M. and Gush, H. P. (1969). *Can. J. Phys.*, **47,** 2115
47. Ketelaar, J. A. A., Colpa, J. P. and Hooge, F. N. (1955). *J. Chem. Phys.*, **23,** 413
48. Colpa, J. P. and Ketelaar, J. A. A. (1958). *Mol. Phys.*, **1,** 14
49. Kiss, Z. J., Gush, H. P. and Welsh, H. L. (1959). *Can. J. Phys.*, **37,** 362
50. Kiss, Z. J. and Welsh, H. L. (1959). *Can. J. Phys.*, **37,** 1249
51. Bosomworth, D. R. and Gush, H. P. (1965). *Can. J. Phys.*, **43,** 729; 751
52. Mactaggart, J. W. and Hunt, J. L. (1969). *Can. J. Phys.*, **47,** 65
53. Kiss, Z. J. and Welsh, H. L. (1959). *Phys. Rev. Lett.,* **2,** 166
54. Heastie, R. and Martin, D. H. (1962). *Can. J. Phys.*, **40,** 122
55. Gebbie, H. A., Stone, N. W. B. and Williams, D. (1963). *Mol. Phys.*, **6,** 215
56. Marteau, P., Granier, R., Vu, H. and Vodar, B. (1967). *Compt. Rend.,* **265,** 685
57. Marteau, P., Vu., H. and Vodar, B. (1968). *Compt. Rend.,* **266,** 1068
58. Trafton, L. M. (1964). *Astrophys. J.,* **140,** 1340
59. Trafton, L. M. (1966). *Astrophys. J.,* **146,** 558
60. Trafton, L. M. (1967). *Astrophys. J.,* **147,** 765
61. Linsky, J. L. (1969). *Astrophys. J.,* **156,** 989
62. Olfe, D. B. (1961). *J. Quant. Spectrosc. Radiat. Transf.,* **1,** 104
63. Patch, R. W. (1971). *J. Quant. Spectrosc. Radiat. Transf.,* **11,** 1311, 1331
64. Futrelle, R. P. (1967). *Phys. Rev. Lett.,* **19,** 479
65. Tanimoto, O. (1965). *Progr. Theoret. Phys.,* **33,** 585
66. Levine, H. B. and Birnbaum, G. (1967). *Phys. Rev.,* **154,** 86
67. Levine, H. B. (1967). *Phys. Rev.,* **160,** 159
68. Okada, K., Kajikawa, T. and Yamamoto, T. (1968). *Progr. Theoret. Phys.,* **39,** 863
69. Sears, V. F. (1968). *Can. J. Phys.*, **46,** 1163
70. Sears, V. F. (1968). *Can. J. Phys.*, **46,** 1501

71. McQuarrie, D. A. and Bernstein, R. B. (1968). *J. Chem. Phys.*, **49**, 1958
72. Brenner, S. L. and McQuarrie, D. A. (1971). *Can. J. Phys.*, **49**, 837
73. Huber, D. L. and Van Vleck, J. H. (1966). *Rev. Mod. Phys.*, **38**, 187
74. Welsh, H. L., Crawford, M. F. and Locke, J. L. (1949). *Phys. Rev.*, **76**, 580
75. Crawford, M. F., Welsh, H. L., MacDonald, J. C. F. and Locke, J. L. (1950). *Phys. Rev.*, **80**, 469
76. Coulon, R., Galatry, L., Robin, R. and Vodar, B. (1955). *J. Phys. Radium*, **16**, 728
77. Coulon, R., Robin, J. and Vodar, B. (1955). *Compt. Rend.*, **240**, 956
78. Coulon, R., Galatry, L., Robin, J. and Vodar, B. (1956). *Discuss. Faraday Soc.*, **22**, 22
79. Gush, H. P., Nanassy, A. and Walsh, H. L. (1957). *Can. J. Phys.*, **35**, 712
80. Watanabe, A. and Welsh, H. L. (1967). *Can. J. Phys.*, **45**, 2859
81. Reddy, S. P. and Lee, W. F. (1968). *Can. J. Phys.*, **46**, 1373
82. Varghese, G. and Reddy, S. P. (1969). *Can. J. Phys.*, **47**, 2745
83. Watanabe, A. (1971). *Can. J. Phys.*, **49**, 1320
84. Mactaggert, J. W. and Welsh, H. L. (1972). *Can. J. Phys.*, in the press
85. Reddy, S. P. and Cho, C. W. (1965). *Can. J. Phys.*, **43**, 794
86. Pai, S. T., Reddy, S. P. and Cho, C. W. (1966). *Can. J. Phys.*, **44**, 2893
87. Poll, J. D. (1960). *Ph.D. Thesis.* (University of Toronto)
88. Welsh, H. L. and Hunt, J. L. (1963). *J. Quant. Spectrosc. Radiat. Transfer.*, **3**, 385
89. Nikitin, E. E. (1959). *Opt. Spektrosk.*, **7**, 744 (*Opt. Spectrosc.*, **7**, 441)
90. Nikitin, E. E. (1960). *Opt. Spektrosk.*, **8**, 264 (*Opt. Spectrosc.*, **8**, 135)
91. Muc, A. M., Reesor, G. F. and Dagg, I. R. (1971). *Can. J. Phys.*, **49**, 1970
92. McKellar, A. R. W. (1971). Private communication
93. Lewis, J. C. and Van Kranendonk, J. (1972). *Can. J. Phys.*, in the press
94. Chapman, S. and Cowling, W. T. (1952). *The mathematical theory of non-uniform gases*, (Cambridge: University Press)
95. Mactaggart, J. W., De Remigis, J. and Welsh, H. L. (1972). *Can. J. Phys.*, in the press
96. Desai, R. C. and Yip, S. (1968). *Phys. Rev.*, **166**, 129; *Advances in Chemical Physics*, Vol. 15 (New York: Interscience Publications)
97. Herzberg, G. (1951). *J. Roy. Astron. Soc. Can.*, **45**, 100
98. Herzberg, G. (1952). *Astrophys. J.*, **115**, 337
99. Kuiper, G. P. (1949). *Astrophys. J.*, **109**, 540
100. Ketelaar, J. A. A. and Rettschnick, R. P. H. (1963). *Z. Physik*, **173**, 101
101. Danielson, R. E. (1966). *Astrophys. J.*, **143**, 949
102. Welsh, H. L. (1969). *J. Atmos. Sci.*, **26**, 835
103. Watanabe, A., Hunt, J. L. and Welsh, H. L. (1971). *Can. J. Phys.*, **49**, 860
104. Spinrad, H. (1963). *Astrophys. J.*, **138**, 1242
105. Giver, L. P. and Spinrad, H. (1966). *Icarus*, **5**, 586
106. Kudian, A., Welsh, H. L. and Watanabe, A. (1967). *J. Chem. Phys.*, **47**, 1190
107. Gordon, R. G. and Cashion, J . K. (1966). *J. Chem. Phys.*, **44**, 1190
108. Cashion, J. K. (1966). *J. Chem. Phys.*, **45**, 1656
109. Cashion, J. K . (1968). *J. Chem. Phys.*, **48**, 94
110. Kudian, A. and Welsh, H. L. (1971). *Can. J. Phys.*, **49**, 230
111. McKellar, A. R. W. and Welsh, H. L. (1972). *Can. J. Phys.*, in the press
112. McKellar, A. R. W. and Welsh, H. L. (1972). *Can. J. Phys.*, in the press
113. Poll, J. D. and Miller, M. S. (1971). *J. Chem. Phys.*, **54**, 2673

3
The Stark Effect

A. D. BUCKINGHAM
University Chemical Laboratory, Cambridge

3.1   INTRODUCTION

3.1.1   Historical

The effect of an electric field on optical spectra was first described by Johannes Stark in 1913[1] and is now known as the Stark effect. He was then working at the Technische Hochschule, Aachen. Stark was a prolific writer; in addition to numerous research papers he wrote at least seven books, including one on National Socialism and the Catholic Church[2].

Figure 3.1   Johannes Stark (15 April 1874–21 June 1957)

The original electric-field effects were seen in the emission spectra of atomic hydrogen and helium in fields of 13, 29 and 31 kV cm^{-1}[1]. The β and γ lines of the Balmer series of atomic hydrogen ($n = 4 \rightarrow n = 2$ and $n = 5 \rightarrow n = 2$) were found to split symmetrically into polarised lines by amounts approximately proportional to the electric field strength E (see Figure 3.2, obtained by Stark at a later date[3]). The displacements in helium were not linear in E and the relative intensities of the components in He were shown to vary with E[1]. Independent work by Lo Surdo[4] in Florence confirmed Stark's observations on H; Lo Surdo employed a small discharge tube and studied the emission from the region near the cathode where there is a large non-uniform field (see Figure 3.3).

It must have been natural to expect an effect after Pieter Zeeman had discovered in 1896 that magnetic fields influence spectra[5]. No doubt the 17-year delay between the discovery of the Zeeman and Stark effects was

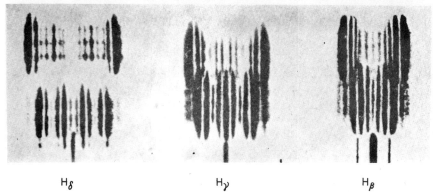

$$H_\delta \qquad\qquad H_\gamma \qquad\qquad H_\beta$$

Figure 3.2 Stark splittings in the Balmer series of the hydrogen atom. This spectrum, attributed to Stark, is reproduced from *J. Franklin Institute*, 1930, **209**, 585. The lowest lines are the field-free spectra (with the H_β line showing Rowland grating ghosts) and the other spectra are presumably for different conditions of polarisation or field strength

due to the rarity of first-order Stark shifts in atomic spectra, in contrast to the abundant first-order Zeeman splittings. An explanation of this difference between electric and magnetic effects is given on page 16.

The first application of quantum-mechanical perturbation theory was to the Stark effect in atomic hydrogen[6, 7]. This was of special interest at the time because the second- and higher-order effects and the intensities differed in quantum mechanics and in the old quantum theory[7, 8].

Earlier reviews of the Stark effect have been published by Stark[9], Foster[10,11], Verleger[12] and Bonch–Bruevich and Khodovoĭ[13].

3.1.2 Significance of the Stark effect

Studies of the Stark effect are of interest primarily because they can yield accurate values of molecular electric dipole moments. The dipole moment is the first moment of the charge distribution in a molecule in a particular vibronic state and gives insight into the structure and properties of the molecule in that state[14]. The effect may also assist in the assignment of complex spectra. The electric field may cause changes in the intensities of spectral lines, and transitions that are normally forbidden may be induced by the field.

Atoms and molecules without permanent dipole moments may exhibit Stark effects. Energy changes due to the field-induced dipole moment, and therefore proportional to E^2, may be measured in favourable cases and interpreted in terms of the polarisability of the molecule in its various vibronic states.

Examples of "new" lines
↓   Blue   Green

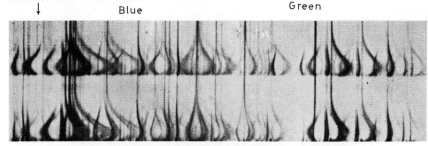

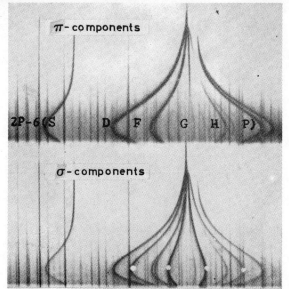

Figure 3.3 The Stark effect in neon and helium in a Lo Surdo cell. The top photograph of the emission of neon was taken by Foster, J. S. and Rowles, W. (*Proc. Roy. Soc. A.*, 1929, **123**, 80) and is reproduced from *J. Franklin Institute*, 1930, **209**, 585. The lower spectrum shows the emission spectrum of He$[6^1(S,D,F,G,H,P) \rightarrow 2^1P]$ near 4144 Å in fields of 0–8.5 MVm^{-1}; it was taken by J. S. Foster and is reproduced from *Can. J. Phys.*, 1959, **37**, 1202
(Photographs by courtesy of the Royal Society and the National Research Council, Canada)

3.1.2.1 Radio-frequency and microwave studies

In the radio-frequency and microwave region, measurements are normally restricted to molecules in the ground electronic state although excited vibrational levels are commonly seen, particularly in molecular-beam electric-resonance experiments[15, 16]. Molecules in metastable states may also be studied; Klemperer and his colleagues have carried out extensive studies on CO in its $a\ ^3\Pi$ state[17-19]. For dipolar asymmetric-rotor molecules it is possible to determine the *components* of the dipole moment along different molecular axes; isotopic substitution is of considerable value as it changes the principal axes of inertia but leaves the charge distribution unaltered. Species of the general formula CH_3COX, where X = H, F and CN, have been studied by Wilson and his colleagues[20-22] and both the *magnitude* and the *components* along molecule-fixed axes of the dipole moment deduced.

By observing the effects of isotopic substitution on molecular rotational magnetic moments it is possible to determine the *sign* of the electric dipole moment[23]. The following signs of dipole moments of molecules in their ground vibronic states have been determined: C^-O^+ [24], H^+I^- [25], H^+F^- [26], Li^+H^- [27], O^-CS^+ [28], O^-CSe^+ [29], $CH_3^+Cl^-$ [30], $CH_3^+CN^-$ [31], $H_2^+CO^-$ [32] and $H^+CF_3^-$ [33]. Most of these signs are in accord with expectation based on rudimentary knowledge of chemical structure. The case of CO is of particular interest, as an approximate Hartree–Fock calculation coupled with an estimate of the effect of correlation yielded approximately the correct magnitude but the wrong sign[34].

Stark effects have been reported in gas-phase electron paramagnetic resonance spectroscopy by Carrington and his colleagues[35] who have measured the dipole moments of a number of unstable free radicals.

3.1.2.2 Optical spectroscopy of gases

In optical spectroscopy the Stark splittings are determined by differences in the interaction of the electric field and the molecule in the appropriate vibronic states. Some early work was carried out on the non-dipolar molecule H_2 [9, 36, 37], which is highly polarisable in some excited electronic states. If first-order Stark splittings, that is, splittings proportional to the field strength, E, are resolved, it is possible to determine both the magnitudes and the relative signs of the dipole moments of the molecule in the appropriate vibronic states. There may be two possible magnitudes, but by studying different rotational lines in the band this ambiguity can be eliminated[38].

Much current work is in progress in this area with the aim of determining the dipole moments of molecules in excited states.

In vibrational spectroscopy, Stark splittings are not normally seen since dipole moment changes are so small (typically of the order of 1 %). However, Maker[39] examined the v_3 fundamental of HCN and identified all M components in the six lines R(2) to P(3); he deduced a dipole moment change $\Delta\mu = +(0.0055 \pm 0.0017)\mu_0$ for the excited v_3 vibrational state of HCN, where μ_0 is the dipole moment of the molecule in the ground state. Maker also studied the v_4 band of CH_3F, v_4 and v_1 of CH_3I, v_1 of NH_3, and the v_1 and v_3 bands of H_2O [39]. These conventional infrared Stark spectra can be useful in making rotational assignments but their scope is limited by the low resolution of conventional spectrometers; the resolution is generally of the same order as the Stark shifts. However, infrared lasers have provided new opportunities and Stark shifts may be used to bring the molecular frequency into coincidence with the laser. By use of this technique, 'Stark modulation spectra' have been observed recently on CH_3—C≡C—H at 3.39 μm [40], on the v_3 band of CH_4 (yielding a dipole moment of 0.024 ± 0.002 D* for the vibrationally excited methane molecule)[41], the v_2 bands of

---

*1D = 1 Debye unit = 10^{-18} e.s.u. = 3.33564×10^{-30} C m. The Debye unit is still in common use and is therefore adopted in this Review.

NH_2D [42, 43], NH_3 and $^{15}NH_3$ [44, 45], and on rotational transitions in D_2O [46] and ND_3 [47] in the far infrared.

The electric field may also affect the intensities of spectral lines. Field-induced atomic lines are well known and are to be found in Figure 3.3. Induced infrared absorption was predicted by Condon[48] who showed that the induced spectrum should exhibit Raman selection rules; this can be appreciated by looking upon the electric field as a static analogue of the optical field causing the Raman transitions. Woodward[49] also proposed the effect. Field-induced vibration–rotation spectra have been recorded in compressed H_2 by Crawford and Dagg[50], Crawford and MacDonald[51], Terhune and Peters[52], and others[52a–e]. Stark modulation spectra of dipolar gases have been recorded in the infrared[39, 39a] and in the ultraviolet on formaldehyde[53, 54] and acrolein vapours[55]. They have also been recorded in the visible region for phenanthrene in a biphenyl crystal at very low temperatures[56].

3.1.2.3   The Kerr effect

The electro-optic Kerr effect, discovered in 1875[57], provided evidence for the existence of the Stark effect[1, 9], and offers another means of studying the interaction of an electric field with molecules in various energy levels. The effect is an induced birefringence proportional to the square of the electric field strength[58]; it leads naturally to electric dichroism, or differential absorbance proportional to the square of the field[59, 60].

Kopfermann and Ladenburg[61] and Bramley[62] studied the Kerr effect near the yellow resonance doublet in sodium vapour. There is a strong effect near the higher frequency line $(3^2P_{\frac{3}{2}} \leftarrow 3^2S_{\frac{1}{2}})$ but none was observed near the other line $(3^2P_{\frac{1}{2}} \leftarrow 3^2S_{\frac{1}{2}})$[61]; this was correctly interpreted in terms of the second-order Stark splitting of the higher frequency line and the absence of any splitting of the other[61]. A magnetic field would split both lines and one would expect a strong Cotton–Mouton effect in the vicinity of both resonance lines[63]. The Kerr effect near the higher frequency resonance line has been predicted to be ~ 2000 larger than that near the other[63, 63a].

The Kerr effect in the infrared in the vicinity of vibrational overtones in liquid nitrobenzene was studied by Charney and Halford[64] who recognised the importance of the technique for determining the polarisation of vibrational transitions. The first high-resolution molecular Kerr spectrum was observed photographically by Freeman and Klemperer[65] in formaldehyde vapour near 3400 Å. More extensive Kerr studies of this molecule have recently been carried out (see Figure 3.4)[60], and the 3821 Å band of propynal ($H—C{\equiv}C—CHO$) has also been photographed[60]. Kerr spectra are generally rather weak but their simplicity can be useful in making rotational assignments.

3.1.2.4   Molecular crystals at low temperatures

Stark effects on the spectra of molecular crystals at low temperatures are being studied extensively by Hochstrasser and his colleagues[56, 66–69]. The

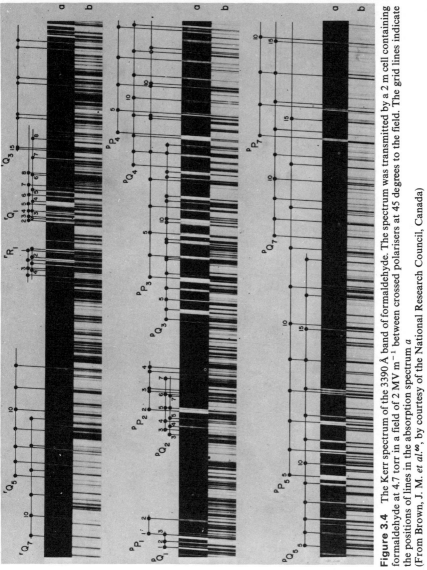

Figure 3.4 The Kerr spectrum of the 3390 Å band of formaldehyde. The spectrum was transmitted by a 2 m cell containing formaldehyde at 4.7 torr in a field of 2 MV m^{-1} between crossed polarisers at 45 degrees to the field. The grid lines indicate the positions of lines in the absorption spectrum *a* (From Brown, J. M. *et al.*[60], by courtesy of the National Research Council, Canada)

sharp lines permit Stark splittings of $c.$ $1-5$ cm^{-1} to be accurately measured. However, there are theoretical difficulties in deducing the 'local' static electric field, leading to uncertainty in the absolute value of the dipole moments of molecules in excited states, although the ratio of the changes in moment in different electronic states should be insensitive to this problem[68]. If the structure is significantly different in the different states, there could be changes in the local field. The Stark effect on colour and impurity centres in crystals has been considered recently by Kaplyanskii, Medvedev and Skvortsov[70]. The effect on dipolar impurities in alkali halide crystals and the interesting topic of 'paraelectric resonance' has been reviewed by Vredevoe, Chilver and Fong[71].

3.1.2.5 Solutions

Electric field effects on the electronic spectra of polar solutes in non-polar solvents have been extensively studied and the results reviewed by Labhart[59, 72] and Liptay[73]. Changes in the positions and intensities of the very broad electronic bands are observed and interpreted to yield values for the change in the dipole moment of the solute on excitation, the direction of the transition moment and components of the transition polarisability (see Section 3.4.8.2)[73]. The effect of the field on the light intensity passing through solutions of large non-polar molecules[74–76] (and polar molecules in rigid organic glasses)[75] have been measured by Stark modulation techniques[72] and the results interpreted in terms of differences in the molecular polarisabilities in the ground and excited electronic states. Bold assumptions about the shapes of these broad bands and about the effects of the environment on the solute are necessary in this work and the results must therefore be treated with caution. However, solution or solid-state studies provide the only practical means of investigating many important non-volatile molecules, and it is also of interest to measure the influence of the environment on molecular dipole moments, polarisabilities and transition moments.

The solvent shifts of electronic absorption bands of polar molecules have been interpreted in terms of a 'solvent Stark effect' to yield changes in the properties of the solute on excitation[77–79]. These results depend on the reaction field model[80, 81] in which the solute molecule polarises its environment, which is usually assumed to form a continuum surrounding some tractable shape for the solute (usually either a sphere or ellipsoid) and to have the bulk properties of the solvent. Such a model can be useful and revealing, but it has obvious limitations.

3.1.2.6 Electric deflexion of molecular beams

Molecules experience a force when in a non-uniform electric field. If a molecule possesses a permanent dipole moment then the energy of some states is lowered and that of others raised by the field. In the former case, the molecule is drawn into regions of stronger field and in the latter it is repelled by the field. If the molecules are non-dipolar, then the energies of all rotational states belonging to the ground vibronic state are lowered by the interaction of the induced moment and the field; this also applies to the lowest state of each multiplicity although excited electronic states may be raised in energy by the field through mixing with lower states, that is, a molecule in an excited state may have a negative polarisability. Hence, if a molecule in its ground vibronic state in a beam in a non-uniform field is deflected into regions of weaker field strength, it must possess a permanent dipole moment. Early electric-deflexion experiments were performed by Kallman and Reiche[82] and Stern[83] (see also Ramsey[15]); the technique has recently been used by Klemperer and his colleagues[84–86] on simple high-temperature species with some very interesting and surprising results. They have shown that lithium halide dimers are planar, Li_2O is linear, the beryllium and magnesium halides are linear, the barium halides are bent, the Group IIB dihalides are linear, the lead dihalides are bent and the dihalides of the transition metals from Mn to Cu are linear[86]. The hexafluorides of S, Se, Mo, Ru, Rh, Te, W, Re, Os, Ir, Pt and U[86a] are non-dipolar, as is XeF_6[86b]; however, IF_7 and ReF_7 behave as non-rigid polar molecules[86c]. A mass spectrometer was used to detect the beams, in some cases after ionisation on a hot tungsten wire[84]. The experiments essentially discriminate between polar and non-dipolar molecules, although care is needed in interpreting the origin of the fragments observed after surface ionisation or electron bombardment in the mass spectrometer. If the permanent dipole is very small, electrostatic deflexion and refocusing of the beam is still possible if there is a first-order Stark effect, as in CH_3D[87]. Electric deflexion may be used to create population inversion between states of opposite parity leading to maser action[87a].

3.1.2.7 Stark effects in rapidly varying fields

If the electric field varies harmonically in time, $E = E_0 \cos 2\pi v_0 t$, new types of Stark effects may be seen if v_0 is much greater than Δv, the half-width at half maximum height, or if v_0 is comparable with an allowed molecular resonant absorption frequency. To understand these effects it is necessary to use Schrödinger's equation including the time

$$(H_0 - \mu_0 E_0 \cos 2\pi v_0 t)\Psi(t) = i\hbar \dot{\Psi}(t)$$

where H_0 is the Hamiltonian operator in the absence of the harmonic field. The problem has been considered theoretically by Blockinzew[88], Autler and Townes[89], Mizushima[90], Bonch-Bruevich and Kodovoĭ[91, 13], Glorieux et al.[92], Macke and Glorieux[93] and Macke[94]; both experimental and theoretical aspects have been described by Townes and Schawlow[95].

If $v_0 \ll \Delta v$, the absorption frequency varies at a frequency v_0 (or $2v_0$ for second-order Stark shifts) between that of the field-free line and that corresponding to a static field E_0; the line appears broadened within the outermost components. If v_0 is much greater than Δv but less than the Stark shift, sidebands are seen at $\pm m v_0$, where m is an integer, and at $\pm 2m v_0$ in the case of second-order Stark effects. The resonant case, $v_0 \approx v_{mn}$, where $h v_{mn} = W_m - W_n$ is the energy difference between the mth and nth levels, was discussed by Autler and Townes[89] and again recently by Macke[94]. Forbidden transitions may be observed in the presence of strong radiation which mixes one of the states through either a one- or two-photon process with a third state[93, 95]; transition-dipole moments may also be measured.

3.2 IONISATION BY ELECTRIC FIELDS

It is interesting to ponder the fact that stationary bound states of atoms and molecules do not exist in a uniform electric field. The metastability is due to a term in the potential energy that varies linearly with the distance z of an electron from the nucleus in the direction of the field E; at large z this term exceeds the attractive potential energy, giving rise to a saddle point in the potential surface. The charged particles may tunnel through the barrier, producing free electrons and positive ions. In a uniform field, the energy approaches minus infinity for infinite z and in such a situation there can be

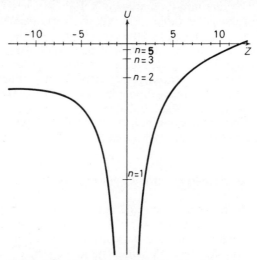

Figure 3.5 The potential energy U of the electron in a H atom as a function of its distance z from a proton in the direction of a uniform field E. The potential is $U = -e^2 r^{-1} (4\pi\varepsilon_0)^{-1} + ezE$ where ε_0 is the permittivity of free space ($4\pi\varepsilon_0 = 1$ e.s.u.). The maximum occurs at $U = -2e^{\frac{3}{2}}E^{\frac{1}{2}}(4\pi\varepsilon_0)^{-\frac{1}{2}}, z = -e^{\frac{1}{2}}E^{-\frac{1}{2}}(4\pi\varepsilon_0)^{-\frac{1}{2}}$, and $U = 0$ at $z = e^{\frac{1}{2}}E^{-\frac{1}{2}}(4\pi\varepsilon_0)^{-\frac{1}{2}}$. The Figure has been drawn for $x = y = 0$ and for a field strength of $\frac{1}{144}$ a.u., that is 3.57×10^9 V m^{-1}; it shows the energy of the unperturbed H atom for principal quantum numbers $n = 1-4$. The distance z is in the atomic units

no discrete spectrum[96]. A two-dimensional representation of the potential of a H atom in a uniform field is illustrated in Figure 3.5. Oppenheimer[97] evaluated the mean lifetime of a H atom in its ground state in an electric field of 100 V m^{-1} and found it to be $10^{10^{10}}$ s; he showed that dissociation of the ground state only becomes appreciable in fields of the order of 10^{12} V m^{-1}. However, as Figure 3.5 discloses, excited states have a much shorter life-expectancy. Early work on this problem has been reviewed by Bethe and Salpeter[98]. Hirschfelder and Curtiss[99] recently looked at the H atom in an excited $n = 5$ state; they used parabolic coordinates to separate the Schrödinger equation in the field and numerically integrated the resulting two one-dimensional equations to obtain the lifetime τ in fields ranging from 8×10^7 to 11×10^7 V m^{-1}; the calculated lifetime varies from 6.97×10^{-8} to 2.54×10^{-14} s[99].

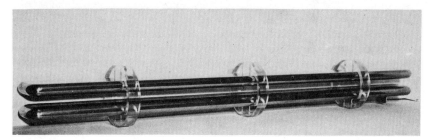

Figure 3.6   A 2 m × 5 cm Stark cell with a gap of 3 cm, no internal spacers and specially rounded edges

(From Buckingham, A. D. et al.[113], by courtesy of the National Research Council, Canada)

3.3   EXPERIMENTAL METHODS

First-order Stark splittings are normally of the order of magnitude of the applied field strength, E, times a molecular dipole moment, μ, ($\mu \sim 1$ D $= 10^{-18}$ e.s.u. $= 3.33564 \times 10^{-30}$ C m); if E is in V m^{-1}, $\mu E \sim 5000E$ Hz. In radio-frequency or microwave spectroscopy the extremely high resolution available means that weak fields ($E < 10^4$ V m^{-1}) can be used. In optical spectroscopy splittings of at least 0.1 cm^{-1} (3 GHz) are desirable and these are normally obtained only in strong fields ($E > 10^6$ V m^{-1}). Stark and Zeeman splittings may conveniently be adjusted to be of the same order of magnitude, that is $\mu E \sim mB$ where m is the magnetic moment ($m \sim 1$ Bohr magnetron $= 9.274 \times 10^{-21}$ e.m.u. $= 9.274 \times 10^{-24}$ C s^{-1} $-$ m^2) and B the magnetic flux density ($\sim 10^4$ gauss $= 1$ T).

The design of Stark cells for use in radio-frequency spectroscopy is discussed by Ramsey[15] and Kusch and Hughes[16]. It is important that the field be uniform in the region in which the radio-frequency power is concentrated. In the most accurate work, dipole moments may be determined to better than 1 part in 10^4 (see, for example, Hebert et al.[100], Muenter[101], de Leeuw and Dymanus[102], Kaiser[103] and Bennewitz et al.[104]).

Stark cells for use in microwave spectroscopy are described in the books by Townes and Schawlow[95] and Gordy, Smith and Trambarulo[105]. In most cases they are suitable only for parallel transitions ($\Delta M = 0$), but Gordy and his colleagues[105-107] have constructed cells which permit both perpendicular ($\Delta M = \pm 1$) and parallel transitions. Lide[108] has also constructed a versatile Stark waveguide for studying perpendicular transitions. Precision cells capable of determining dipole moments to c. 1 part in 10^4 have been constructed by Gordy[106-107] and Laurie[109, 110] and their colleagues. A gas-phase electron-resonance cavity which allows the application of a parallel electric field both for dipole moment determination and Stark modulation has been described by Carrington, Levy and Miller[111].

In optical spectroscopy the high fields required have been produced[38, 39, 44, 112] in cells with small gaps of the order of the mean free path of the gas. Another possible method of avoiding electric breakdown is to use a wide gap (~ 30 mm) and to add a gas (commonly SF_6) to suppress discharges[113] (see Figure 3.6). This approach produces accurately known fields but can suffer from pressure broadening by the foreign gas. An alternative approach is to use the intense fields in the dark space near the cathode of a Lo Surdo cell[10, 12, 114–116] (see Figure 3.7) and to determine the field strength at a particular height above the cathode by use of a calibrant gas (see Figure 3.3). These Lo Surdo-type experiments yield emission spectra and the Balmer lines of the H atom provide a convenient standard for calibration of the field.

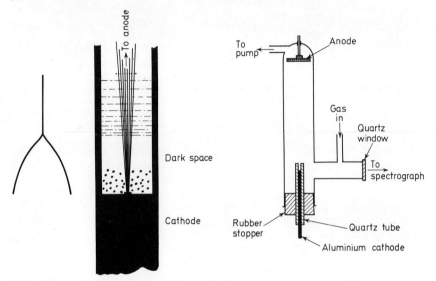

Figure 3.7 The Lo Surdo discharge tube. The sketch on the left shows the region in the Crookes dark space where the electric field is strong (the 'inverted Y' represents a typical emission spectrum in the non-uniform field—see Figure 3.3) and was copied by Foster[10] from a sketch by Lo Surdo, by courtesy of the Franklin Institute and the National Research Council, Canada. The diagram on the right is from Thomson and Dalby[116]

The infrared laser technique of using the Stark effect to bring a molecular frequency into coincidence with a particular laser line (see Section 3.1.2.2) offers the promise of improved resolving power. However, it suffers from the limitation that there must be a nearby laser line. Recently, Oka and Shimizu[117] have observed infrared-microwave two-photon transitions in $^{15}NH_3$. A tunable high-powered microwave source (20 W at 23 GHz) provided power to a 1 m absorption cell containing $^{15}NH_3$ at $1-10^{-2}$ torr in the 10 µm radiation from a CO_2/N_2O gas laser fitted with an adjustable grating for selecting individual vibration–rotation lines. The laser frequency was c. 300 MHz lower than that associated with the transition in $^{15}NH_3$ and under these conditions an intense two-photon absorption was seen; its width was substantially smaller than that of a Doppler-broadened line, presumably because of the non-linearity of the process and the narrowness of the two sources of radiation. Substantially higher resolution (half-width at half-height $\Delta v \sim 100\,\text{kHz}$) can be achieved by use of 'optical–optical double resonance' techniques[118–120], and these have recently been applied to the v_3 band of CH_3F at 9.5 μm by Brewer[121]. The accurately known Stark field was varied until the molecule was brought into resonance with the small difference frequency of two independent but similar CO_2 lasers. The inter-action is non-linear and an extremely narrow line (300 times sharper than the Doppler-broadened line) results from a mechanism similar to that leading to the Lamb dip[122]. Brewer[121] obtained dipole moments of the CH_3F molecule in its ground and v_3 excited (C—F stretch) vibrational states of 1.8596 ± 0.0010 D and 1.9077 ± 0.0010 D respectively; the rotational quantum numbers were $J = 12$, $K = 2$ in both states. Very high resolution infrared Stark spectra have also been observed in NH_2D[42] and in the v_3 band of CH_4[123], yielding a dipole moment of 0.0200 ± 0.0001 D for the vibrationally excited CH_4 molecule (a lower resolution experiment yielded 0.024 ± 0.002 D[41]).

These new non-linear techniques offer exciting prospects in optical spectroscopy and in particular for Stark-effect studies.

3.4   QUANTUM MECHANICAL BACKGROUND

3.4.1   The Schrödinger equation

To understand the Stark effect in a static field we require knowledge of the eigenvalues and eigenfunctions of Schrödinger's equation for the system in a uniform electric field E,

$$H(r, p, E)\Psi_n(r, E) = W_n(E)\Psi_n(r, E) \qquad (3.1)$$

where $H(r, p, E)$ is the Hamiltonian operator for the system in the field and is dependent on the positions r and momenta p of the nuclei and electrons and $\Psi_n(r, E)$ is the eigenfunction associated with the energy level $W_n(E)$.

$$H(r, p, E) = H_0(r, p) - \mu.E \qquad (3.2)$$

where $H_0(r, p)$ is the Hamiltonian when $E = 0$ and μ is the electric dipole moment operator,

$$\mu = \sum_i e_i r_i \qquad (3.3)$$

Even in the case of the H atom, it is not possible to solve equation (3.1) directly, and various approximate procedures must be employed. The standard approach is to use perturbation theory, treating $-\mu.E$ as a perturbation to H_0 to obtain the energy as a power series in E (see Section 3.4.2). However, procedures which do not involve an expansion in powers of E have been used in certain cases, as in the application of the Wentzel–Kramers–Brillouin (WKB) method to the H atom in a strong field[98, 124], and in the 'method of finite perturbations' in which a particular value of E is taken and the variation method used to determine the 'best' wave function in that field[125-129].

3.4.2 Perturbation theory

Normally the field strength E can be varied and the spectrum studied as a function of E. Quantum-mechanical perturbation theory is therefore an appropriate means of determining the effect of the field on the energy levels and wave functions. The interaction with the field, $-\mu.E$, is treated as a perturbation to H_0, and

$$\Psi_n(E) = \Psi_n^{(0)} + \Psi_n^{(1)} + \Psi_n^{(2)} + \dots$$
$$W_n(E) = W_n^{(0)} + W_n^{(1)} + W_n^{(2)} + \dots$$

where the superscript in parenthesis indicates the dependence on E of that term in the series (thus $\Psi_n^{(1)}$ and $W_n^{(1)}$ are the first-order corrections to the wave function and energy and are linear in E). Since Schrödinger's equation (3.1) is true for all E, it entails

$$(H_0 - W_n^{(0)})\Psi_n^{(0)} = 0 \qquad (3.4)$$

$$(H_0 - W_n^{(0)})\Psi_n^{(1)} - \mu.E\Psi_n^{(0)} = W_n^{(1)}\Psi_n^{(0)} \qquad (3.5)$$

$$(H_0 - W_n^{(0)})\Psi_n^{(2)} - (\mu.E + W_n^{(1)})\Psi_n^{(1)} = W_n^{(2)}\Psi_n^{(0)} \qquad (3.6)$$

If $\Psi_n^{(0)}$ is m-fold degenerate, then the function $\Psi_n^{(0)}$ in equations (3.5) and (3.6) must be one of the m-independent linear combinations of the degenerate set in which $H_0 - \mu.E$ is diagonal. The zeroth and first order energies are

$$W_n^{(0)} = \langle \Psi_n^{(0)} | H_0 | \Psi_n^{(0)} \rangle \qquad (3.7)$$

$$W_n^{(1)} = -E\langle \Psi_n^{(0)} | \mu_z | \Psi_n^{(0)} \rangle \qquad (3.8)$$

where we have chosen the axis of quantisation z as the direction of E. The normalised wave function to first order in E is

$$\Psi_n^{(0)} + \Psi_n^{(1)} = \Psi_n^{(0)} + E \sum_{j \neq n} \frac{\langle \Psi_j^{(0)} | \mu_z | \Psi_n^{(0)} \rangle}{W_j^{(0)} - W_n^{(0)}} \Psi_j^{(0)} \qquad (3.9)$$

and the second-order energy is

$$W_n^{(2)} = -\langle \Psi_n^{(0)} | \mu_z E | \Psi_n^{(1)} \rangle$$
$$= -E^2 \sum_{j \neq n} \frac{|\langle \Psi_j^{(0)} | \mu_z | \Psi_n^{(0)} \rangle|^2}{W_j^{(0)} - W_n^{(0)}} \qquad (3.10)$$

where the summations in equations (3.9) and (3.10) over all the unperturbed states except $\Psi_n^{(0)}$ include an integration over the continuum states of the system. The dipole selection rules restrict the total angular momentum quantum numbers of $\Psi_j^{(0)}$ to be the same as those of $\Psi_n^{(0)}$ or to differ from them by ± 1.

If $W_n^{(1)}$ is non-zero, there is said to be a first-order Stark splitting of the nth level; if $W_n^{(1)} = 0$, $W_n^{(2)} \neq 0$ the level exhibits a second-order Stark effect.

Equations (3.8) and (3.10) for $W_n^{(1)}$ and $W_n^{(2)}$ are adequate provided the

Stark splittings and shifts are small compared to the energy difference $W_j^{(0)} - W_n^{(0)}$ between the unperturbed states. If this is not so, the energy levels and wave functions may be obtained through diagonalisation of the matrix of the full Hamiltonian H. In this case the energy may not vary either linearly or quadratically with E, although for very large E an approximate first-order Stark effect may exist. Thus in the case of the H atom in the $n = 2$ state, spin–orbit coupling splits the states by 0.365 cm^{-1} into two quadruplets (see Figure 3.8). The upper ($j = \frac{3}{2}$) are 2s-2p hybrids. If a field E exists such

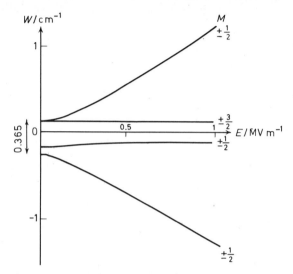

Figure 3.8 The hydrogen atom in the $n = 2$ state in a field E. At $E = 0$, spin–orbit coupling and the Lamb shift splits the states into a lowest $^2P_{\frac{1}{2}}$ state, followed by the $^2S_{\frac{1}{2}}$ and $^2P_{\frac{3}{2}}$ states. In strong fields the levels nearly independent of E are the π states ($l_z = \pm 1$) and these are split by spin–orbit coupling into levels at $\pm\frac{1}{3}$ of the zero-field spin–orbit splitting, that is, ± 0.122 cm^{-1}

that the Stark energy is small compared to the spin–orbit splitting, the lower $j = \frac{1}{2}$ levels show a first-order splitting into two doublets, whereas the upper $j = \frac{3}{2}$ levels experience only second-order effects due to field-dependent mixing with the $j = \frac{1}{2}$ states. If the field strength is very large, then the zero-field splitting can be neglected and there is a first-order splitting of the states $(2s \pm 2p_\sigma)/\sqrt{2}$ but only a very small second-order effect due to mixing with states of other n on the $2p_\pi$ states (spin–orbit coupling splits these states, as shown in Figure 3.8). Actually the $j = \frac{1}{2}$ quadruplet of states is split when $E = 0$ by the Lamb shift into a lower $^2P_{\frac{1}{2}}$ pair and an upper $^2S_{\frac{1}{2}}$ pair. The splitting is only 1058 MHz = 0.0353 cm^{-1} and will have a significant influence on the Stark effect only in fields under about 5×10^4 V m^{-1}; in extremely weak fields there is no first-order splitting of the $j = \frac{1}{2}$ levels but only a second-order effect due to mixing of the $^2S_{\frac{1}{2}}$ and $^2P_{\frac{1}{2}}$ levels by the field. These H-atom effects are well described by Bethe and Salpeter[98], and are illustrated in Figure 3.8.

The case of two nearly degenerate states that are mixed by the field is considered in Section 3.4.10.

3.4.3 Parity

It is of interest to examine the parity of some appropriate operators and eigenfunctions. The parity operator $\mathscr{P}$ inverts the coordinates of all particles of the system, which in our case means the nuclei and electrons of the molecule, but not the fixed charges producing the external field E. For example, under $\mathscr{P}$

$$r \to -r, \, p \to -p, \, H_0 \to H_0 \text{ and } \mu \to -\mu \tag{3.11}$$

The eigenvalues of $\mathscr{P}$ are ± 1, since $\mathscr{P}^2 = 1$.

An operator or function f is said to be of *even* parity if, under $\mathscr{P}$, f $\to$ +f and of *odd* parity if f $\to$ −f. Now $\mathscr{P}$ and H_0 commute, that is

$$\mathscr{P}H_0 - H_0\mathscr{P} = 0 \tag{3.12}$$

since H_0 is determined by p_i^2 and $|\,r_i - r_j\,|$ and is therefore unaffected by inversion of all coordinates. Thus the eigenfunctions $\Psi_n^{(0)}$ of H_0 are also eigenfunctions of $\mathscr{P}$ and are therefore of even or odd parity. Since μ is of odd parity

$$\mu_n^{(0)} = \langle \Psi_n^{(0)} \, | \, \mu \, | \, \Psi_n^{(0)} \rangle = 0 \tag{3.13}$$

that is, no eigenstate of definite parity can have a permanent dipole moment and hence a first-order Stark effect. It is of interest to compare μ with the magnetic moment operator m,

$$m = \sum_i (e_i/2m_i)(r_i \times p_i + g_i s_i) \tag{3.14}$$

where $r_i \times p_i$ and s_i are the orbital and spin angular momentum operators of the ith particle. Since, under $\mathscr{P}$

$$m \to +m \tag{3.15}$$

the parity of m is even, and molecules in states $\Psi_n^{(0)}$ of either even or odd parity may possess permanent magnetic moments.

The eigenfunctions associated with different components Mh of the total angular momentum Jh are of the same parity. For a single particle in a central potential

$$\mathscr{P}Y_{l,m}(\theta,\phi) = Y_{l,m}(\pi - \theta, \pi + \phi) = (-1)^l Y_{l,m}(\theta, \phi) \tag{3.16}$$

and the parity of any state l,m is even if l is even and odd if l is odd. Thus M-degeneracy does not lead to mixed parity and therefore to first-order Stark effects. In any case M remains a good quantum number in any strength of field if the axis of quantisation is chosen as the direction of E.

Hence first-order Stark effects occur only if there exist degenerate or nearly degenerate levels of opposite parity. Thus diatomic molecules in Σ states, for example HCl and CO, show only a second-order Stark effect (there may be a first order Zeeman effect arising from the rotational magnetic moment, since m is of even parity) as do all asymmetric rotor molecules when the asymmetry splitting is large compared to the Stark shift.

3.4.4 Induced dipole moments

The total Hamiltonian $H = H_0 - \mu.E$ does *not* commute with $\mathscr{P}$, since E is not changed by the operation of inversion, so the field destroys parity and if $\Psi_n(E)$ is an eigenfunction of H

$$\mu_n = \langle \Psi_n(E) | \mu | \Psi_n(E) \rangle \neq 0 \qquad (3.17)$$

and induced molecular moments exist. In weak fields the induced dipole is linearly dependent on E and we write

$$\mu_n = \alpha_n.E \qquad (3.18)$$

where the second-rank tensor α_n is called the polarisability of the molecule in the state $\Psi_n^{(0)}$. In strong fields it may be necessary to include higher powers of E and in general

$$\mu_n = \mu_n^{(0)} + \alpha_n.E + \tfrac{1}{2}\beta_n:E^2 + \tfrac{1}{6}\gamma_n \vdots E^3 + \cdots \qquad (3.19)$$

and $\beta_n, \gamma_n, \ldots$ are the first, second, ... hyperpolarisabilities[130, 131].

3.4.5 The hydrogen atom

If $\Psi_n^{(0)}$ is degenerate with functions of opposite parity, the electric field E mixes them and the stationary states are those linear combinations which diagonalise H. If the zero-field splittings due to spin–orbit coupling and the Lamb shift (see Section 3.4.2) are negligible, as in strong fields, the stationary states of the hydrogen atom in the excited $n = 2$ states are the mixed parity-states

$$\Psi_\pm = (\Psi_{2s} \pm \Psi_{2p_z})/\sqrt{2} \qquad (3.20)$$

together with the unperturbed $2p_\pi$ states. The dipole moments in the mixed states are

$$\mu_\pm = \langle \Psi_\pm | \mu_z | \Psi_\pm \rangle = \pm \langle \Psi_{2s} | \mu_z | \Psi_{2p_z} \rangle = \pm 3ea = \pm 7.625 \text{ D} \qquad (3.21)$$

The $2p_\pi$ states are also degenerate with the 2s when $E = 0$ but are unaffected except that in strong fields they are lowered slightly through mixing with higher nd_π states, just as the ground 1s state experiences a second-order lowering through mixing with higher np_z states. The states $\Psi_\pm$ experience first-order Stark shifts, $-\mu_\pm E$. The angular momentum quantum number m remains intact in the presence of the field provided the axis of quantisation is chosen to be the direction of the field (the z axis).

The eigenstates and dipole moments for $n = 3$ in the field are

$$\Psi_{\sigma 0} = \frac{1}{\sqrt{3}} \Psi_{3s} - \frac{2}{\sqrt{3}} \Psi_{3d_{z^2}}, \qquad \mu_{\sigma 0} = 0$$

$$\Psi_{\sigma \pm} = \frac{1}{\sqrt{3}} \Psi_{3s} \pm \frac{1}{\sqrt{2}} \Psi_{3p_z} + \frac{1}{\sqrt{6}} \Psi_{3d_{z^2}}, \qquad \mu_{\sigma \pm} = \pm 9ea$$

$$\Psi_{\pi \pm} = (\Psi_{3p_\pi} \pm \Psi_{3d_\pi})/\sqrt{2}, \qquad \mu_{\pi \pm} = \pm \tfrac{9}{2}ea \qquad (3.22)$$

$$\Psi_\delta = \Psi_{3d_\delta}, \qquad \mu_\delta = 0$$

The general case is best treated by separating the Schrödinger equation in parabolic coordinates[6, 7, 8, 98]. The states with principal quantum number n are split into $2n-1$ energetically distinct levels regularly and symmetrically displaced about the unperturbed energy:

$$W_{nkm}(E) = -e^2(2n^2a)^{-1} + \tfrac{3}{2}n(k_1 - k_2)eaE + O(E^2) \tag{3.23}$$

where the parabolic quantum numbers k_1 and k_2 are zero or positive integers less than $n - |m|$ such that

$$n = k_1 + k_2 + 1 + |m| \tag{3.24}$$

The intensities are governed by the usual polarisation restrictions

$$\begin{aligned} \Delta m &= 0 \quad \text{(parallel polarisation)} \\ \Delta m &= \pm 1 \text{ (perpendicular polarisation)} \end{aligned} \tag{3.25}$$

but there is no selection rule for $k_1 - k_2$, although the transitions which involve a change of sign of $k_1 - k_2$ are weak[6]. Equation (3.23) is generally in good agreement with experiment. Higher order terms in E^2 and E^3 have been evaluated[8, 98].

In weak fields it is necessary to consider the effects of spin–orbit coupling and possibly for $n = 2$, the Lamb shift as well. The appropriate theory is straightforward[98] and has been described briefly in Section 3.4.2.

It should be mentioned that the electric field does not remove all the $2n^2$ degeneracy of the hydrogen atom. In fact, each level in the field is at least doubly degenerate in the absence of a magnetic field, in accordance with Kramers' theorem[132, 133].

Hochstrasser and Zewail[133a] observed a first-order Stark splitting of the degenerate $^1E''(\pi^* \leftarrow n)$ states of s-triazine in a single crystal at 4.2 K when the field is in the plane of the molecule. The effect is of special interest since it arises from a similar source to that in the hydrogen atom, that is, from a field-induced mixing of degenerate orbitals; it is the first example of a first-order Stark effect in a non-polar aromatic molecule.

3.4.6 Other atoms

For atoms with more than one electron the l degeneracy of the one-electron states is lifted by electron–electron repulsion and there is in general no first-order Stark effect. There is a second-order effect which in the case of an atom in an S state can be expressed as

$$W_n(E) = W_n^{(0)} - \tfrac{1}{2}\alpha_n E^2 \tag{3.26}$$

where α_n is the mean polarisability of the atom in the quantum state Ψ_n. In general, the second-order energy may be expressed as[8]

$$W_{nJM}(E) = W_{nJ}^{(0)} - (A_{nJ} - B_{nJ}M^2)E^2 \tag{3.27}$$

where J and M are the total angular momentum quantum numbers. However, it is often preferable to use irreducible spherical tensors and write[134, 135]

$$W_{nJM}(E) = W_{nJ}^{(0)} - \tfrac{1}{2}\alpha_{nJ}E^2 + \tfrac{1}{2}\alpha'_{nJ}[(J^2 + J - 3M^2)J^{-1}(2J-1)^{-1}]E^2 \tag{3.28}$$

where α_{nJ} is the mean, or scalar, polarisability and α'_{nJ} the tensor polarisability of the atom in the states Ψ_{nJ}. If $J = 0$ or $\frac{1}{2}$, $\alpha'_{nJ} = 0$ in agreement with the well-known fact that the total angular momentum quantum number of a system must be equal to or greater than 1 if it is to have anisotropic second-rank tensor properties such as a quadrupole moment or tensor polarisability α'_{nJ}. For $M = \pm J$ the effective polarisability is $\alpha_{nJ} + \alpha'_{nJ}$.

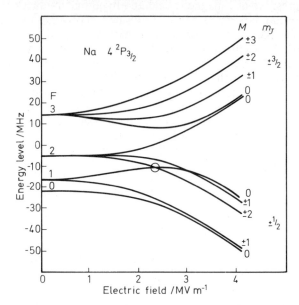

Figure 3.9 Calculated energy levels for the $4\ ^2P_{\frac{3}{2}}$ state of the sodium atom in an electric field. The zero-field splittings arise from the magnetic and quadrupole interaction of the nuclear spin ($I = \frac{3}{2}$) with the electronic angular momentum ($J = \frac{3}{2}$), and depend on the total angular momentum F. M is the component of the total angular momentum and M (but not F) is a good quantum number. Since the $4\ ^2P_{\frac{1}{2}}$ level is lower by 1.69×10^5 MHz, J is approximately conserved. Levels of different M may cross; a crossing of $M = \pm 2$ and $M = 0$ occurs near 2.3 MV m^{-1}.
(From Schmeider, R. W. et al.[136], by courtesy of the American Institute of Physics)

The tensor polarisability α'_{nJ} can be measured in atomic-beam electric-resonance experiments in which radio-frequency power causes $\Delta M = \pm 1$ transitions in a uniform electric field or by level-crossing spectroscopic studies. In recent applications of the latter technique[135, 136] the changes in the resonance fluorescence intensity scattered by an atomic beam in a uniform field when two energy levels cross (see Figure 3.9) was detected to yield accurate values for α'_{nJ} for the second $^2P_{\frac{3}{2}}$ states of K, Rb and Cs. Schmeider, Lurio and Happer[136] have listed calculated and experimental values of both α_{nJ} and α'_{nJ} for the first three $^2P_{\frac{3}{2}}$ states of the five alkali metal atoms Li to Cs.

Atoms in S states must have isotropic polarisabilities, but if they have spin angular momentum, as in metastable helium in its lowest ^3S state, there is presumably a small tensor polarisability since the total angular momentum quantum number is 1 and there is a non-vanishing spin–orbit coupling[134]. Player and Sanders[134a] obtained $\alpha' = 5.07 \pm 0.25 \times 10^{-28}$ cm^3 for He in its lowest ^{3}S state by observing Stark splittings in an atomic-beam resonance experiment.

The general question of the existence of an electric dipole moment of an atom in the presence of a magnetic field and hyperfine interactions has been considered by Sandars[137]; he has incorporated a possible intrinsic dipole moment of the electron. An intrinsic moment would be enhanced by a factor of the order of 10^2 in a highly polarisable atom like Cs or Xe. A recent experiment on Xe in the metastable ^{3}P$_2$ state indicates that the dipole moment of the electron is smaller than $(3 \pm 10) \times 10^{-16}$D [138].

The Stark effect on atoms in a magnetic field was studied extensively by Foster[139]. The isotropic second hyperpolarisabilities γ (see equation (3.18)) of the inert gas atoms He through to Xe have been measured through the Kerr effect[140]. In the case of He several calculations of γ using the variation principle have been performed[126, 127, 141, 142]; there is a small discrepancy with the experimental value[140].

3.4.7 Symmetric rotor molecules

Non-planar symmetric rotor molecules, in which the component $K\hbar$ of angular momentum about the symmetric rotor axis is non-zero, are degenerate, and in the presence of the electric field, are effectively of mixed parity[143, 144] and may exhibit first-order Stark effects. The quantum number K is a signed number so the angular momentum $K\hbar$ about the axis of symmetry may be positive or negative. The states $\pm K$ are degenerate but there is no tendency for the field to mix these states of opposite angular momentum, for $\langle J,K,M|\mu|J',-K,M'\rangle = 0$ except for $K = 0$ or $\pm\frac{1}{2}$ by the well-known dipole selection rule restricting changes in J,K,M to 0 or ± 1. In the case of nearly rigid non-planar symmetric tops, for example CH$_3$F, there is the additional near-degeneracy arising from the inversion motion of the nuclei through the plane perpendicular to the symmetric top axis. This degeneracy doubles the number of states, each of which has a well-defined parity, and leads to first-order Stark splittings when $K \neq 0$. However, this inversion motion in a molecule-fixed frame is not a 'feasible' transformation[144a] in a rigid molecule like CH$_3$F, and it could be said that the parity operation is not appropriate so that Ψ_{JKM} does not have parity for a rigid non-planar symmetric rotor. The situation can be appreciated through considerations of non-rigid molecules such as NH$_3$ (see Section 3.4.7.1).

3.4.7.1 Inversion motion in molecules like NH$_3$

The vibrational motion associated with the movement of the N atom through the plane of the three H atoms, from one equilibrium configuration to the other, causes a doubling of the rotational lines with a doublet splitting equal to the inversion frequency for a full cycle from one equilibrium form to the other and back. The vibrational wave functions describing the lower and upper inversion states are symmetric and antisymmetric with respect to the inversion coordinate, and for particular rotational quantum numbers J, K, M, the two inversion levels are of opposite parity. If the splitting is vanishingly small, as for a rigid symmetric rotor, the states may be considered to have mixed parity and therefore exhibit first-order Stark effects (see Figure 3.10). If the interaction with the field is small compared to the inversion doubling, as in NH_3 in weak electric fields, there is only a second-order Stark effect, although its magnitude is large because of extensive mixing of the inversion pairs[145, 146].

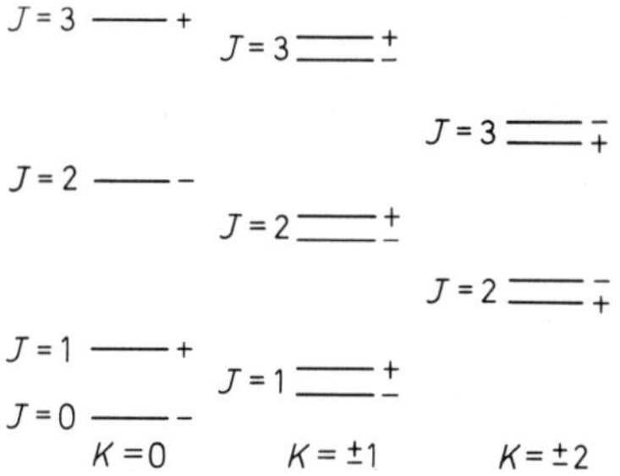

Figure 3.10 Energy levels for the rotation–inversion motion of a non-rigid symmetric-top molecule like NH_3. The + and − labels denote the parity of the states. The electric dipole selection rule restricts transitions to + ⟷ −. There is no splitting of the $K = 0$ states of NH_3 because of nuclear spin symmetry[143]

The dipole selection rules restrict changes in the angular momentum quantum numbers to 0 or ± 1 in a matrix element of $-\boldsymbol{\mu} . \boldsymbol{E}$. Hence a small table of non-vanishing matrix elements can be given. The unperturbed symmetric rotor states are written in the form

$$\Psi_n^{(0)} = \psi_n \Psi_{JKM} = \psi_n \Theta_{JKM}(\theta) e^{iK\chi} e^{iM\phi}$$

where θ, ϕ, χ are the Euler angles describing the rotation of the molecule in space-fixed axes and ψ_n is the 'internal' wave function of the molecule. If there is no interaction of the rotation with the electronic and vibrational degrees of freedom, ψ_n is independent of the rotational quantum numbers, but in the most general case ψ_n depends on J and K.

The matrix elements of the dipole moment operator are proportional to the 'molecular dipole moment' $\mu_n = \langle \psi_n | \mu | \psi_n \rangle$, which is the usual permanent moment relative to molecule-fixed axes, or to the transition moment $\langle \psi_n | \mu | \psi_j \rangle$. These 'internal' moments must be multiplied by direction cosine matrix elements. The non-vanishing elements are given in Table 3.1 which includes elements in $\Delta M = \pm 1$ although these do not enter the perturbation equations (3.8)–(3.10). The phases of the wave functions are identical to those of Townes and Schawlow[147] and are consistent with Condon and Shortley[148].

Table 3.1   Values of the direction cosine matrix element factors

Factor	$J' = J+1$	$J' = J$	$J' = J-1$
$\phi_{J,J'}$	$\frac{1}{4}(J+1)^{-1}(2J+1)^{-\frac{1}{2}}(2J+3)^{-\frac{1}{2}}$	$\frac{1}{4}J^{-1}(J+1)^{-1}$	$\frac{1}{4}J^{-1}(2J-1)^{-\frac{1}{2}}(2J+1)^{-\frac{1}{2}}$
$(\phi_a)_{JK,J'K}$	$2[(J+1)^2 - K^2]^{\frac{1}{2}}$	$2K$	$2[J^2 - K^2]^{\frac{1}{2}}$
$(\phi_b \text{ or } \pm i\phi_c)_{JK,J'K\pm 1}$	$\mp(J+1\pm K)^{\frac{1}{2}}(J+2\pm K)^{\frac{1}{2}}$	$(J\mp K)^{\frac{1}{2}}(J+1\pm K)^{\frac{1}{2}}$	$\pm(J\mp K)^{\frac{1}{2}}(J-1\mp K)^{\frac{1}{2}}$
$(\phi_z)_{JM,J'M}$	$2[(J+1)^2 - M^2]^{\frac{1}{2}}$	$2M$	$2[J^2 - M^2]^{\frac{1}{2}}$
$(\phi_x \text{ or } \mp i\phi_y)_{JM,J'M\pm 1}$	$\mp(J+1\pm M)^{\frac{1}{2}}(J+2\pm M)^{\frac{1}{2}}$	$(J\mp M)^{\frac{1}{2}}(J+1\pm M)^{\frac{1}{2}}$	$\pm(J\mp M)^{\frac{1}{2}}(J-1\mp M)^{\frac{1}{2}}$

[The a, b, c and x, y, z axes are molecule-fixed and space-fixed right-handed frames with a along the molecular symmetry axis and z is the axis of space quantisation. The matrix element of the cosine of the angle between the a, b, or c axis (α) and the x, y, or z axis (i) is the product $\phi_{J,J'}(\phi_\alpha)_{JK,J'K'}(\phi_i)_{JM,J'M'}$.]

3.4.7.2   The perturbed energy

The perturbed energy of a symmetric rotor molecule in a uniform field is

$$W_{jJKM} = W_{jJK}^{(0)} + W_{jJKM}^{(1)} + W_{jJKM}^{(2)} + \cdots \tag{3.30}$$

where

$$W_{jJK}^{(0)} = W_j^{(0)} + B_j J(J+1) + (X_j - B_j)K^2 \tag{3.31}$$

$$W_{jJKM}^{(1)} = -\mu_j E \frac{KM}{J(J+1)} \tag{3.32}$$

$$W_{jJKM}^{(2)} = \frac{\mu_j^2 E^2}{2B_j}\left[\frac{(J^2-K^2)(J^2-M^2)}{J^3(2J-1)(2J+1)} - \frac{[(J+1)^2-K^2][(J+1)^2-M^2]}{(J+1)^3(2J+1)(2J+3)}\right](J\neq 0) \tag{3.33a}$$

$$W_{j000}^{(2)} = -\frac{\mu_j^2 E^2}{6B_j} \tag{3.33b}$$

where $W_j^{(0)}$ is the vibronic energy, $X_j = A_j$ for a prolate and C_j for an oblate symmetric top, and $A_j \geqslant B_j \geqslant C_j$ are the rotational constants (in energy units).

The spectral line arising from the transition $\Psi_{jJ'K'M'} \leftarrow \Psi_{nJKM}$ therefore, has a Stark displacement

$$\Delta W = E\left[\frac{\mu_n KM}{J(J+1)} - \frac{\mu_j K'M'}{J'(J'+1)}\right] + W_{jJ'K'M'}^{(2)} - W_{nJ'KM}^{(2)} + O(E^3) \tag{3.34}$$

For a linear molecule in a Σ state, or for a symmetric top with $K = 0$, there is no first-order Stark effect and

$$\Delta W = \frac{\mu_j^2 E^2}{2B_j}\frac{J^2+J-3M^2}{J(J+1)(2J-1)(2J+3)} + O(E^4) \tag{3.35}$$

The terms in E^4 and E^6 have been evaluated and a continued-fraction formula valid for all fields deduced by W. E. Lamb[16, 95]. This has been used to tabulate the energy levels of a linear molecule in a field[16]. The results are shown in Figure 3.11 for the unperturbed states $J = 0-3$ (see Kusch and Hughes[16] and Maker[39]). Notice the quadratic dependence on E for small E and the maximum in the energy of the first excited state with $M = 0$—this state is initially repelled by the ground state and ultimately lowered in energy by mixing with higher states with $M = 0$. States having different M values may cross (see Figure 3.9).

Equations (3.32) and (3.33) give the Stark energy of a rigid symmetric rotor without hyperfine interactions. Various additional contributions to the energy are now considered.

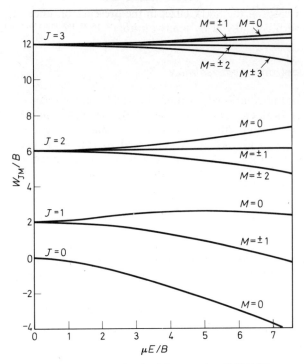

Figure 3.11 The energy levels of the rotor in a field E. The permanent dipole moment is μ and the rotation constant is B. Notice how the lowest state with a particular value of M enjoys a second-order lowering of its energy and that the second state with $M = 0$ is initially raised in energy but is ultimately depressed by mixing with higher states having $M = 0$

3.4.7.3   Effects due to electronic and nuclear spin

The rigid rotor energy levels may be split by a variety of field-independent interactions. For example, in linear molecules having electronic angular momentum there may be Λ-type doubling, and if the total electron spin S is not zero there are spin splittings[149]. There may be hyperfine interactions involving the coupling of the nuclear magnetic moments (if $I \geqslant \frac{1}{2}$) with the rotation of the molecule, with the electronic angular momentum and with themselves, and of the nuclear quadrupole moments (if $I \geqslant 1$) with the local electric field gradient in the rotating molecule[150].

The Stark effect in these molecules is treated theoretically by setting up an appropriate matrix of the full Hamiltonian

$$H = H_{\text{rotation}} + H_{\text{interaction}} - \boldsymbol{\mu}.\boldsymbol{E} \qquad (3.36)$$

The component of the total angular momentum in the direction of E, $M\hbar$, is a constant of the motion, so the matrix is evaluated and diagonalised for each value of M. Trial values of the effective dipole moment may have to be used and the actual value deduced through a comparison with observed spectra.

Particular cases of the Stark effect in the presence of other small interactions have been considered by many authors[15, 16, 35, 103, 113, 136, 145, 151, 152].

3.4.7.4 Electronic polarisability

Distortion of the rotational motion of a rigid rotor molecule leads to the second-order Stark energy in equation (3.33). The full second-order energy in equation (3.10) includes distortion of the internal wave function. Normally the internal states are far apart compared to rotational energies, so the additional energy for the internal state ψ_j may be represented by the effective Hamiltonian $-\frac{1}{2}\boldsymbol{\alpha}_j : \boldsymbol{E}^2$, where $\boldsymbol{\alpha}_j$ is the polarisability of the fixed molecule in the state ψ_j For symmetric rotors this simplifies to

$$H_{\text{polarisability}} = -\tfrac{1}{2}\alpha_j E^2 - \tfrac{1}{3}(\Delta\alpha_j)(\tfrac{3}{2}\cos^2\theta - \tfrac{1}{2})E^2 \qquad (3.37)$$

where α_j is the mean polarisability and $\Delta\alpha_j = (\alpha_{||} - \alpha_{\perp})_j$ is the difference in the polarisability parallel and perpendicular to the symmetric top axis, and θ is the angle between this axis and $\boldsymbol{E}$. To order E^2, the extra energy is

$$\Delta W_{jJKM} = -\tfrac{1}{2}\alpha_j E^2 - \tfrac{1}{3}\Delta\alpha_j E^2 \, \frac{(J^2+J-3K^2)}{J(J+1)} \, \frac{(J^2+J-3M^2)}{(2J-1)(2J+3)} (J \geqslant 1) \qquad (3.38a)$$

$$\Delta W_{j000} = -\tfrac{1}{2}\alpha_j E^2 \qquad (3.38b)$$

If $\alpha_{||}$ and $\alpha_{\perp}$ are positive (which is true for the ground internal state ψ_n), ΔW_{jJKM} is negative for all rotational states ψ_{JKM}, in contrast with the second-order dipolar energy in equation (3.33a) which may be of either sign (see Section 3.1.2.6).

If $\Delta\alpha = 10^{-24}$ cm^3, then $\tfrac{1}{3}\Delta\alpha E^2 = 0.5$ MHz for $E = 3 \times 10^6$ V m^{-1} = 10^2 e.s.u. Thus polarisability contributions to Stark shifts in microwave or radio-frequency experiments are normally rather small, though interesting[110, 153].

3.4.7.5 Non-rigidity

The non-rigidity of a molecule causes the dipole moment to differ in different vibrational and rotational states and may introduce additional terms into the expression for the energy in the field[154]. It is convenient to expand the dipole function μ, that is the dipole moment obtained by averaging over the electronic coordinates for fixed nuclear positions, as a power series in the normal vibrational coordinates. For a diatomic molecule the dipole and the potential are expanded in powers of $\xi = (r - r_e)/r_e$ where r_e is the equilibrium separation of the nuclei:

$$\mu(\xi) = \mu_e + \mu'_e \xi + \tfrac{1}{2}\mu''_e \xi^2 + \cdots \qquad (3.39)$$

The expectation value of the dipole moment for the particular vibration–rotation state ψ_{vJ} may be obtained by perturbation theory and is[155, 156]

$$\mu_{vJ} = \mu_e + (v + \tfrac{1}{2})(B_e/\omega_e)(\mu_e'' - 3a\mu_e') + 4(J^2 + J)(B_e/\omega_e)^2\mu_e' + \cdots \quad (3.40)$$

where B_e and ω_e are the rotational and vibrational constants and the anharmonic constant a is the ratio of the coefficient of the cubic to that of the quadratic term in the potential energy function. In general, μ_e, μ_e', μ_e'', ... are of comparable magnitude, a is very approximately -2 and B_e/ω_e a small number of the order of 10^{-2}–10^{-3}. Hence the relative variation of the dipole moment with v is of the order of 1% and that with J much less except for very high J. For $H^{35}Cl$ in the $v = 0$, $J = 1$ and $v = 0$, $J = 2$ states, Kaiser[103] found a dipole of 1.1085 ± 0.0005 D; for $v = 1$, $J = 1$, 1.1390 ± 0.0010 D, and for $v = 2$, $J = 1$, 1.1685 ± 0.0010 D; for $D^{35}Cl$ in the $v = 0$, $J = 1$ state, the dipole is 1.1033 ± 0.0005 D[103]. These results were combined with absolute infrared intensity data to yield the following dipole parameters for HCl[103]:

$$\mu_e = 1.0933 \pm 0.0005 \text{ D}, \quad \mu_e' = +1.18 \pm 0.03 \text{ D}, \quad \mu_e'' = +0.26 \pm 0.18 \text{ D}.$$

The dipole function for DCl was found to differ slightly from that for HCl and this was attributed to a breakdown of the Born–Oppenheimer approximation. The breakdown of this approximation is solely responsible for the dipole of HD which was found by Trefler and Gush[157] using pure-rotation intensity measurements to be 0.58×10^{-3} D; this value is in approximate agreement with a calculation of Blinder[158] (H^+D^- 0.567×10^{-3} D) but not with the supposedly more accurate evaluation of Kolos and Wolniewicz[159] (1.54×10^{-3} D).

Except for extremely non-rigid molecules like NH_3[44] and HNCO and HN_3[107], the dependence of the dipole moment on the rotational quantum numbers J and K is barely detectable. This is unfortunate since the sign and magnitude of μ_e'/μ_e could be determined from such measurements (see equation (3.40)). Symmetrical non-linear molecules like CH_4 may exhibit moments in excited vibrational[41] or rotational states[160, 161, 161a].

In some molecules, as in CH_3D[87, 162], there is, even within the Born–Oppenheimer approximation, a dipole moment due to isotopic substitution; Wofsy, Muenter and Klemperer[87] found in a molecular-beam electric-resonance experiment dipoles of 5.6409×10^{-3} and 5.6794×10^{-3} D for CH_3D in the $J = 1$, $K = 1$ and $J = 2$, $K = 2$ states. The dipole agrees reasonably well with a simple calculation of Gangemi[163]. The microwave spectrum of acetylene-d_1, H—C≡C—D, was studied by Muenter and Laurie[153] who observed the $J = 1 \leftarrow 0$ transition in a strong field and deduced approximate values for both the μ and $\Delta\alpha$. Deuterium isotope effects on dipole moments have been studied by Lide[164] and Muenter and Laurie[109].

3.4.8 Intensities

3.4.8.1 Unperturbed intensities

The usual dipole selection rules limit changes in the rotational quantum numbers to 0 or ± 1. The relative intensities of the possible transitions in a symmetric rotor can be obtained from Table 3.1 and are of the form $|\langle \psi_j | \mu | \psi_n \rangle|^2$ times the square of the appropriate factors in Table 3.1. The

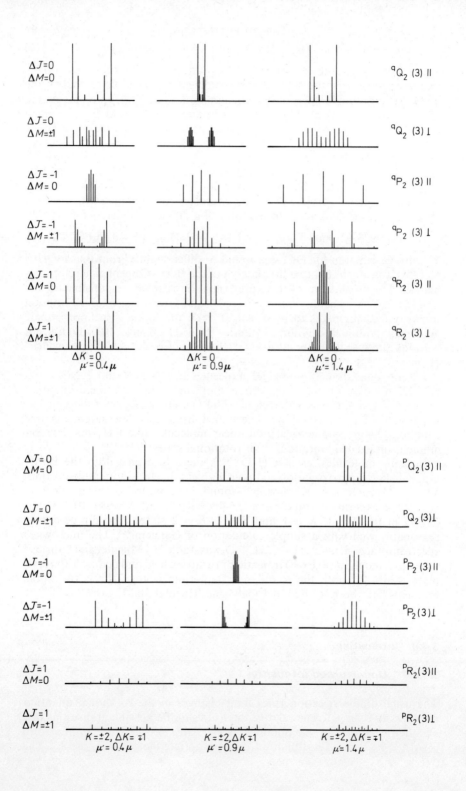

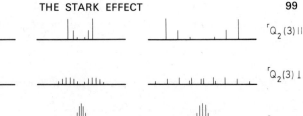
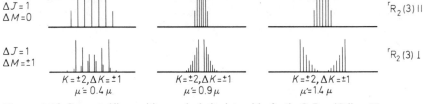

Figure 3.12 Computed line positions and relative intensities for the Q, P and R lines $\Psi_{j,J'K'M'} \leftarrow \Psi_{n,J=3,K=\pm2,M}$ in a symmetric-top molecule for various ratios of the permanent dipole moments $\mu_j/\mu_n = \mu'/\mu$. There is no P branch when $J = 3$, $K = \pm2$ and $\Delta K = \pm1$. The two innermost ${}^rQ_2$ (3) lines in perpendicular polarisation for $\mu'/\mu = 0.4$ are accidentally doubly degenerate, the energies for $M' = \mp3 \leftarrow M = \mp2$ and $M' = \pm2 \leftarrow M = \pm1$ coinciding. The regular spacing of the 12 lines in the ${}^qQ_2$ (3) perpendicular transition, and of the 10 lines in the ${}^pP_2$ (3) perpendicular transition, when $\mu'/\mu = 1.4$, is a consequence of this particular dipole moment ratio.

transition moment $\langle\psi_j|\mu|\psi_n\rangle$ determines whether the electric dipole transition is polarised parallel ($\Delta K = 0$) or perpendicular ($\Delta K = \pm1$) to the axis of the symmetric rotor. For pure rotational transitions $\Delta K = 0$.

For the Q branch in parallel polarisation ($\Delta M = 0$) most of the intensity is in the lines having maximum Stark displacement, and they therefore exhibit strong Stark-modulation signals; in perpendicular polarisation ($\Delta M = \pm1$) the strongest lines tend to have the least Stark displacements and hence weak Stark-modulation signals. In the P and R branches, the position is reversed and the displaced lines have maximum intensity in perpendicular polarisation. Figure 3.12 illustrates the situation for various relative values of the permanent molecular dipole moments in the lower and upper internal states. The clarity of the distinction between Q branch lines on the one hand and P and R lines on the other is seen to vary and to depend on the relative dipole moments. In the case of the ${}^qQ_2$ (3) line, the Stark modulation signal is stronger in perpendicular polarisation when the dipole moments are approximately equal in the two states; this is generally true for ${}^qQ$ lines if $\mu \approx \mu'$, since the Stark splittings are then significant only if $M' \neq M$.

Stark modulation spectra of HCN and other gases in the infrared[39] and on formaldehyde[53, 54] and acrolein[55] in the ultraviolet exhibit the above qualitative features and their relative simplicity can be useful in assigning complicated spectra. The magnetic dipole character of some vibrational bands in the 3500 Å system of formaldehyde[165, 166] has been directly demonstrated by Lombardi, Freeman and Klemperer[167] using the Stark effect. Since the electric and magnetic vectors of an electromagnetic wave are at right angles to each other, the relative intensities of the individual Stark lines in a magnetic dipole transition are opposite to those in an electric dipole transition, that is, in parallel polarisation $\Delta M = 0$ for an electric dipole transition and ±1 for a magnetic dipole transition, and in perpendicular polarisation $\Delta M = \pm 1$ for an electric and 0 for a magnetic dipole transition.

3.4.8.2 Perturbed wave functions and induced intensities

As indicated in equation (3.9), the field distorts the rotational motion of the molecule. In Rayleigh–Schrödinger perturbation theory this is described as a mixing with neighbouring rotational states. The first-order perturbed symmetric-rotor functions are

$$\Psi_{n,J,K,M} = \Psi^{(0)}_{n,J,K,M} - \frac{\mu_n E}{2B_n}\left[\frac{(J^2-K^2)^{\frac{1}{2}}(J^2-M^2)^{\frac{1}{2}}}{J^2(2J-1)^{\frac{1}{2}}(2J+1)^{\frac{1}{2}}}\;\Psi^{(0)}_{n,J-1,K,M} - \right.$$
$$\left. \frac{[(J+1)^2-K^2]^{\frac{1}{2}}[(J+1)^2-M^2]^{\frac{1}{2}}}{(J+1)^2(2J+1)^{\frac{1}{2}}(2J+3)^{\frac{1}{2}}}\;\Psi^{(0)}_{n,J+1,K,M}\right] \qquad (3.41)$$

provided the mixing with other internal states $\Psi_j(j \neq n)$ is insignificant and if the permanent dipole is along the axis of the symmetric top. If the molecule is an asymmetric top whose spectrum approximates that of a symmetric top (as at high K values and for $J \sim K$), it may happen that the dipole has a non-vanishing component at right angles to the top axis; in that case equation (3.9) must be used for the perturbed wave function; the first-order perturbed wave function then incorporates the unperturbed states $K \pm 1$.

The perturbed wave function (equation (3.41)) allows transitions $\Delta J = -2$ or $+2$, that is, O and S branches. For linear dipolar molecules in Σ states, or for symmetric rotors with $K = 0$, there is also an induced Q branch. These electric field induced lines have intensity proportional to the square of E[168].

The theory of field-induced Σ–Σ and Π–Σ transitions in diatomic molecules has been discussed[168]. The induced intensities summed over all M for O and S branch lines in perpendicular polarisation $(\Delta M \pm 1)$ are $\frac{3}{4}$ those for parallel polarisation[168]. The line shapes and intensities to be expected in Stark modulation spectroscopy have been described[39a, 54].

For molecules in states with $K = 0$ (e.g. a diatomic molecule in a Σ state) the molecular-beam electric-resonance experiment[15, 16], in which $\Delta J = 0$ and $\Delta M = \pm 1$, *depends on* electric-field-induced intensity, for there is no unperturbed transition dipole moment (see Table 3.1).

The external field also distorts the internal states of the molecule by mixing into ψ_n some ψ_k if $\langle n|\mu|k\rangle \neq 0$. If $\mu_z = \mu_\alpha a_{z\alpha}$ where $a_{z\alpha}$ is a direction cosine,

$$\Psi_{n,J,K,M} = \Psi^{(0)}_{n,J,K,M} - E\sum_{J_n,K_n,k\neq n}\frac{\langle k|\mu_\alpha|n\rangle}{W^{(0)}_{n,J,K} - W^{(0)}_{k,J_n,K_n}}$$
$$\langle J_n,K_n,M|a_{z\alpha}|J,K,M\rangle\Psi^{(0)}_{k,J_n,K_n,M} + \ldots \qquad (3.42)$$

If the internal state k is far removed in energy from n, the denominator in equation (3.42) can be approximated by $W^{(0)}_n - W^{(0)}_k$. The intensity of the corresponding induced transition polarised in the i direction in space is proportional to

$$|\langle\Psi_{n,J,K,M}|\mu_i|\Psi_{j,J',K',M'}\rangle|^2 = E^2|\langle J,K,M|\alpha^{(nj)}_{iz}|J',K',M'\rangle|^2 \qquad (3.43)$$

where $\qquad \alpha_{\alpha\beta}^{(nj)} = \sum_{k \neq j} \dfrac{\langle n|\mu_\alpha|k\rangle\langle k|\mu_\beta|j\rangle}{W_k^{(0)} - W_j^{(0)}} + \sum_{k \neq n} \dfrac{\langle n|\mu_\beta|k\rangle\langle k|\mu_\alpha|j\rangle}{W_k^{(0)} - W_n^{(0)}}$ (3.44)

is the *transition polarisability tensor*[168–170].

If $j = n$, $\alpha_{\alpha\beta}^{(nn)} = \alpha_{\beta\alpha}^{(nn)}$ is the usual polarisability tensor. For true symmetric rotors (molecules possessing a threefold or higher axis of symmetry), there are at most three independent components of $\alpha^{(nj)}$, namely,

$$\alpha_{33}^{(nj)} = \alpha_{\|}^{(nj)}, \; \alpha_{11}^{(nj)} = \alpha_{22}^{(nj)} = \alpha_{\perp}^{(nj)}, \; \alpha_{12}^{(nj)} = -\alpha_{21}^{(nj)}$$

(the last vanishes for a molecule with a plane of symmetry at right angles to the principal axis[171]). Hence if **1** is the unit vector along the principal axis of the molecule

$$\alpha_{\alpha\beta}^{(nj)} = \alpha_{\perp}^{(nj)}\delta_{\alpha\beta} + (\alpha_{\|}^{(nj)} - \alpha_{\perp}^{(nj)})l_\alpha l_\beta + \tfrac{1}{2}(\alpha_{12}^{(nj)} - \alpha_{21}^{(nj)})\varepsilon_{\alpha\beta\gamma}l_\gamma \qquad (3.45)$$

where $\delta_{\alpha\beta}$ is the Kronecker delta $(= 1$ if $\alpha = \beta$, $= 0$ if $\alpha \neq \beta)$ and $\varepsilon_{\alpha\beta\gamma} = 1$ or -1 if $\alpha\beta\gamma$ is an even or odd permutation of x, y, z and is zero otherwise. When combined with equation (3.43), the first term of (3.45) generates a Q branch in parallel polarisation $(\Delta J = \Delta K = 0, \Delta M = 0)$, the second O, P, Q, R, S branches in parallel and perpendicular polarisation $(\Delta J = -2, -1, 0, 1, 2; \Delta K = 0; \Delta M = 0, \pm 1)$ and the final term P, Q and R branches in perpendicular polarisation $(\Delta J = -1, 0, 1; \Delta K = 0; \Delta M = \pm 1)$.

If ψ_n and ψ_j are vibrational states of the same electronic state, it is convenient to expand $\alpha^{(nj)}$ as a Taylor series about the equilibrium configuration in the normal coordinates. If $j = n$ there may be electric-field-induced pure-rotational transitions and if $j \neq n$, the vibrational fundamentals are by far the most intense transitions. Measurements of the intensities of the Q and S branches in the infrared spectrum of compressed H_2 in strong fields[51, 52a, 52d] has provided values for the two components of the transition polarisability for the vibrational fundamental. These may be converted into the first derivatives of $\alpha_{\|}$ and $\alpha_{\perp}$ with respect to the internuclear distance at the equilibrium separation. The width of the field-induced lines has been studied as a function of pressure[52a–e]; the results provide information relevant to Raman line shapes[52b]. In the case of H_2, the experimental transition polarisabilities are probably less accurate than those deduced from accurate *ab initio* calculations by Kołos and Wolniewicz[172], but electric-field-induced infrared spectra of inactive vibrations in other molecules (e.g. N_2, F_2, CO_2, CH_4) should lead to interesting new information.

An external electric field gradient E' distorts the rotational states through interaction with the molecular quadrupole moment; the intensities of allowed transitions $(\Delta J = 0, \pm 1)$ should show a *linear* dependence on E' and on the permanent quadrupole moment of the molecule in the upper and lower states.

3.4.8.3 Induced Raman lines

The electric field may also affect Raman scattering intensities. The perturbed wave function in equation (3.42) indicates that polar molecules acquire N, O, P, Q, R, S, T branches $(\Delta J = -3, -2, -1, 0, 1, 2, 3)$ with intensities proportional to E^2.

Tetrahedral molecules, such as CH_4 and CF_4, are of special interest since they normally exhibit no pure rotational transitions either in microwave (or far infrared) or Raman spectroscopy, since there is no change in the dipole moment or polarisability as a molecule rotates. In the presence of an electric field E_z the polarisability tensor is changed (see equation (3.19)) to

$$\pi_{n_{\alpha\beta}} = \alpha_{n_{\alpha\beta}} + \beta_{n_{\alpha\beta z}}E_z + \tfrac{1}{2}\gamma_{n_{\alpha\beta zz}}E_z^2 + \cdots \qquad (3.46)$$

where β and γ are first and second hyperpolarisabilities[130, 131]. For tetrahedral molecules there is only one independent component of β, namely $\beta_{123} = \beta$ where the 1, 2, 3 axes are those of a cube in which the four equivalent nuclei are at $(1, 1, 1), (1, -1, -1), (-1, 1, -1), (-1, -1, 1)$. Hence in the field there is a small rotating dipole proportional to βE_z^2 and this would give rise to a very weak far infrared (or microwave) absorption proportional to E_z^4 in which $\Delta J = 0, \pm 1, \pm 2, \pm 3$. There is also a rotating polarisability proportional to βE_z and this yields induced Raman lines ($\Delta J = 0, \pm 1, \pm 2, \pm 3$) whose intensities are proportional to $\beta^2 E_z^2$. Typical magnitudes for $\beta E/\alpha$ might be 10^{-4}–10^{-3} so unfortunately these interesting induced effects, which would yield pure rotational spectra of tetrahedral molecules and a value for β, are likely to be elusive.

Electric fields have been used by Hauchecorne, Kerhervé and Mayer[173] to measure second hyperpolarisabilities γ by inducing second-harmonic scattering from gases and liquids in intense laser beams.

3.4.9 Stark effects in asymmetric rotors

Stark effects in asymmetric rotors are usually second order since the states are of definite parity when $E = 0$. The theory has been thoroughly considered by Golden and Wilson[174]. If near-degeneracies do not occur, the Stark energy is given by equation (3.10) and involves matrix elements of the dipole operator for many rotational states; these are not simple, as for a symmetric rotor, but are well known[95].

The second-order energy is of the form

$$W_{J\tau M}(E) = \tfrac{1}{2}(A+C)J(J+1) + \tfrac{1}{2}(A-C)E_\tau + (F_{J\tau} + G_{J\tau}M^2)E^2 \qquad (3.47)$$

where τ is the usual asymmetric-rotor quantum number (an integer between $-J$ and J), A and C are the largest and smallest of the three rotational constants A, B, C and the functions $E_\tau, F_{J\tau}$ and $G_{J\tau}$ have been tabulated[95, 174]

The Hamiltonian for a rigid rotor in an electric field may be written as

$$H = \tfrac{1}{2}(B+C)P^2 + [A - \tfrac{1}{2}(B+C)]P_a^2 + \tfrac{1}{2}(B-C)(P_b^2 - P_c^2) - \mu_z E_z \qquad (3.48)$$

where P_a, P_b, P_c are the rotational angular momenta (in units of $\hbar$) about the three principal moments of inertia and $P^2 = P_a^2 + P_b^2 + P_c^2$ is the square of total angular momentum. If A is nearly equal to B (as in a near-oblate rotor), the analogous equation

$$H = \tfrac{1}{2}(A+B)P^2 + [C - \tfrac{1}{2}(A+B)]P_c^2 + \tfrac{1}{2}(A-B)(P_a^2 - P_b^2) - \mu_z E_z \qquad (3.49)$$

is more appropriate. If the electric field is not too strong, so that J, in addition to M, remains a good quantum number, the eigenvalues and eigenfunctions may easily be obtained by using the $2J+1$ symmetric-rotor K-states as the basis for the calculation.

The non-vanishing matrix elements are (for the near-prolate case)

$$\langle JKM \mid P^2 \mid JKM \rangle = J(J+1)$$

$$\langle JKM \mid P_a^2 \mid JKM \rangle = K^2$$

$$\langle JKM \mid P_b^2 - P_c^2 \mid J, K \pm 2, M \rangle = \tfrac{1}{2}[\{J(J+1) - K(K \pm 1)\} \\ \{J(J+1) - (K \pm 1)(K \pm 2)\}]^{\frac{1}{2}}$$

$$\langle JKM \mid \mu_z \mid JKM \rangle = \mu_a KM / J(J+1)$$

$$\langle JKM \mid \mu_z \mid J, K \pm 1, M \rangle = \tfrac{1}{2}(\mu_b \mp i\mu_c)M\sqrt{[(J \mp K)(J+1 \pm K)]} / J(J+1)$$

with analogous expressions for a near-oblate rotator.

The theory of the Stark effect in asymmetric rotors has been discussed by Lombardi[175] who shows how Stark patterns in optical spectra can be used to obtain all three components μ_a, μ_b, μ_c of the molecular dipole moment in the excited state and its direction relative to the ground state.

3.4.10 Mixing of nearly degenerate wave functions

If two states ψ_{nJKM} and $\psi_{jJ'K'M'}$ are nearly degenerate the electric field mixes them extensively if $\langle nJKM \mid \mu_z \mid jJ'K'M' \rangle \neq 0$, that is if $J' = J$ or $J \pm 1$, $K' = K$ or $K \pm 1$, $M' = M$ and $\langle n \mid \mu \mid j \rangle \neq 0$. There is no mixing of M states by the field (if the axis of quantisation is chosen to be in the direction of the field) so the eigenvalues and eigenfunctions can be obtained from the solutions of the secular equations. These have a solution only if

$$\begin{vmatrix} W_{nJKM}(E) - W & -\langle nJKM \mid \mu_z \mid jJ'K'M \rangle E \\ -\langle jJ'K'M \mid \mu_z \mid nJKM \rangle E & W_{jJ'K'M}(E) - W \end{vmatrix} = 0 \quad (3.50)$$

where
$$W_{nJKM}(E) = W_{nJK}^{(0)} - \langle nJKM \mid \mu_z \mid nJKM \rangle E$$
$$W_{jJ'K'M}(E) = W_{jJ'K'}^{(0)} - \langle jJ'K'M \mid \mu_z \mid jJ'K'M \rangle E.$$

The two eigenvalues and eigenfunctions are

$$W_{\pm} = \tfrac{1}{2}[W_{nJKM}(E) + W_{jJ'K'M}(E)] \pm C \quad (3.51)$$

$$\left. \begin{array}{l} \Psi_{+} = \cos\theta\,\Psi_{nJKM} + \sin\theta\,\Psi_{jJ'K'M} \\ \Psi_{-} = -\sin\theta\,\Psi_{nJKM} + \cos\theta\,\Psi_{jJ'K'M} \end{array} \right\} \quad (3.52)$$

where the positive quantity

$$C = \tfrac{1}{2}\{[(W_{nJKM}(E) - W_{jJ'K'M}(E)]^2 + 4 \mid \langle nJKM \mid \mu_z \mid jJ'K'M \rangle \mid^2 E^2\}^{\frac{1}{2}}$$

and the angle θ is between 0 and π such that

$$\left. \begin{array}{l} C \cos 2\theta = \tfrac{1}{2}[W_{nJKM}(E) - W_{jJ'K'M}(E)] \\ C \sin 2\theta = -\langle nJKM \mid \mu_z \mid jJ'K'M \rangle E \end{array} \right\} \quad (3.53)$$

If the energy difference $W_{nJK}^{(0)} - W_{jJ'K'}^{(0)}$ is large compared to the Stark energies, $\cos 2\theta \sim 1$ and mixing is slight. If there is no first-order Stark effect on either state, as when the unperturbed states have definite parity,

$$\cos 2\theta = \frac{W_{nJK}^{(0)} - W_{jJ'K'}^{(0)}}{[(W_{nJK}^{(0)} - W_{jJ'K'}^{(0)})^2 + 4|\langle nJKM|\mu_z|jJ'K'M\rangle|^2 E^2]^{\frac{1}{2}}} \qquad (3.54)$$

If the original states are degenerate, $\cos 2\theta = 0$, $\theta = \frac{\pi}{2}$, and the stationary states are equal mixtures of the two basis states of opposite parity. Equation (3.51) shows a smooth change from a second-order to a first-order Stark effect as E increases.

If there are more than two states that are mixed significantly by the field, it is necessary to obtain the roots of a larger secular determinant. Each state has the same value of M, and one seeks the eigenvalues and eigenfunctions of the Hamiltonian matrix:

$$\begin{bmatrix} W_1 - \mu_1 E & -\mu_{1\,2}E & -\mu_{1\,3}E \ldots\ldots \\ -\mu_{2\,1}E & W_2 - \mu_2 E & -\mu_{2\,3}E \ldots\ldots \\ -\mu_{3\,1}E & -\mu_{3\,2}E & W_3 - \mu_3 E \ldots\ldots \end{bmatrix}$$

Normally these would be obtained numerically for each value of M, using trial values for the unknown molecular dipole moment. The stationary states are field-dependent sums of the eigenfunctions of the molecule when $E = 0$.

This mixing of states can lead to the acquisition of intensity by transitions that are normally forbidden. For example, if the molecule is centrosymmetric and if ψ_1 and ψ_2 are of g and u symmetry, then transitions from other g states to state 1, and from u states to state 2, have transition dipole moments proportional to E and intensities proportional to E^2. Electric-field-induced transitions are most intense when the transition from which the intensity is acquired has a similar frequency to the induced line. When a near-degeneracy occurs, the spectra show anomalous Stark effects and analysis of these lines is capable of yielding values for the dipole moment of the molecule in the perturbing state, the transition dipole moment between the interacting states and the separation of the unperturbed energy levels.

Huang and Lombardi[176] have seen an interesting electric-field perturbation in the 0–0 band of the $\pi^* \leftarrow \pi(^1B_2 \leftarrow {}^1A_1)$ transition of benzonitrile near 2738 Å. They observed strong non-linear Stark effects in fields as low as $50 \, kV \, m^{-1}$, determined the dipole moment of the perturbing state of $^1B_2$ symmetry $(1.35 \pm 0.3 \, D)$ and the interaction moment $(1.0 \pm 0.5 \, D)$, and argued that the most probable perturbing electronic state is the previously unobserved excited $^1B_1$ state arising from a $\pi^* \leftarrow n$ transition from the ground $^1A_1$ state. They used a simplified theory in which only the strongest lines were incorporated.

3.5 SOME RECENT EXPERIMENTAL RESULTS

In this final Section some recent results of Stark effect observations are collected and briefly discussed.

Table 3.2   The dipole moments of some diatomic hydrides

Molecule	State	Dipole moment/D	Reference
DH	$X\ ^1\Sigma^+\ (v = 0)$	0.000585 ± 0.000017	157
^{7}LiH	$X\ ^1\Sigma^+\ (v = 0, J = 1)$	5.8820 ± 0.0004	177, 178
		$(\mathrm{Li^+H^-})$	27
	$X\ ^1\Sigma^+\ (v = 1, J = 1)$	5.9905 ± 0.0004	177, 178
^{7}LiD	$X\ ^1\Sigma^+\ (v = 0, J = 1)$	5.8677 ± 0.0005	177, 178
^{6}LiH	$X\ ^1\Sigma^+\ (v = 0, J = 1)$	5.8836 ± 0.0012	177, 178
BeH	$X\ ^2\Sigma^+\ (r = 2.538\ \mathrm{a.u.})$	$0.18\ (\mathrm{Be^+H^-})$	179
BH	$X\ ^1\Sigma^+\ (v = 0)$	1.27 ± 0.21	180
	$A\ ^1\Pi\ (v = 0)$	0.58 ± 0.04	180
CH	$X\ ^2\Pi_{\frac{1}{2}}\ (J = \frac{1}{2})$	1.46 ± 0.06	181
		$(\mathrm{C^-H^+})$	182
NH	$X\ ^3\Sigma^-$	$1.63\ (\mathrm{N^-H^+})$	182, 183
	$A\ ^3\Pi$	1.31 ± 0.03	184
	$c\ ^1\Pi$	1.70 ± 0.07	184
	$a\ ^1\Delta$	1.49 ± 0.06	184
OH	$X\ ^2\Pi_{\frac{1}{2}}\ (v = 0, J = \frac{7}{2})$	1.666 ± 0.010	185
	$X\ ^2\Pi_{\frac{1}{2}}\ (v = 0, J = \frac{1}{2})$	1.72 ± 0.03	186
	$X\ ^2\Pi_{\frac{1}{2}}\ (v = 1, J = \frac{1}{2})$	1.66 ± 0.05	186
	$A\ ^2\Sigma^+\ (v = 0)$	1.98 ± 0.08	116a
FH	$X\ ^1\Sigma^+\ (v = 0, J = 1)$	1.8264 ± 0.0026	187
		$(\mathrm{F^-H^+})$	26
SH	$X\ ^2\Pi_{\frac{3}{2}}\ (J = \frac{3}{2})$	0.62 ± 0.01	35, 188
SeD	$X\ ^2\Pi_{\frac{3}{2}}\ (J = \frac{3}{2})$	0.483 ± 0.003	188
^{35}ClH	$X\ ^1\Sigma^+\ (v = 0, J = 1)$	1.1085 ± 0.0005	103
	$X\ ^1\Sigma^+\ (v = 1, J = 1)$	1.1390 ± 0.0010	103
	$X\ ^1\Sigma^+\ (v = 2, J = 1)$	1.1685 ± 0.0010	103
^{35}ClD	$X\ ^1\Sigma^+\ (v = 0, J = 1)$	1.1033 ± 0.0005	103
	$X\ ^1\Sigma^+\ (v = 1, J = 1)$	1.1256 ± 0.0010	103
^{79}BrH	$X\ ^1\Sigma^+\ (v = 0)$	0.8280 ± 0.0006	189
^{81}BrH	$X\ ^1\Sigma^+\ (v = 0)$	0.8282 ± 0.0006	189
IH	$X\ ^1\Sigma^+\ (v = 0)$	0.4477 ± 0.0005	189
		$(\mathrm{I^-H^+})$	25

The values quoted are from experiment with the exception of BeH and NH which were deduced for the experimental internuclear distance by quantum mechanical calculations[179, 182, 183]. The symbol X represents the ground electron configuration, A, B, C, ... denote excited states of the same multiplicity as the ground state and a, b, c, ... those of different multiplicity.

3.5.1   Diatomic hydrides

In Table 3.2 the dipole moments of diatomic hydrides in various electronic, vibrational and rotational states are listed. These numbers are primarily of interest in that they reflect the distribution of electric charge in the molecule in a particular stationary state. However, they may also be important astrophysically, for they determine absolute probabilities for pure rotational transitions and can therefore be used to determine the concentration of the species in interstellar space or in comets[181, 184]. Dalby in Vancouver has recently measured the dipole moments of several hydrides and other molecules with a cell of the Lo Surdo type. He has observed fields up to $3.2 \times 10^7\ \mathrm{V\ m^{-1}}$ in the dark space near the cathode of the discharge tube, and these strong fields have enabled him to measure second-order Stark effects and hence dipole moments of molecules in Σ states[116a, 180].

The dipole moments of the molecules XH are shown in Figure 3.13 as a function of the atomic number of X for the series Li to F. *Ab initio* calculations approaching Hartree–Fock accuracy have been performed by Cade and Huo[183] on the first- and second-row diatomic hydrides; the computed dipole moments for the equilibrium internuclear distances are rather larger

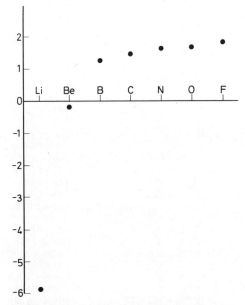

Figure 3.13 The dipole moments of the diatomic hydrides XH in their ground states, where X runs from Li to F. The dipoles are in Debye and are positive in the sense X^-H^+.

in magnitude than the experimental values. Accurate molecular orbital wave functions have also been computed for LiH by Kahalas and Nesbet[190] ($\mu_e = -5.89$ D) and for FH by Clementi[191] ($\mu_e = 1.98$ D, $\mu'_e = 2.9$ D) and Nesbet[192] ($\mu_e = 2.01$ D, $\mu'_e = 1.80$ D—the subscript e denotes the experimental equilibrium internuclear distance). Cade and Huo[183] found that the sign of the dipole moment changed from X^+H^- to X^-H^+ between Groups II and III (i.e. between BeH and BH, and MgH and AlH) in both rows.

3.5.2 Open-shell diatomic molecules

Some Stark effect studies have been performed recently on molecules in metastable or reactive states using molecular-beam electric resonance, microwave spectroscopy or the interesting new techniques of gas-phase electron paramagnetic resonance[35]. Table 3.3 lists results for various diatomic species other than hydrides. Some computed moments have been listed by Byfleet, Carrington and Russell[188]. Self-consistent-field wave functions near the Hartree–Fock limit have been calculated by O'Hare and Wahl for several diatomic fluorides and their positive and negative ions, and dipole and quadrupole moments were evaluated[195, 197, 199, 206]. These calculations provide insight into the charge distribution in a series of similar species; for example, they point out that the trends of the dipole and quadrupole moments of CO, NO and O_2 are continued smoothly through to FO[199].

The dipole moments of the diatomic fluorides CF through to F_2 in their ground states are plotted as a function of atomic number in Figure 3.14.

Table 3.3 The dipole moments of some open-shell diatomic molecules

Molecule	State	Dipole moment D	Reference
CN	$X\ ^2\Sigma^+$	1.45 ± 0.08	116
	$B\ ^2\Sigma^+$	-1.15 ± 0.08	116
CO	$X\ ^1\Sigma^+$	0.111 ± 0.005	193 (106)
		$(C^- O^+)$	24
	$a\ ^3\Pi(\Omega = 2, v = 0, J = 2)$	1.37485 ± 0.0003	18
	$a\ ^3\Pi(\Omega = 2, v = 0, J = 6)$	1.37548 ± 0.0003	18
	$a\ ^3\Pi(\Omega = 2, v = 3, J = 2)$	1.37802 ± 0.0003	18
	$a\ ^3\Pi(\Omega = 1, v = 0, J = 1)$	1.37454 ± 0.0003	18
CF	$X\ ^2\Pi_\frac{3}{2}(J = \frac{3}{2})$	0.65 ± 0.05	194
		$(C^- F^+)$	195
NO	$X\ ^2\Pi_\frac{1}{2}(J = \frac{1}{2})$	0.15872 ± 0.00002	196
NF	$X\ ^3\Sigma^-$	$-0.41\ (N^+F^-)$	197
	$a\ ^1\Delta$	0.37 ± 0.06	198
	$b\ ^1\Sigma^+$	$0.08\ (N^-F^+)$	197
OF	$X\ ^2\Pi$	$0.361\ (O^+F^-)$	199
CS	$X\ ^1\Sigma^+\ (v = 0)$	1.958 ± 0.005	200
	$X\ ^1\Sigma^+(v = 1)$	1.936 ± 0.010	200
	$A\ ^1\Pi(v = 0)$	0.63 ± 0.04	201
	$a'\ ^3\Sigma^+(v = 10)$	1.7 ± 0.4	201
NS	$X\ ^2\Pi$	1.82 ± 0.02	188, 202
OS	$X\ ^3\Sigma^-$	1.55 ± 0.02	203
	$a\ ^1\Delta$	1.32 ± 0.04	188, 204
FS	$X\ ^2\Pi$	0.87 ± 0.05	188
OCl	$X\ ^2\Pi$	1.239 ± 0.010	188, 205
OBr	$X\ ^2\Pi$	1.61 ± 0.04	188
OI	$X\ ^2\Pi$	2.45 ± 0.05	188
OSe	$a\ ^1\Delta$	2.01 ± 0.06	188
FSe	$X\ ^2\Pi$	1.52 ± 0.05	188

All values are from experiment, except those for NF in the Σ states, and for OF, which were deduced from accurate molecular orbital wave functions[197, 199].

3.5.3 Stark effects on $\pi^* \leftarrow n$ transitions

A number of Stark effect studies have been made on high resolution spectral lines in the near ultraviolet resulting from $\pi^* \leftarrow n$ transitions. The change in the molecular dipole moment on excitation is particularly revealing and can provide a useful test of approximate wave functions and of general ideas on charge migration. Table 3.4 shows the molecules and the transitions which have been studied either in the gas phase or in crystals at low temperatures.

In the simple molecular orbital description of the electronic structure of these molecules, the non-bonding electron n is localised on the N,O or S atom whereas the π^* orbital is associated with the unsaturated bonds. The change in dipole moment on excitation in formaldehyde has the expected direction but is smaller in magnitude than that required (~ 3 D) to transfer an electron from the O atom to the mid-point of the CO bond. This is hardly surprising.

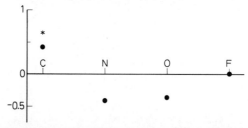

Figure 3.14 The dipole moments of the diatomic fluorides XF in their ground states, where X runs from C to F. The dipoles are in Debye and are positive in the sense X^-F^+. The star denotes Carrington and Howard's experimental value for CF[194] and the dots are results calculated by O'Hare and Wahl[195, 197, 199]

The observed dipoles provide a very valuable check on any approximate wave function. The minimal-basis self-consistent-field wave function of Foster and Boys[218] yields approximately the correct change of dipole[38], although the absolute values are too low; they find that the n orbital contains a significant part of the two H 1s orbitals and is therefore partly de-

Table 3.4 The dipole moments of molecules in their ground and $\pi^* \leftarrow n$ singlet and triplet excited electronic states (in Debye units)

Molecule	μ_0	$\mu_{1_{\pi^*n}}$	$\mu_{3_{\pi^*n}}$	Reference
Formaldehyde	2.33 ± 0.02	1.56 ± 0.07	1.29 ± 0.03	38, 113
Formyl fluoride	0.595 ± 0.006	1.66 ± 0.1	—	207, 208
	(μ_a)	(μ_a)		
Propynal	2.39 ± 0.04	0.7 ± 0.2	—	209, 210
	(μ_a)	(μ_a)		
Difluoro-diazirine	~ 0	1.5 ± 0.2	—	211
Benzonitrile	4.18 ± 0.08	1.35 ± 0.3	—	176
Pyridine	2.19 ± 0.04	$-1.0 \pm 0.1(B)$	—	69
Pyrimidine	2.2	$1.6 \pm 0.1(B)$	$1.2(B)$	212, 213
			0.4	212
Quinoxaline	0.49	$-0.12 \pm 0.05(N)$	—	213, 214
		$-0.16 \pm 0.05(D)$	—	213
Phthalazine	4.8	$1.75 \pm 0.10(N)$	—	214a, 215
Benzophenone	2.98	1.46 ± 0.1	1.79 ± 0.1	67
p,p'-Dichlorobenzophenone	1.6	0.31 ± 0.06	0.92 ± 0.05	213a, 216

The relative signs of the moments in the ground and excited states are determined by the Stark effect observations. In formyl fluoride and propynal in their excited states, only the components of the dipole along the near-symmetric top axis (μ_a) were determined. The first five molecules were studied in the gas phase and the last six in crystals at low temperature. The letters B, D and N in parentheses denote that the Stark measurements were observed in single crystals of benzene (B), durene (D) and naphthalene (N); the absence of a symbol attached to a crystal dipole moment indicates that the value was obtained in the pure crystal. Stark splittings in a crystal yield the magnitude of the change in the dipole moment on excitation; the values quoted for the dipole of a molecule in an excited state in a crystal are based on the assumptions that μ_0 is accurately known and that $\Delta\mu$ is negative. Except where specific references are given the dipole moments of the molecules in their ground states are from the compilation of Nelson, Lide and Maryott[217].

localised. The more elaborate configuration-interaction wave-function of Whitten and Hackmeyer[219] yields dipoles and dipole differences that are too large and it erroneously predicts that the triplet dipole is larger than the singlet. Actually both *ab initio* wave functions were evaluated for the geometry appropriate to the ground state, whereas a significant lengthening of the CO bond occurs and the molecular becomes non-planar[113]. The c component of the dipole moment of H_2CO in the excited A_2 states was not measured since the rapid inversion motion, which causes a splitting of $130 \ cm^{-1}$ in the $^1A_2$ state and $40 \ cm^{-1}$ in $^3A_2$, causes its Stark effect to be of the second order and small. The difference in the dipole moment of formaldehyde in the $^3A_2$ and $^1A_2$ states, $\Delta\mu_a = -0.27\pm0.10$ D, is of interest. It has been rationalised by supposing that the 'relaxation' of the electrons to a lower energy distribution in the triplet state will normally be such that it tends to reduce the existing charge asymmetry[113]. This point has recently been examined by Liebman[220]. Dixon, using semi-empirical wave functions, predicted the correct value of $\Delta\mu_a$, although his spin constants are not accurate[113, 221]. The larger dipole moment change in propynal[209] is consistent with the greater distance of the centroid of the π^* orbital from the O atom. The change in formyl fluoride, coupled with the shift of the $\pi^* \rightarrow n$ absorption to higher frequency, presents problems for the simple molecular orbital description[207]. The dipoles are believed to have the sense of O^-F^+ so the increase on excitation could be explained by the n orbital having a substantial amount of F character; however, since the n orbital on the F atom has a lower energy than that on O, this interaction would cause a red shift. F contributions to the π^* orbital would cause a blue shift but a decrease in dipole moment[207]. A compromise mixing of F atomic states with both the n and π orbitals might explain both facts.

In interpreting the Stark spectra of crystals one must remember the problems associated with the 'local field' (see Section 3.1.2.4). In the triplet state of pyrimidine there is a dependence of the dipole on the crystalline environment[212, 213], but in the quinoxaline singlet almost the same dipole is deduced in naphthalene and in durene[212, 213]. The triple-state dipole is smaller than the singlet in pyrimidine, but this is not so for the benzophenones. The larger change of dipole moment in phthalazine compared to quinoxaline[213] is of interest and is consistent with the bigger movement of the electron from the n to the π^* orbital expected on naive grounds.

Stark studies of $\pi^* \leftarrow \pi$ transitions have also been carried out at high resolution. Lombardi and his colleagues have measured Stark effects in the rotational fine structure of the ultraviolet spectra of aromatic molecules. They found increases in the dipole moment on excitation of 0.30 ± 0.07 D for fluorobenzene[222], 0.25 for chlorobenzene[223], 0.2 ± 0.2 for phenol[224], 0.85 ± 0.15 for aniline[224], 0.44 ± 0.10 for p-fluorophenol[225], 0.82 ± 0.09 for p-fluoroaniline[225], 0.36 ± 0.10 for p-chlorofluorobenzene[226], $\mu_a = 0.20\pm0.05$ D for m-chlorofluorobenzene[227] and 0.13 D for styrene[228], and discovered an interesting excited-state interaction effect in benzonitrile (see Section 3.4.10)[176]. The singlet and triplet $\pi^* \leftarrow \pi$ transitions in phenanthrene have been studied in a biphenyl single crystal at 1.8 K by Hochstrasser and Noe[56], who observed Stark modulation of the fluorescence and phosphorescence; they deduced dipole moment changes of 0.34 ± 0.05 D for the singlet and $\leqslant 0.12$ D for the triplet. They also observed azulene in naphthalene at

4.2 K and deduced dipole changes of -1.21 ± 0.05 and -1.10 ± 0.05 D for the first $(^1B_1)$ and second $(^1A_1)$ $\pi^* \leftarrow \pi$ excited singlet states[68].

A correlation between the dipole and structural changes occurring on $\pi^* \leftarrow \pi$ excitation has been proposed and explored by Lombardi[229].

3.5.4   Determination of polarisability anisotropies

Very accurate measurement of second-order Stark shifts can yield the anisotropy in the polarisability, $\Delta\alpha$ (see Section 3.4.7.4). The results are most accurate if the dipole moment is small. A number of slightly polar molecules have been studied by Laurie and his colleagues. The $J = 1 \leftarrow 0$ transition of H—C$\equiv$C—D was observed by Muenter and Laurie[230] in fields of 8.2–11 MV m^{-1}; by assuming that the permanent moment is equal to the difference in the moments of CH_3—C$\equiv$C—D and CH_3—C$\equiv$C—H (0.012 D) they deduced that $\Delta\alpha = 1.85 \pm 0.05 \times 10^{-24}$ cm^3. The ratio of the $\Delta\alpha$ to the μ^2/B contribution to the second-order Stark shift is in this case $-\Delta\alpha/2\mu^2 B^{-1} = -1.27$; this would increase if higher energy transitions were studied. Scharpen, Muenter and Laurie[110, 231] have made very accurate measurements on OCS, NNO and $CD_3C\equiv C$—H, using a microwave spectrometer incorporating a special parallel-plate absorption cell capable of sustaining large fields that are uniform in the absorption region. They obtain $\Delta\alpha(OCS) = 4.67 \pm 0.16 \times 10^{-24}$ cm^3, $\Delta\alpha(N_2O) = 3.222 \pm 0.046 \times 10^{-24}$ cm^3 and $\Delta\alpha(CD_3C_2H) = 2.7 \pm 0.6 \times 10^{-24}$ cm^3. They also found $\mu(N_2O) = 0.160830 \pm 0.000016$ D. $\mu(CD_3C_2H) = 0.78780 \pm 0.00013$ D and evidence of a small decrease in the dipole of propyne on increasing K. These polarisability anisotropies are those appropriate to static electric fields and include the atomic polarisability[154], unlike the optical methods for measuring $\Delta\alpha$. English and MacAdam[232] obtained $\Delta\alpha = 0.2897 \times 10^{-24}$ cm^3 by observing the Stark effect in a molecular-beam magnetic-resonance experiment on D_2 in the $J = 1$ rotational state. Muenter[232a] has found $\Delta\alpha(HF) = 0.22 \pm 0.02 \times 10^{-24}$ cm^3 using the molecular-beam electric-resonance method.

Polarisability anisotropies of atoms have been measured in level-crossing experiments[135–136] (see Section 3.4.6). Angel, Sandars and Woodgate[233] determined the quadrupole moment of the Al atom in the $^2P_{\frac{3}{2}}$ state by observing resonance frequency shifts of c. 100 Hz in a field gradient of 4×10^8 V m^{-2} in an atomic-beam apparatus. The second-order Stark effect complicates the experiment but can be separated by taking advantage of the fact that the quadrupole interaction changes sign with the voltage, unlike the second-order Stark effect. The atomic quadrupole moment, defined as $-e\langle JJ|\Sigma_i(\frac{3}{2}z_i^2 - \frac{1}{2}r_i^2)|JJ\rangle$, was found to be 2.53 ± 0.15 e.a.$^2 = 3.40 + 0.20 \times 10^{-26}$ e.s.u. $= 11.35 \pm 0.67 \times 10^{-40}$ C m^2, which is approximately 10% lower than the Hartree–Fock value[233, 234].

Eisenthal and Rieckhoff[235–237] claim to have measured the optical polarisability anisotropy of naphthalene $-d_8$ in a triplet excited state in a glass matrix at 77 K by observing the change in the refractive index induced in the glass by an intense mercury–xenon arc source which populated the long-lived $^3B_{1u}$ state of naphthalene by intersystem crossing from the excited $^1B_{1u}$

state. Polarisability anisotropies of naphthalene in the same glass have also been determined by Kuball et al.[238, 239] through a Kramers–Kronig transformation of the triplet–triplet absorption coefficient; they believe that the anisotropy determined by Eisenthal and Rieckhoff is too small. This is not strictly an application of the Stark effect, but it is of interest since it provides another source of knowledge of polarisability changes on excitation.

3.5.5 Tabulations of dipole moments

The dipole moments of many free molecules in their ground electronic states have been accurately determined through Stark-effect measurements, principally in the microwave and radio-frequency regions or through dielectric polarisation studies. Dielectric measurements may also be made on dilute solutions in non-polar solvents and 'solution dipole moments' deduced. These are useful, though they may differ somewhat from gas-phase values[14]. There exist valuable compilations of dipole moments, in particular those of Nelson, Lide and Maryott[217] and McClellan[216]. The tabulated moments determined through Stark effect measurements relative to OCS should be multiplied by 1.0038 to take account of the re-determinations of μ(OCS) by Muenter[101] and de Leeuw and Dymanus[102].

3.6 CONCLUSION

The previous Sections describe some of the progress that has been made since Stark's and Lo Surdo's initial observations on the H atom in 1913. The most important application in chemistry is undoubtedly to the determination of electric dipole moments of molecules in their ground and excited states. Dipole moments reflect the electron distribution in the molecule, and differences in values from state to state provide the basis for our 'feeling' for charge migration and for physical and chemical properties. The moment $\mu_n^{(0)}$ is determined by the eigenfunction $\Psi_n^{(0)}$; it therefore provides an important test of the accuracy of approximate wave functions and a close link to quantum chemical calculations.

There are other important uses of the Stark effect—the determination of polarisabilities, the assignment of spectral lines, field-induced spectra, the separation and alignment of molecules, Stark modulation and tuning. There will surely be new ones, for electric interactions are so fundamental and experimental techniques are developing so fast.

Acknowledgements

The author is particularly grateful to Dr. D. A. Ramsay for assistance in preparing this review. He also acknowledges the help of R. G. Brewer, J. M. Brown, A. Carrington, M. G. Clark, F. W. Dalby, W. Gordy, R. M. Hochstrasser, W. Klemperer, V. W. Laurie, D. R. Lide, J. R. Lombardi, J. S. Muenter, T. Oka and of the Medical and Technical Publishing Company.

References

1. Stark, J. (1913). *Sitz. Akad. Wiss. Berlin*, **47**, 932
2. Stark, J. (1931). *Nationalsozialismus und Katholische Kirche*. (Munchen: Verlag Frz. Eher Nachf.)
3. See Plate I in Foster, J. S. (1930). *J. Franklin Inst.*, **209**, 585
4. Lo Surdo, A. (1913). *Atti Reale Accademia Lincei*, **22**, (Part 2), 664
5. See the *Proceedings of the Zeeman Centennial Conference* published in 1967, *Physica*, **33**, 1–293
6. Schrödinger, E. (1926). *Ann. Physik*, **80**, 457
7. Epstein, P. S. (1926). *Phys. Rev.*, **28**, 695
8. Condon, E. U. and Shortley, G. H. (1935). *Theory of Atomic Spectra*, Chap. 17. (Cambridge: Cambridge University Press)
9. Stark, J. (1927). *Handbuch der Experimentalphysik*. Vol. 21, 399 (Wien, W. and Harms, F. Eds.). (Leipzig: Akademische Verlagsgesellschaft)
10. Foster, J. S. (1930). *J. Franklin Institute*, **209**, 585
11. Foster, J. S. (1939). *Reports Progress in Physics*, **5**, 233
12. Verleger, H. (1939). *Ergeb. exakt Naturwiss.*, **18**, 99
13. Bonch-Bruevich, A. M. and Khodovoǐ, V. A. (1967). *Usp. Fiz. Nauk*, **93**, 71 [(1968). *Soviet Phys. Uspekhi*, **10**, 637]
14. Buckingham, A. D. (1970). *Physical Chemistry. An Advanced Treatise*, Vol. 4, 349. (Eyring, H., Henderson, D. and Jost, W., Eds.). (New York: Academic Press)
15. Ramsey, N. F. (1956). *Molecular Beams*, Chap. 10. (Oxford: Oxford University Press)
16. Kusch, P. and Hughes, V. W. (1959). *Handbuch der Physik*, Vol. 37, Part I, 1 (Flugge, S. Ed.). (Berlin: Springer)
17. Freund, R. S. and Klemperer, W. (1965). *J. Chem. Phys.*, **43**, 2422
18. Stern, R. C., Gammon, R. H., Lesk, M. E., Freund, R. S. and Klemperer, W. (1970). *J. Chem. Phys.*, **52**, 3467
19. Gammon, R. H., Stern, R. C., Lesk, M. E., Wicke, B. G. and Klemperer, W. (1971). *J. Chem. Phys.*, **54**, 2136; Gammon, R. H., Stern, R. C. and Klemperer, W. (1971). *J. Chem. Phys.*, **54**, 2151
20. Kilb, R. W., Lin, C. C. and Wilson, E. B. (1957). *J. Chem. Phys.*, **26**, 1695
21. Pierce, L. and Krisher, L. C. (1959). *J. Chem. Phys.*, **31**, 875
22, Krisher, L. C. and Wilson, E. B. (1959). *J. Chem. Phys.*, **31**, 882
23. Townes, C. H., Dousmanis, G. C., White, R. L. and Schwarz, R. F. (1955). *Discuss. Faraday Soc.*, **19**, 56
24. Rosenblum, B., Nethercot, A. H. and Townes, C. H. (1958). *Phys. Rev.*, **109**, 400
25. Burrus, C. A. (1959). *J. Chem. Phys.*, **30**, 976; **31**, 1270
26. Ramsey, N. F. (1961). *American Scientist*, **49**, 509
27. Lawrence, T. R., Anderson, C. H. and Ramsey, N. F. (1963). *Phys. Rev.*, **130**, 1865
28. Flygare, W. H., Hüttner, W., Shoemaker, R. L. and Foster, P. D. (1969). *J. Chem. Phys.*, **50**, 1714
29. Shoemaker, R. L. and Flygare, W. H. (1970). *Chem. Phys. Lett.*, **6**, 576
30. Vanderhart, D. L. and Flygare, W. H. (1970). *Molec. Phys.*, **18**, 77
31. Pochan, J. M., Shoemaker, R. L., Stone, R. G. and Flygare, W. H. (1970). *J. Chem. Phys.*, **52**, 2478
32. Hüttner, W., Lo, M.-K. and Flygare, W. H. (1968). *J. Chem. Phys.*, **48**, 1206
33. Flygare, W. H. and Benson, R. C. (1971). *Molec. Phys.*, **20**, 225
34. Nesbet, R. K. (1964). *J. Chem. Phys.*, **40**, 3619
35. Carrington, A., Levy, D. H. and Miller, T. A. (1967). *J. Chem. Phys.* **47**, 3801
36. MacDonald, J. K. L. (1929). *Proc. Roy. Soc. A.*, **123**, 103; (1931). ibid., **131**, 146
37. Foster, J. S., Jones, D. C. and Neamtan, S. M. (1937). *Phys. Rev.*, **51**, 1029
38. Freeman, D. E. and Klemperer, W. (1966). *J. Chem. Phys.*, **45**, 52
39. Maker, P. D. (1961). *Stark Effects in the Near Infrared Spectra of Simple Polyatomic Molecules*, Ph.D. Thesis, University of Michigan
39a Willson, P. L. and Edwards, T. H. (1971). *J. Chem. Phys.*, **55**, 5191
40. Feld, M. S., Parks, J. H., Schlossberg, H. R. and Javan, A. (1966). *Physics of Quantum Electronics.*, 567 (Kelly, P. L., Lax, B. and Tennenwald, P. E., Eds.). (New York: McGraw-Hill)
41. Uehara, K., Sakurai, K. and Shimoda, K. (1969). *J. Phys. Soc. Jap.*, **26**, 1018

42. Brewer, R. G., Kelly, M. J. and Javan, A. (1969). *Phys. Rev. Lett.*, **23,** 559
43. Brewer, R. G. and Swalen, J. D. (1970). *J. Chem. Phys.*, **52,** 2774
44. Shimizu, F. (1969). *J. Chem. Phys.*, **51,** 2754; (1970). ibid., **52,** 3572; **53,** 1149
45. Shimizu, T. and Oka, T. (1970). *Phys. Rev. A.*, **2,** 1177
46. Duxbury, G. and Jones, R. G. (1971). *Molec. Phys.*, **20,** 721
47. Duxbury, G. and Jones, R. G. (1971). *Chem. Phys. Lett.*, **8,** 439
48. Condon, E. U. (1932). *Phys. Rev.*, **41,** 759
49. Woodward, L. A. (1950). *Nature (London)*, **165,** 198
50. Crawford, M. F. and Dagg, I. R. (1953). *Phys. Rev.*, **91,** 1569
51. Crawford, M. F. and MacDonald, R. E. (1958). *Can. J. Phys.*, **36,** 1022
52. Terhune, R. W. and Peters, C. W. (1959). *J. Mol. Spectrosc.*, **3,** 138
52a. Foltz, J. V., Rank, D. H. and Wiggins, T. A. (1966). *J. Mol. Spectrosc.*, **21,** 203
52b. Brannon, P. J., Church, C. H. and Peters, C. W. (1968). *J. Mol. Spectrosc.*, **27,** 44
52c. Hunt, R. H., Barnes, W. L. and Brannon, P. J. (1970). *Phys. Rev. A.*, **1,** 1570
52d. Buijs, H. L. and Gush, H. P. (1971). *Can. J. Phys.*, **49,** 2366
52e. Robertson, C. W., Hunt, R. H. and Brannon, P. J. (1971). *J. Chem. Phys.*, **55,** 4149
53. Buckingham, A. D. and Ramsay, D. A. (1965). *J. Chem. Phys.*, **42,** 3721
54. Bridge, N. J., Haner, D. A. and Dows, D. A. (1968). *J. Chem. Phys.*, **48,** 4196
55. Bridge, N. J., Haner, D. A. and Dows, D. A. (1966). *J. Chem. Phys.*, **44,** 3128
56. Hochstrasser, R. M. and Noe, L. J. (1970). *Chem. Phys. Lett.*, **5,** 489
57. Kerr, J. (1875). *Phil. Mag.* [4], **50,** 337, 446
58. Kerr, J. (1880). *Phil. Mag.* [5], **9,** 157
59. Labhart, H. (1967). *Advan. Chem. Phys.*, **13,** 179
60. Brown, J. M., Buckingham, A. D. and Ramsay, D. A. (1971). *Can. J. Phys.*, **49,** 914
61. Kopfermann, H. and Ladenburg, R. (1925). *Ann. Physik.* [4], **78,** 659
62. Bramley, A. (1928). *J. Franklin Inst.*, **205,** 539
63. Bogaard, M. P., Buckingham, A. D. and Orr, B. J. (1967). *Molec. Phys.*, **13,** 533
63a. Bogaard, M. P. and Orr, B. J. (1968). *Molec. Phys.*, **14,** 557
64. Charney, E. and Halford, R. S. (1958). *J. Chem. Phys.*, **29,** 221
65. Freeman, D. E. and Klemperer, W. (1964). *J. Chem. Phys.*, **40,** 604
66. Hochstrasser, R. M. and Noe, L. J. (1968). *J. Chem. Phys.*, **48,** 514; Hochstrasser, R. M. and Lin, T.-S. (1968). *J. Chem. Phys.*, **49,** 4929; (1970). ibid., **52,** 1624
67. Hochstrasser, R. M. and Noe, L. J. (1971). *J. Molec. Spectrosc.*, **38,** 175
68. Hochstrasser, R. M. and Noe, L. J. (1969). *J. Chem. Phys.*, **50,** 1684
69. Hochstrasser, R. M. and Michaluk, J. W. (1971). *J. Chem. Phys.*, **55,** 4668
70. Kaplyanskii, A. A., Medvedev, V. N. and Skvortsov, A. P. (1970). *Optics and Spectrosc.*, **29,** 481
71. Vredevoe, L. A., Chilver, C. R. and Fong, F. K. (1971). *Progr. Solid State Chem.*, **5,** 431
72. Labhart, H. (1961). *Chimia*, **15,** 20
73. Liptay, W. (1969). *Angew. Chem.*, **81,** 195, (English edition, **8,** 177)
74. Malley, M., Feher, G. and Mauzerall, D. (1968). *J. Molec. Spectrosc.*, **25,** 544
75. Seibold, K., Navangul, H. and Labhart, H. (1969). *Chem. Phys. Lett.*, **3,** 275
76. Varma, C. A. G. O. and Oosterhoff, L. J. (1971). *Chem. Phys. Lett.*, **8,** 1
77. Lippert, E. (1955). *Z. Naturforsch.*, **10a,** 541; (1957). *Z. Electrochem.*, **61,** 962
78. Liptay, W. (1965). *Z. Naturforsch.*, **20a,** 1441
79. Baur, M. E. and Nicol, M. (1966). *J. Chem. Phys.*, **44,** 3337
80. Onsager, L. (1936). *J. Amer. Chem. Soc.*, **58,** 1486
81. Linder, B. (1967). *Advan. Chem. Phys.*, **12,** 225
82. Kallmann, H. and Reiche, F. (1921). *Z. Physik.*, **6,** 352
83. Stern, O. (1922). *Physik. Z.*, **23,** 476
84. Wharton, L., Berg, R. A. and Klemperer, W. (1963). *J. Chem. Phys.*, **39,** 2023
85. Büchler, A., Stauffer, J. L., Klemperer, W. and Wharton, L. (1963). *J. Chem. Phys.*, **39,** 2299
86. Büchler, A., Stauffer, J. L. and Klemperer, W. (1964). *J. Amer. Chem. Soc.*, **86,** 4544
86a. Kaiser, E. W., Muenter, J. S., Klemperer, W., Falconer, W. E. and Sunder, W. A. (1970). *J. Chem. Phys.*, **53,** 1411
86b. Falconer, W. E., Büchler, A., Stauffer, J. L. and Klemperer, W. (1968). *J. Chem. Phys.*, **48,** 312
86c. Kaiser, E. W., Muenter, J. S., Klemperer, W. and Falconer, W. E. (1970). *J. Chem. Phys.*, **53,** 53

87. Wofsy, S. C., Muenter, J. S. and Klemperer, W. (1970). *J. Chem. Phys.*, **53**, 4005
87a. Townes, C. H. and Schawlow, A. L. (1955). *Microwave Spectroscopy*, 432. (New York: McGraw-Hill)
88. Blokhintzev, D. (1933). *Phys. Z. U.S.S.R.*, **4**, 501
89. Autler, S. H. and Townes, C. H. (1955). *Phys. Rev.*, **100**, 703
90. Mizushima, M. (1964). *Phys. Rev.*, **133**, A414
91. Bonch-Bruevich, A. M. and Khodovoi, V. A. (1965). *Usp. Fiz. Nauk.*, **85**, 3 [*Soviet Phys. Uspekhi*, **8**, 1]
92. Glorieux, P., Legrand, J., Macke, B. and Messelyn, J. (1972). *J. Quant. Spectrosc. Radiative Transfer*, **12**, 731
93. Macke, B. and Glorieux, P. (1971). *J. Physique*. In the press
94. Macke, B. (1972(. *Nuovo Cimento*, **8B**, 321
95. Townes, C. H. and Schawlow, A. L. (1955). *Microwave Spectroscopy*, Chap. 10. (New York: McGraw-Hill)
96. Weyl, H. (1910). *Math. Ann.*, **68**, 220
97. Oppenheimer, J. R. (1928). *Phys. Rev.*, **31**, 66
98. Bethe, H. A. and Salpeter, E. E. (1957). *Quantum Mechanics of One- and Two-Electron Atoms*, Chap. IIIb. (Berlin: Springer-Verlag)
99. Hirschfelder, J. O. and Curtiss, L. A. (1971). *J. Chem. Phys.*, **55**, 1395
100. Hebert, A. J., Lovas, F. J., Melendres, C. A., Hollowell, C. D., Story, T. L. and Street, K. (1968). *J. Chem. Phys.*, **48**, 2824
101. Muenter, J. S. (1968). *J. Chem. Phys.*, **48**, 4544
102. de Leeuw, F. H. and Dymanus, A. (1970). *Chem. Phys. Lett.*, **7**, 288
103. Kaiser, E. W. (1970). *J. Chem. Phys.*, **53**, 1686
104. Bennewitz, H. G., Haerten, R., Klais, O. and Müller, G. (1971). *Chem. Phys. Lett.*, **9**, 19
105. Gordy, W., Smith, W. V. and Trambarulo, R. F. (1953). *Microwave Spectroscopy*, Chap. 1. (New York: Wiley)
106. Bhattacharya, B. N. and Gordy, W. (1960). *Phys. Rev.*, **119**, 144
107. White, K. J. and Cook, R. L. (1967). *J. Chem. Phys.*, **46**, 143
108. Lide, D. R. (1964). *Rev. Sci. Instr.*, **35**, 1226
109. Muenter, J. S. and Laurie, V. W. (1966). *J. Chem. Phys.*, **45**, 855
110. Scharpen, L. H., Muenter, J. S. and Laurie, V. W. (1970). *J. Chem. Phys.*, **53**, 2513
111. Carrington, A., Levy, D. H. and Miller, T. A. (1967). *Rev. Sci. Instr.*, **38**, 1183
112. Lombardi, J. R. (1969). *J. Chem. Phys.*, **50**, 3780
113. Buckingham, A. D., Ramsay, D. A. and Tyrrell, J. (1970). *Can. J. Phys.*, **48**, 1242
114. Lo Surdo, A. (1914). *Phys. Z.*, **15**, 122
115. Foster, J. S. (1927). *Proc. Roy. Soc. A.*, **114**, 47
116. Thomson, R. and Dalby, F. W. (1968). *Can. J. Phys.*, **46**, 2815
116a. Scarl, E. A. and Dalby, F. W. (1971). *Can. J. Phys.*, **49**, 2825
117. Oka, T. and Shimizu, T. (1971). *Appl. Phys. Lett.*, **19**, 88
118. Schlossberg, H. R. and Javan, A. (1966). *Phys. Rev. Lett.*, **17**, 1242
119. Notkin, G. E., Rautian, S. G. and Feoktistov, A. A. (1967). *Soviet Phys. J.E.T.P.*, **25**, 1112
120. Feld, M. S. and Javan, A. (1969). *Phys. Rev.*, **177**, 540
121. Brewer, R. G. (1970). *Phys. Rev. Lett.*, **25**, 1639; (1971). *Proceedings of the Esfakan, Iran, Symposium on Fundamental and Applied Laser Physics*, to be published
122. Lamb, W. E. (1964). *Phys. Rev.*, **134**, A1429
123. Luntz, A. C. and Brewer, R. G. (1971). *J. Chem. Phys.*, **54**, 3641
124. Lanczos, C. (1930). *Z. Physik*, **65**, 431
125. Cohen, H. D. and Roothaan, C. C. J. (1965). *J. Chem. Phys.*, **43**, S34
126. Cohen, H. D. (1965). *J. Chem. Phys.*, **43**, 3558; (1966). ibid., **45**, 10
127. Grasso, M. N., Chung, K. T. and Hurst, R. P. (1968). *Phys. Rev.*, **167**, 1 (and references therein)
128. Schweig, A. (1967). *Chem. Phys. Lett.*, **1**, 163, 195
129. Pople, J. A., McIver, J. W. and Ostlund, N. S. (1968). *J. Chem. Phys.*, **49**, 2960
130. Buckingham, A. D. and Pople, J. A. (1955). *Proc. Phys. Soc. A.*, **68**, 905
131. Buckingham, A. D. and Orr, B. J. (1967). *Quart. Rev. Chem. Soc.*, **21**, 195
132. Kramers, H. A. (1930). *Proc. Koninkl. Ned. Akad. Wetenschap.*, **33**, 959
133. Wigner, E. P. (1959). *Group Theory and its Application to the Quantum Mechanics of Atomic Spectra*. (New York: Academic Press)
133a. Hochstrasser, R. M. and Zewail, A. H. (1971). *Chem. Phys. Lett.*, **11**, 157

134. Angel, J. R. P. and Sandars, P. G. H. (1968). *Proc. Roy. Soc. A.,* **305,** 125
134a. Player, M. A. and Sanders, P. G. H. (1969). *Phys. Lett.,* **30A,** 475
135. Khadjavi, A., Lurio, A. and Happer, W. (1968). *Phys. Rev.,* **167,** 128
136. Schmieder, R. W., Lurio, A. and Happer, W. (1971). *Phys. Rev. A.,* **3,** 1209
136a. Franken, P. A. (1961). *Phys. Rev.,* **121,** 508
137. Sandars, P. G. H. (1968). *J. Phys. B.,* **1,** 499, 511
138. Player, M. A. and Sandars, P. G. H. (1970). *J. Phys. B.,* **3,** 1620
139. Foster, J. S. and Pounder, E. R. (1947). *Proc. Roy. Soc. A.,* **189,** 287
140. Buckingham, A. D. and Dunmur, D. A. (1968). *Trans. Faraday Soc.,* **64,** 1776
141. Sitz, P. and Yaris, R. (1968). *J. Chem. Phys.,* **49,** 3546
142. Buckingham, A. D. and Hibbard, P. G. (1968). *Symposia Faraday Soc.,* **2,** 41
143. Townes, C. H. and Schawlow, A. L. (1955). *Microwave Spectroscopy,* Chap. 3. (New York: McGraw-Hill)
144. Herzberg, G. (1966). *Electronic Spectra and Electronic Structure of Polyatomic Molecules.* (New York: Van Nostrand)
144a. Longuet-Higgins, H. C. (1963). *Molec. Phys.,* **6,** 445
145. Jauch, J. M. (1947). *Phys. Rev.,* **72,** 715
146. Coles, D. K., Good, W. E., Bragg, J. K. and Sharbaugh, A. H. (1951). *Phys. Rev.,* **82,** 877
147. Townes, C. H. and Schawlow, A. L. (1955). *Microwave Spectroscopy,* Chap. 4. (New York: McGraw-Hill)
148. Condon, E. U. and Shortley, G. H. (1935). *Theory of Atomic Spectra,* Chap. 3. (Cambridge: Cambridge University Press)
149. Raynes, W. T. (1964). *J. Chem. Phys.,* **41,** 3020
150. Townes, C. H. and Schawlow, A. L. (1955). *Microwave Spectroscopy,* Chaps. 6 and 8. (New York: McGraw-Hill)
151. Low, W. and Townes, C. H. (1949). *Phys. Rev.,* **76,** 1295
152. Buckingham, A. D. and Stephens, P. J. (1964). *Molec. Phys.,* **7,** 481
153. Muenter, J. S. and Laurie, V. W. (1964). *J. Amer. Chem. Soc.,* **86,** 3901
154. Wharton, L. and Klemperer, W. (1963). *J. Chem. Phys.,* **39,** 1881
155. Schlier, C. (1961). *Fortshr. Physik,* **9,** 455
156. Buckingham, A. D. (1962). *J. Chem. Phys.,* **36,** 3096
157. Trefler, M. and Gush, H. P. (1968). *Phys. Rev. Lett.,* **20,** 703
158. Blinder, S. M. (1961). *J. Chem. Phys.,* **35,** 974
159. Kolos, W. and Wolniewicz, L. (1966). *J. Chem. Phys.,* **45,** 944
160. Fox, K. (1971). *Phys. Rev. Letters,* **27,** 233
161. Watson, J. K. G. (1971). *J. Molec. Spectrosc.,* **40,** 536. See also Oka, T., Shimizu, F. O., Shimizu, T. and Watson, J. K. G. (1971). *Astrophys. J.,* **165,** L15
161a. Ozier, I. (1971). *Phys. Rev. Letters,* **27,** 1329
162. Ozier, I., Ho, W. and Birnbaum, G. (1969). *J. Chem. Phys.,* **51,** 4873
162a. Hirota, E. and Matsumura, C. (1971). *J. Chem. Phys.,* **55,** 981
163. Gangemi, F. A. (1963). *J. Chem. Phys.,* **39,** 3490
164. Lide, D. R. (1960). *J. Chem. Phys.,* **33,** 1519
165. Sidman, J. W. (1958). *Chem. Revs.,* **58,** 689
166. Callomon, J. H. and Innes, K. K. (1963). *J. Molec. Spectrosc.,* **10,** 166
167. Lombardi, J. R., Freeman, D. E. and Klemperer, W. (1967). *J. Chem. Phys.,* **46,** 2746
168. Dows, D. A. and Buckingham, A. D. (1964). *J. Molec. Spectrosc.,* **12,** 189
169. Buckingham, A. D. and Dows, D. A. (1963). *Discuss. Faraday Soc.,* **35,** 48
170. Liptay, W., Dumbacher, B. and Weisenberger, H. (1968). *Z. Naturforsch.,* **23a,** 1601
171. Birss, R. R. (1963). *Reports Progress Phys.,* **26,** 307
172. Kolos, W. and Wolniewicz, L. (1967). *J. Chem. Phys.,* **46,** 1426
173. Hauchecorne, G., Kerhervé, F. and Mayer, G. (1971). *J. Phys.,* **32,** 47
174. Golden, S. and Wilson, E. B. (1948). *J. Chem. Phys.,* **16,** 669
175. Lombardi, J. R. (1968). *J. Chem. Phys.,* **48,** 348
176. Huang, K.-T. and Lombardi, J. R. (1971). *J. Chem. Phys.,* **55,** 4072
177. Wharton, L., Gold, L. P. and Klemperer, W. (1960). *J. Chem. Phys.,* **33,** 1255
178. Rothstein, E. (1969). *J. Chem. Phys.,* **50,** 1899
179. Chan, A. C. H. and Davidson, E. R. (1968). *J. Chem. Phys.,* **49,** 727
180. Thomson, R. and Dalby, F. W. (1969). *Can. J. Phys.,* **47,** 1155
181. Phelps, D. H. and Dalby, F. W. (1966). *Phys. Rev. Lett.,* **16,** 3
182. Huo, W. M. (1968). *J. Chem. Phys.,* **49,** 1482

183. Cade, P. E. and Huo, W. M. (1966). *J. Chem. Phys.,* **45,** 1063
184. Irwin, T. A. R. and Dalby, F. W. (1965). *Can. J. Phys.,* **43,** 1766
185. Powell, F. X. and Lide, D. R. (1965). *J. Chem. Phys.,* **42,** 4201
186. Phelps, D. H. and Dalby, F. W. (1965). *Can. J. Phys.,* **43,** 144
187. Weiss, R. (1963). *Phys. Rev.,* **131,** 659
188. Byfleet, C. R., Carrington, A. and Russell, D. K. (1971). *Molec. Phys.,* **20,** 271
189. van Dijk, F. A. and Dymanus, A. (1970). *Chem. Phys. Lett.,* **5,** 387
190. Kahalas, S. L. and Nesbet, R. K. (1963). *J. Chem. Phys.,* **39,** 529
191. Clementi, E. (1962). *J. Chem. Phys.,* **36,** 33
192. Nesbet, R. K. (1962). *J. Chem. Phys.,* **36,** 1518
193. Burrus, C. A. (1958). *J. Chem. Phys.,* **28,** 427
194. Carrington, A. and Howard, B. J. (1970). *Molec. Phys.,* **18,** 225
195. O'Hare, P. A. G. and Wahl, A. C. (1971). *J. Chem. Phys.,* **55,** 666
196. Neumann, R. M. (1970). *Astrophys. J.,* **161,** 779
197. O'Hare, P. A. G. and Wahl, A. C. (1971). *J. Chem. Phys.,* **54,** 4563
198. Curran, A. H., MacDonald, R. G., Stone, A. J. and Thrush, B. A. (1971). *Chem. Phys. Lett.,* **8,** 451, and personal communication
199. O'Hare, P. A. G. and Wahl, A. C. (1970). *J. Chem. Phys.,* **53,** 2469
200. Winnewisser, G. and Cook, R. L. (1968). *J. Molec. Spectrosc.,* **28,** 266
201. Field, R. W. and Bergeman, T. H. (1971). *J. Chem. Phys.,* **54,** 2936
202. Amano, T., Saito, S., Hirota, E. and Morino, Y. (1969). *J. Molec. Spectrosc.,* **32,** 97
203. Powell, F. X. and Lide, D. R. (1964). *J. Chem. Phys.,* **41,** 1413
204. Saito, S. (1970). *J. Chem. Phys.,* **53,** 2544
205. Amano, T., Saito, S., Hirota, E., Morino, Y., Johnson, D. R. and Powell, F. X. (1969). *J. Molec. Spectrosc.,* **30,** 275
206. O'Hare, P. A. G. and Wahl, A. C. (1970). *J. Chem. Phys.,* **53,** 2834
207. Lombardi, J. R., Campbell, D. and Klemperer, W. (1967). *J. Chem. Phys.,* **46,** 3482
208. Le Blanc, O. H., Laurie, V. W. and Gwinn, W. D. (1960). *J. Chem. Phys.,* **33,** 598
209. Freeman, D. E., Lombardi, J. R. and Klemperer, W. (1966). *J. Chem. Phys.,* **45,** 58
210. Howe, J. A. and Goldstein, J. H. (1955). *J. Chem. Phys.,* **23,** 1223
211. Lombardi, J. R., Klemperer, W., Robin, M. B., Basch, H. and Kuebler, N. A. (1969). *J. Chem. Phys.,* **51,** 33
212. Lin, T.-S. (1969). Ph.D. Thesis, University of Pennsylvania
213. Hochstrasser, R. M. (1971). Personal communication
213a. Hochstrasser, R. M. and Michaluk, J. W. (1972). *J. Molec. Spectrosc.,* **42,** 197 and personal communication
214. Hurley, J. and Le Fèvre, R. J. W. (1967). *J. Chem. Soc. B,* 824
214a. Hochstrasser, R. M. and Wiersma, D. A. (1972). *J. Chem. Phys.,* **56,** 528
215. Crossley, J. and Walker, S. (1968). *Can. J. Chem.,* **46,** 2369
216. McClellan, A. L. (1963). *Tables of Experimental Dipole Moments.* (San Francisco: W. H. Freeman and Co.)
217. Nelson, R. D., Lide, D. R. and Maryott, A. A. (1967). *Selected Values of Electric Dipole Moments for Molecules in the Gas Phase.* (Washington: National Bureau of Standards, NSRDS-NBS 10)
218. Foster, J. M. and Boys, S. F. (1960). *Rev. Mod. Phys.,* **32,** 303
219. Whitten, J. L. and Hackmeyer, M. (1969). *J. Chem. Phys.,* **51,** 5584
220. Liebman, J. F. (1971). *Molec. Phys.,* **21,** 563
221. Dixon, R. N. (1967). *Molec. Phys.,* **12,** 83; **13,** 77
222. Huang, K.-T. and Lombardi, J. R. (1970). *J. Chem. Phys.,* **52,** 5613
223. Wu, C. Y. and Lombardi, J. R. (1971). *J. Chem. Phys.,* **55,** 1997
224. Lombardi, J. R. (1969). *J. Chem. Phys.,* **50,** 3780
225. Huang, K.-T. and Lombardi, J. R. (1969). *J. Chem. Phys.,* **51,** 1228
226. Janiak, M. J., Hartford, A. and Lombardi, J. R. (1971). *J. Chem. Phys.,* **54,** 2449
227. Wu, C. Y. and Lombardi, J. R. (1971). *J. Chem. Phys.,* **54,** 3659
228. Parker, H. and Lombardi, J. R. (1971). *J. Chem. Phys.,* **54,** 5095
229. Lombardi, J. R. (1970). *J. Amer. Chem. Soc.,* **92,** 1831
230. Muenter, J. S. and Laurie, V. W. (1964). *J. Amer. Chem. Soc.,* **86,** 3901
231. Scharpen, L. H., Muenter, J. S. and Laurie, V. W. (1967). *J. Chem. Phys.,* **46,** 2431
232. English, T. C. and MacAdam, K. B. (1970). *Phys. Rev. Lett.,* **24,** 555
232a. Muenter, J. S. (1972). *J. Chem. Phys.,* **56,** 5409, 45

233. Angel, J. R. P., Sandars, P. G. H. and Woodgate, G. K. (1967). *J. Chem. Phys.,* **47,** 1552
234. Watson, R. E. and Freeman, A. J. (1961). *Phys. Rev.,* **123,** 521
235. Eisenthal, K. B. and Rieckhoff, K. E. (1968). *Phys. Rev. Lett.,* **20,** 309
236. Eisenthal, K. B. and Rieckhoff, K. E. (1970). *Chem. Phys. Lett.,* **6,** 441
237. Eisenthal, K. B. and Rieckhoff, K. E. (1971). *J. Chem. Phys.,* **55,** 3317
238. Kuball, H.-G., Euing, W. and Karstens, T. (1970). *Ber. Bunsengesellsch. Phys. Chem.,*
     74, 316
239. Kuball, H.-G., Klett, R. and Euing, W. (1971). *Chem. Phys. Lett.,* **11,** 454

4
Phosphorescence–Microwave Multiple Resonance Spectroscopy

M. A. EL-SAYED
University of California, Los Angeles

4.1   INTRODUCTION

Active studies on the relationship between phosphorescence and the lowest triplet state started in the 1940s. At present, the properties of the triplet state are being studied by a large number of scientists in different research disciplines, e.g. optical spectroscopy, magnetic resonance, theoretical chemists and physical-organic photochemists. Molecules reaching the lowest triplet state are very popular among researchers in all these fields, however, the same molecules are recently becoming very unpopular among scientists working in dye lasers due to the optical losses these molecules introduce because of their triplet → triplet absorptions.

During the past five years, the two research disciplines of optical spectroscopy and magnetic resonance have merged when it became evident that at low temperatures, microwave radiation of resonance frequencies with the zero-field (z.f.) transitions of the lowest triplet state could affect the phosphorescence intensity as well as the spectrum. Quantitative information can then be obtained from these phosphorescence–microwave multiple resonance experiments from which the magnetic, the radiative and the non-radiative properties of the triplet state can be determined.

This Review is divided into six Sections. Sections 4.1, 4.2 and 4.3 give a historical review of the phosphorescence and the triplet state, as well as the important historical development of phosphorescence–microwave double resonance (p.m.d.r.). These Sections are then followed by a description of the important properties of the triplet state. Section 4.4 describes the state of spin alignment, which is important to attain before a successful p.m.d.r. experiment can be carried out. This Section contains the important basic equations that predict the changes in the phosphorescence intensity upon the application of microwave radiation of resonance frequencies. Most of these equations are derived during the course of writing this Review. Section 4.5 gives a brief description of the experimental methods used in this field, as well as references to more detailed descriptions of these methods. The last three Sections describe the application of p.m.d.r. to the study of the magnetic (Section 4.6), optical (Section 4.7) and non-radiative (Section 4.8) properties of the lowest triplet state. Table 4.1 gives a descriptive summary of the important review articles in this field.

It is recommended that readers in this field follow the order of presentation given here. Some of them might even omit Section 4.3. For readers from other fields, we recommend that they read Section 4.3 before Section 4.2, or leave out Section 4.2 altogether.

Table 4.1 Summary of relevant reviews

Topic	Reference
Triplet state, pre-multiple resonance era	17
Optical pumping in atomic and ionic systems	32
M.O.D.R., experimental	75
Spin alignment and spin levels	88
P.M.D.R., general	99
Optical pumping of the triplet state; steady-state equations	59
The spectroscopy of the individual z.f. levels	52
Experimental methods in p.m.d.r.	60
P.M.D.R. and the intersystem-crossing process	91

4.2 HISTORICAL DEVELOPMENT

4.2.1 Pre-multiple resonance history

4.2.1.1 Phosphorescence and the triplet state

The phosphorescence emission of organic molecules in rigid media was first observed[1] in 1895. The first identification of a spectroscopic phenomenon with the triplet level of an organic molecule was made by Sklar in 1937[2], when he observed a weak absorption system in the 3400 Å region of benzene and assigned it to a singlet–triplet intercombination on the basis of its feeble intensity. Prior to this, Jablónski[3] had suggested an energy-level scheme to explain the phosphorescence emission as the direct radiative decay of a 'metastable state' that is situated somewhat below the energy of the 'normal', 'labile', or fluorescent (excited singlet) state. The relation between the wavelengths of Sklar's absorption and the known phosphorescence spectrum of benzene was evidently not noticed. In 1941, Lewis, Lipkin and Magel[4] suggested that the characteristic absorption spectra of phosphorescing molecules (T–T absorption) might be due to triplet excitation, but the authors were unable to distinguish between this possibility and some type of molecular tautomerism. In 1943–1944, Terenin[5] pointed out the similarity between phosphorescence and the well-characterised multiplicity-forbidden transitions of atomic systems and suggested the involvement of a triplet biradical. Independently in 1944, Lewis and Kasha[6] published their widely cited paper identifying phosphorescence as radiative inter-combination between the lowest triplet excited state and the (singlet) ground state. Lewis and Calvin[7,8] demonstrated the existence of unpaired electrons in the phosphorescent state. Weissman and Lipkin[9] established the electric dipole nature of phosphorescence (in fluorescein), eliminating the possibility that the long-lived emission is a result of a (less probable) electric quadrupole or magnetic dipole transition. For an electric dipole emission to arise from a triplet → singlet transition, a spin–orbital coupling process has to be invoked for intercombinations involving the metastable state. That spin–orbit coupling is responsible for the emission was shown by the observed large, heavy-atom substituent perturbation experiments of McClure[10] on aromatic hydrocarbons. This again supported the triplet character of the phosphorescent state. It was Hutchison and Mangum's characterisation[11,12] of the triplet state of naphthalene by e.s.r. studies that provided the final proof for assigning the phosphorescent state to a triplet state. Furthermore, it was soon realised[13] that fluorescence results from a transition from the lowest excited singlet state to the ground state, whereas phosphorescence originates from a transition between the lowest triplet state and the ground state.

4.2.1.2   The triplet state during the 1950–1965 period

(a) *Polarisation era* — It seems that the period 1944–1950 was spent identifying the electronic transition that gives rise to the phosphorescence emission. Triplet → singlet transitions should not give rise to electric dipole radiation unless spin–orbit perturbation is involved. It was, thus, natural that researchers in this field turned their attention to identify the spin–orbit coupling schemes responsible for the radiative power of the lowest triplet state. Different theoretical approaches using different schemes were used. The long phosphorescence lifetime characteristics of aromatic hydrocarbons were found[14] to be due to the vanishing contribution of the one- and two-centre terms resulting from the spin–orbit interaction between singlet and triplet π,π^* states in planar molecules. An actual calculation[15] of the spin–orbit interaction between π,π^* states was done later which gave a lifetime of 190 s for the radiative lifetime of benzene. Earlier, spin–orbit interaction between singlet σ,π^* and the lowest triplet π,π^* states was considered and a radiative lifetime of several hundred seconds was calculated[16]. At the time, the experimental value of the radiative lifetime determined by different methods gave different values and no experimental distinction between the two theoretical spin–orbit coupling schemes could be made on the basis of the value of the lifetime. However, should the phosphorescence arise from spin–orbit interaction with σ,π^* singlet states, then it should be polarised parallel to the $S_0 \rightarrow S_{\sigma,\pi^*}$ transition, i.e. perpendicular to the molecular plane of aromatic molecules. Spin–orbit coupling between π,π^* states would result in an in-plane polarised emission.

A vast amount of experimental studies was done in the 1960s on the polarisation measurements[17] of the emission of a number of molecules, which led to a great deal of understanding of the spin–orbit coupling schemes responsible for the phosphorescence emission. The method of photoselection, as well as absolute polarisation measurements, has been used in these studies. The following important conclusions have been made[17]:

(i) In aromatic hydrocarbons, phosphorescence results because of the spin–orbit interaction between the lowest triplet π,π^* states and the relatively high energy, radiatively weak σ,π^* singlet state.

(ii) In nitrogen heterocyclic molecules, e.g. quinoxaline, pyrazine, or carbonyl compounds, e.g. benzaldehyde, benzophenone, the spin–orbit coupling between the n,π^* and π,π^* states determines the phosphorescence characteristics of the lowest triplet states of these molecules.

(iii) Internal heavy-atom effect (i.e. the effect of decreasing the radiative lifetime of aromatic hydrocarbons by substituting a halogen for one of its hydrogen atoms) results from at least two different types of spin–orbit perturbations[18]. The spin–orbit interaction involves the halogen states and mixes both the σ,π^* as well as the π,π^* singlet states with the emitting π,π^* triplet state. The second mixing (between S_{π,π^*} and T_{π,π^*}) involves vibronic coupling (along the normal modes involving the halogen) in addition to the spin–orbit coupling. This mixing results in the so called subspectrum II which has a false origin separated from the true origin (the O,O band) by the energy of a non-totally symmetric halogen mode.

(iv) Host crystal states could[19a] mix with the guest molecular states in mixed crystals. This leads to a change of the polarisation characteristic[19b] of the phosphorescence of the latter.

(b) *E.S.R. studies*—A great deal of research was carried out in the period 1960–1970 on the e.s.r. spectra of the triplet state. These studies could constitute a review by themselves. The author is not working in this field and thus does not feel that he is qualified to review it. Only topics and developments relevant to the main body of this Review will be discussed below. For detailed reviews, the reader is referred to Reference 17b.

The two parallel unpaired electrons of the triplet states can have three different orientations with respect to an applied magnetic field. Thus, a change in the spacing of the z.f. levels would occur. The transition between these different levels gives rise to the e.s.r. spectrum of the triplet state. The strength of the (Zeeman) interaction between the magnetic field and the two unpaired electrons (i.e. the energy spacing between the triplet sublevels), depends on the angle between the field and the magnetic moment of the two parallel unpaired spins. The splitting of the three Zeeman triplet components of the triplet state is expected to be sensitive to the angle between the molecule and the magnetic field. This requires that, in order to observe a limited number of transitions between these Zeeman levels, the molecules to be studied should all have similar orientation with respect to the magnetic field. The molecule to be studied must then be dissolved substitutionally in a host lattice and the single crystal is oriented in a specific orientation with respect to the field in the e.s.r. cavity.

The number of molecules studied in single crystals so far is limited in number (~ 20). The small number is undoubtedly due to the limitation of finding proper hosts of known crystal structures and proper crystal properties. In the period 1960–1970, two important advances in the field of triplet-state spectroscopy were made. The first was the recognition[20–23] that e.s.r. of the triplet state could be observed in a randomly oriented sample, e.g. triplet naphthalene in a glass. The second was the development of the magnetophotoselection techniques[24–26]. This method could be classified as the first double-resonance method in this field. It involves the excitation of a randomly oriented sample with polarised monochromatic u.v. light while recording the e.s.r. spectra. An assignment of the e.s.r. lines or the u.v. absorption could be made from the observed changes in the relative intensities of the e.s.r. line as the polarisation or the wavelength of the exciting light is changed.

It should be pointed out that from the e.s.r. spectrum of the triplet state, the z.f. splittings of the triplet state can be calculated. These quantities are useful in identifying the electron distribution in the triplet state, from which a unique assignment of its symmetry can be made. Unfortunately, molecules with short triplet lifetimes, e.g. molecules with n,π^* lowest triplet state (carbonyls and some N-heterocyclic molecules) as well as halogenated aromatic molecules, can not be studied by e.s.r. because of their low steady-state concentrations. Fortunately, the z.f. splittings of these molecules are most conveniently determined by optical detection as will be discussed below.

4.2.2 Historical development of microwave–optical multiple resonance techniques

Historically, this field can be related to the optical detection of magnetic resonance transitions of simple atoms and ions. Research in optical detection of radio-frequency transitions was first made in 1925 on gases by Fermi and Rasetti[27] as well as by Breit and Ellet[28]. This was done by observing a change in the polarisation of resonance fluorescence upon a change in frequency of an applied alternating magnetic field. The first determination of magnetic transitions in ionic solids was not made[29–31] until 1959. For a review of the field, the reader is referred to the book by Bernhein[32].

In 1967, the optical detection of the e.s.r. transitions of the triplet state was made[33–35]. This was done at 4.2 K, where the difference in the Boltzmann population of the z.f. levels is large enough so that microwave radiation can have a measurable effect on the emission intensity. The observation and assignment of the state of spin alignment, in which the population of the z.f. levels is not Boltzmann, was first detected[36] in 1966 by observing e.s.r. emission, rather than absorption lines. In actuality, the state of spin alignment was first detected[37] but not assigned, in 1963 by observing the non-exponential behaviour of the phosphorescence decay whose origin was not then understood. Attributing the observed decay non-exponentiality to the presence of a state of spin alignment was made[38] in 1967. The first optical detection of z.f. transitions was made[39] in 1968. The first application of phosphorescence–microwave double resonance (p.m.d.r.) techniques to the understanding of the optical spectroscopy of the triplet state was made[40] in 1969. In the same year, the first optical detection of electron–nuclear double resonance[41, 42] (e.n.d.o.r.) was made. A great deal of advance has been accomplished in 1970, e.g. the use of the delay (p.d.m.r.) techniques[43] in determining the relative rates of the intersystem crossing (ISC) process to the different z.f. levels, the optical detection of electron–electron double resonance[44] (e.e.d.o.r.), the determination of the relative ISC rates using steady-state saturation methods[45], the optical determination of the polarisation of the z.f. magnetic transitions[46] and the optical detection[47] of the level anticrossing in the triplet state. Up to the date of writing this Review in 1971, a number of interesting reports on the phosphorescence–microwave double resonance techniques has appeared. The detection of phosphorescence modulation due to the coherent coupling of the triplet spin system to a strong microwave field was made[48]. Using p.m.d.r. techniques to determine the coherent length of triplet excitations was also reported[49]. The use of pulsed excitation-pulsed (or fast sweep of) microwave to determine relative intersystem crossing rates has been made[50]. The detection of the state of spin alignment at 77 K using e.s.r. and short pulsed excitation was also made[51].

4.3   RELEVANT PROPERTIES OF THE TRIPLET STATE

4.3.1   Physical picture of the origin of the zero-field splitting in aromatic molecules

Let us discuss the triplet state of an atom in an electronic state with an electronic motion, and thus a distribution, that is independent of the spatial direction around the nucleus, e.g. the electronic distribution in a ^3S state. In the absence of a magnetic field, one may select three perpendicular planes (e.g. MN, LN and ML) and may quantise the spin motion of the two unpaired electrons in these three perpendicular planes, so that the component of the spin angular momentum in the direction perpendicular to the plane chosen is zero. If the two electrons are spinning in the ML plane, the component of the spin angular momentum along N is zero, and the z.f. level is to be called the τ_N level. Similarly, the τ_M and τ_L z.f. levels correspond to the magnetic levels of the triplet state in which the two unpaired electrons are spinning in the NL and MN planes, respectively, with a zero component of spin angular momentum along the M and L axes, respectively. It is obvious that for the ^3S state of an atom in zero field, the average distance of the two parallel spins in the three different planes is the same. The repulsive magnetic dipolar interaction between the two similar magnetic dipoles of the two unpaired electrons is thus independent of the plane in which the two electrons select to spin. Thus the three levels, τ_N, τ_M and τ_L, in Figure 4.1 should all have the same energy.

If the electronic distribution is disc-like, as in the π-system of benzene, the average distance of the two unpaired electrons would be the same in the two planes perpendicular to the disc but different from that in the disc plane. In the plane of the benzene ring (call it the ML plane), the electron density of the π electrons is zero (if σ,π coupling is neglected). It is thus expected that the repulsive dipolar interaction is less than in the spherical atom, thus stabilising the τ_N z.f. level. The three z.f. levels then split into the pattern shown in the middle of Figure 4.1. If the disc is elongated in the plane along L (e.g. in the π system of naphthalene), the average distance between the two

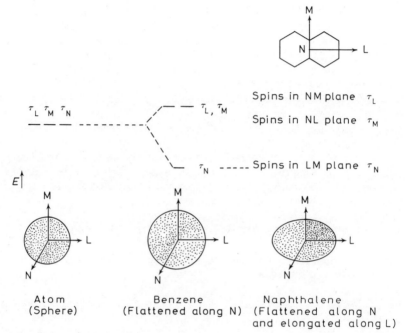

Figure 4.1   The origin of zero-field splitting in the triplet state of symmetric aromatic molecules. Since the electronic distribution in these molecules is not spherical, the average distance and spin density (and thus the magnetic dipolar interactions) will depend on the plane in which the two electron spins are placed

(Reprinted from El-Sayed, M. A. (1971). *Accounts Chem. Res.*, **4**, 23, copyright 1971, by the American Chemical Society. Reprinted by permission of the copyright holder)

spins in the planes perpendicular to the elongated disc is no longer equal and one observes the z.f. pattern shown in the right-hand portion of Figure 4.1.

The three z.f. triplet levels of the triplet state of aromatic molecules are thus split in the absence of magnetic field. The origin of the splitting, as discussed above, is the anisotropy of the spin–spin interaction in molecules. For some molecules in certain states, spin–orbit interactions could lead to further contributions to the splittings. We have neglected this interaction in the above discussion since we are considering molecules in which the unpaired electron distribution is largely localised on atoms of low atomic number and whose z.f. splittings are largely due to spin–spin interaction as in the π,π^* states of aromatic hydrocarbons. The value of the z.f. splitting varies

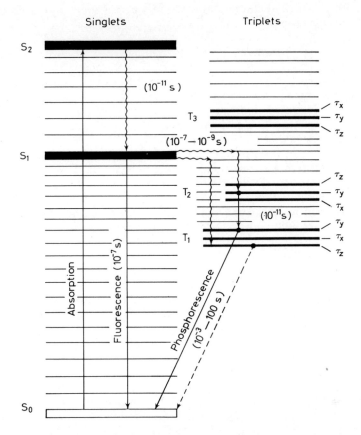

Figure 4.2 Jablónski-type diagram. The zero-field splittings (between τ_x, τ_y and τ_z) are magnified due to their importance in the discussion in this Article. The processes shown in the triplet manifold are those for systems in which the spin–lattice relaxation processes among τ_x, τ_y and τ_z are slower than the $T_2 \rightarrow T_1$ non-radiative processes or the $T_1 \rightarrow S_0$ relaxation processes

(Reprinted from El-Sayed, M. A. (1971). *Accounts Chem. Res.*, **4**, 23, copyright 1971, by the American Chemical Society. Reprinted by permission of the copyright holder)

from 0.1 cm^{-1} in aromatic π,π^* states, 0.3^{-1} cm^{-1} for the n,π^* states of some aromatic carbonyls, up to a few wave numbers as in O_2.

4.3.2 Production of the triplet state by optical excitation at low temperatures

Figure 4.2 shows the different processes that take place following the excitation of a molecule by light absorption. $S_x \leadsto S_1$ and $T_y \leadsto T_1$ are non-radiative (internal conversion) processes and $S_1 \leadsto T_y$, as well as $T_1 \overset{\leadsto}{\to} S_0$, are known as intersystem crossing processes ($\leadsto$ is non-radiative, while $\to$ is a radiative process). T indicates the spatial (orbital) function, whereas τ_i indicates the spin function of the triplet state.

When a molecule changes its spin state from the lowest excited singlet state (S_1) to one of the triplet states, T_k, it does so by exchanging its spin angular momentum with its orbital angular momentum. This leads not only to a change in the total spin angular momentum from $S = 0$ (singlet state) to $S = 1$ (triplet state), but also to a preferential spin direction in the molecular framework, depending on the symmetry of the spatial distribution of the excited electrons in S_1 and T_k. Different molecules might proceed to different triplet levels with different rates, depending on the magnitude of the spin–orbit interaction involved[17]. Thus, it is possible that $S_1 \leadsto T_2$ (in Figure 4.2) leads to the spin direction τ_y (y could coincide with any of the molecular axes L, M, or N), whereas $S_1 \leadsto T_1$ could lead to the spin direction τ_z. Depending on the molecule, these processes take place in $\sim 10^{-7}$–10^{-9} s. The molecule that selects to cross over to the triplet state via the $S_1 \leadsto T_2$ route loses its $E_{T_2} - E_{T_1}$ vibration electronic energy in $\sim 10^{-11}$ s, and ends up in T_1 with the same spin direction, τ_y, as long as the temperature is low so that the spin–lattice relaxation time between the different spin directions is longer than 10^{-11} s. Those molecules initially crossing to the T_1 electronic vibrational levels with energy comparable to S_1 lose their vibration energy in $\sim 10^{-11}$ s and end up in the τ_z spin z.f. level. Since the rate (or rate constant) of the $S_1 \leadsto T_1$ process would, in general, be different from that for the $S_1 \leadsto T_2$ process, the number of molecules reaching the τ_z level per unit time would be different from those reaching the τ_y z.f. level of the lowest triplet state. Molecules in the different z.f. levels of T_1 could return to S_0 by the phosphorescence process and/or by non-radiative processes with different rate constants.

4.4 SPIN ALIGNMENT IN THE TRIPLET STATE

4.4.1 Production of the state of spin alignment

If one defines the degree of spin alignment as a measure of the deviation from the Boltzmann population distribution, then it is obvious that the first requirement for the existence of the triplet state in a state of spin alignment is that the spin–lattice relaxation (SLR) be absent or very slow compared with the rates for the other processes. The actual value could be related to the population ratios of the z.f. levels. In the absence of the SLR processes, the value of this ratio would depend on the mode of preparation of the triplet state. Three extreme cases could be realised. In the first case, the system is being continuously excited and decays continuously, i.e. the system is in a steady state. By equating the rates of pumping any z.f. level with its rate of decay, one obtains the following ratios (equation (4.1)):

$$(n_j/n_i)_t = K_j k_i/K_i k_j \tag{4.1}$$

where n, k and K are the population, decay constant and pumping constants, respectively of the z.f. levels i and j.

On the other hand, if after reaching the steady state, the excitation is cut off, then the above ratio decays according to the following equation:

$$(n_j/n_i)_t = (K_j k_i/K_i k_j) e^{-(k_j - k_i)t} \tag{4.2}$$

Thus, if the value of equation (4.1) is not different from unity, i.e. the degree of spin alignment is small, larger values for the spin alignment could be created by letting[43] the system, whose decay constants k_i and k_j are different, decay. If the K values are different, a large value of spin alignment could be obtained before the system reaches its steady state. This is accomplished experimentally by using pulsed excitation[50]. The population ratio is simply equal to the ratio of the pumping rates if the rate of pumping the z.f. level is much faster than its decay rate (a situation which is true at short times after short pulsed excitation)[50], i.e.

$$n_j/n_i = K_j/K_i \tag{4.3}$$

These are different situations from which the state of spin alignment could be formed. In order for the triplet state to absorb microwave radiation, $n_j - n_i \neq 0$ or $n_j/n_i \neq 1$. Thus, for a successful phosphorescence–microwave double resonance experiment, the triplet state should be in a state of spin alignment.

4.4.2 Detection of spin alignment in zero field

An accurate method for detecting the state of spin alignment is by observing microwave absorption in excess of that expected from a Boltzmann population distribution. The observation of a stimulated microwave emission instead of absorption is better yet at proving the existence of the spin alignment. Due to the difficulties in quantitatively measuring changes in microwave intensity, these methods have not as yet been attempted in z.f. The observation of microwave emission, instead of absorption of e.s.r. lines was the first signal of the presence of the state of spin alignment[36].

In z.f., there are a number of optical methods that can very easily detect the state of spin alignment. The large effects that microwaves have on the phosphorescence spectrum[40], intensity[39], lifetime[52] and polarisation[53], give strong indications of the presence of a spin alignment state. Some of these will be discussed in detail below. Optical spectroscopy, without microwave radiation, can also be used to detect the presence of the state of spin alignment[54]. At temperatures for which the SLR processes have comparable

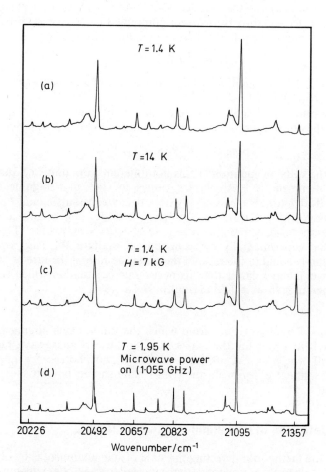

Figure 4.3 Effects of different types of perturbations on the phosphorescence spectrum of 2,3-dichloroquinoxaline in durene at 1.4 K; (a) at this low temperature, the lowest triplet state is in a spin aligned state, (b) this shows the effect of increasing the temperature from 1.4 to 14 K, (c) this shows the effect of applying a 7 kG magnetic field on the spectrum of a polycrystalline sample and (d) this shows the effect of saturating the two radiating zero-field levels with microwave radiation

(From Tinti, D. S. and El-Sayed, M. A.[52] by courtesy of the American Institute of Physics)

rates to the unimolecular decay processes, the decay curve becomes sensitive to temperature. The decay constant of the phosphorescence at temperatures for which the SLR are fast is related to the unimolecular decay constants k_i of the individual z.f. levels by equation (4.4)[55]:

$$k = \Sigma\, k_i\, X_i \qquad\qquad (4.4)$$

where X_i is the fraction of the total population of the z.f. level i. This equation was first tested for quinoxaline[38] and later for pyrazine[56]. The different decay components observed at low temperatures for pyrazine are found to have different polarisation characteristics[57] indicating that all three z.f. levels in this molecule are at least partially radiative.

For systems whose emission is composed of progressions with false origins, different progressions could arise from different z.f. levels. If the population ratio of the different z.f. levels is changed by temperature, decay, pulsed excitation (time-resolved spectroscopy) or magnetic fields, the relative intensity of the different progressions is expected to change. The effect of the first and last type perturbations has been demonstrated[54] and is shown in Figure 4.3. Higher temperatures increase the rates of the SLR processes, which tends to decrease the degree of spin alignment. An applied magnetic field mixes the z.f. functions, thereby equalising the radiative and pumping rates of the different[57a] levels. An applied magnetic field also increases the average energy spacings between the levels and increases the SLR rates[38]. If the application of a magnetic field or increasing the temperature changes the relative band intensities in the spectrum or changes the non-exponential decay into an exponential one, then the system must exist in a state of spin alignment prior to the application of the perturbation.

4.4.3 Effect of microwave radiation; basic equations

If the system under study has its triplet state in a state of spin alignment, then the application of microwave radiation would change the degree of spin alignment. More important, microwave radiation of resonance frequencies should change the number of radiating molecules in a certain level, thus causing a change in the intensity of the emission from that level. The ratio of the intensity in the presence of resonant microwaves to that in the absence of the microwaves depends on the degree of spin alignment (which depends on the excitation mode), the manner by which the microwave is applied (pulsed, rapidly or slowly swept) and the microwave power used. Microwaves could be applied slowly causing a new steady state to be established. On the other hand, microwaves could be swept rapidly as compared to the decay times of the levels. Depending on the power of the microwaves, the population of the two levels could be equalised (saturation) or even completely inverted[58] upon the fast passage of the microwaves. Slow sweep or the continuous exposure of the microwave can only be applied to a system in a steady state, i.e. when the degree of spin alignment is given by equation (4.1). For simplicity, let us assume that the total emission intensity of the phosphorescence, or the emission of a vibronic band whose intensity is being monitored, originates from the τ_i z.f. level. Furthermore, if the rate of sweep is very slow to ensure the satisfaction of the steady-state requirements, saturation of the z.f. levels is the maximum that can be hoped for when using microwaves of sufficient powers. If I^v and I are the phosphorescence intensity with microwaves (that saturates the $\tau_i \leftrightarrow \tau_j$ transition) and without microwaves, respectively, then I^v/I is given by[59]:

$$I^v/I = n_i^v/n_i = [(K_i + K_j)/K_i] \cdot [k_i/(k_i + k_j)] \tag{4.5}$$

The sweeping of the microwaves across resonance can be used in all three extreme cases whose population ratios are given by equations (4.1)–(4.3). The sweep rate across resonance should be fast compared with the decay rate of the z.f. levels. Two types of effects can be discussed: (i) if the microwaves have enough power to saturate the z.f. transition being swept (i.e. to equalise the population of the corresponding z.f. levels) or (ii) the microwaves invert the population of the z.f. levels involved.

If the microwaves saturate the $\tau_i \leftrightarrow \tau_j$ transition, then the intensity ratio of the emission from z.f. level τ_i is given by:

$$I^v/I = \left(\frac{n_i + n_j}{2}\right)/n_i = \tfrac{1}{2}(1 + n_j/n_i) \tag{4.6}$$

Depending on the mode of excitation, n_j/n_i would be related to the rate constants involving the τ_i and τ_j levels as given by equations (4.1)–(4.3). Thus, for a system that is in a steady state before sweeping the microwaves:

$$I^v/I = \tfrac{1}{2}(1 + K_j k_i/K_i k_j) \tag{4.7}$$

If the steady state is allowed to decay and the microwaves are swept at time t after excitation is cut off, then:

$$(I^v/I)_t = \tfrac{1}{2}(1 + (K_j k_i/K_i k_j) e^{-(k_i - k_j)t}) \tag{4.8}$$

On the other hand, if the state of spin alignment is formed by pulsed excitation and the microwave radiation is swept shortly after excitation to saturate the $\tau_i \leftrightarrow \tau_j$ transition, then

$$I^v/I = \tfrac{1}{2}(1 + K_j/K_i) \tag{4.9}$$

Equations have been derived[60] for systems in which the emission originates from two z.f. levels.

If the microwave radiation is swept in a very short time with enough power to completely invert the population of the τ_i and τ_j levels, then:

$$I^v/I = n_j/n_i \tag{4.10}$$

Substituting equations (4.1)–(4.3) in equation (4.10), one obtains:

$$I^v/I = K_j k_i/K_i k_j \tag{4.11}$$

and

$$(I^v/I)_t = (K_j k_i/K_i k_j) e^{-(k_j - k_i)t} \tag{4.12}$$

and

$$I^v/I = K_j/K_i \tag{4.13}$$

for the steady state, decay, and pulsed excitation experiments, respectively. It is thus obvious that the signal-to-noise ratio obtained in an optical-detection method for a given system would greatly depend on the mode of excitation, mode of exposure to the microwaves and mode of detection. The difference between the signals obtained for different molecules, using the same technique, would depend on the rate constants of pumping and decay, i.e. on the degree of spin alignment. In general, the steady-state excitation method (thus using equation (4.7) or (4.11)) most useful when one of the z.f. levels being irradiated with microwaves has a larger K but smaller k than the other z.f. levels. If the two intersystem crossing rates are comparable, but the decay constants are very different, then the decay technique (thus equations (4.8) or (4.12)) is more appropriate to use. On the other hand, if the opposite is true, i.e. the decay constants are comparable but the intersystem crossing rates are different, the pulsed method (thus equation (4.9) or (4.13)) is most useful.

4.5 EXPERIMENTAL TECHNIQUES

For a recent detailed examination of the different experimental methods in this field, the reader is referred to Reference 60. A summary of these methods is given below.

There are several methods by which microwave resonance can be optically detected. The methods that are utilised in our laboratory, in order of increasing complexity of the attendant electronics are: (i) continuous wave (c.w.) operation; (ii) modulation of the microwaves with lock-in detection at the modulation frequency (lock-in); (iii) cut-off of the exciting light followed by a sweep of a microwave frequency region after a specified delay time (delay); and (iv) sweep of microwave frequency region after subjecting the sample to a short-lived light excitation pulse (pulse). Each of these methods incorporates certain advantages; which one is most useful for a given molecular system will depend on the rate constants for population (intersystem crossing to) and depopulation (radiationless or phosphorescent decay) of the individual triplet sublevels as shown by the equations given in the previous Section.

Optical detection of microwave transitions involves irradiating the sample with light of sufficient energy to populate the triplet sublevels, exposing the sample to microwave radiation and observing changes in the phosphorescence emission intensity as the microwave frequency is scanned. Population of the triplet sublevels is most commonly achieved, as discussed above, by optical pumping of the molecule to its first excited singlet state, followed by intramolecular intersystem crossing to the triplet manifolds; but any other pathways leading to the triplet state (singlet–triplet absorption, host–guest energy transfer, etc.) can equally well suffice, and microwave resonance studies can yield important information about such pathways[59].

In order for optical detection of microwave resonances to be feasible for a given molecular system, three important conditions must be satisfied: (i) the triplet state in question must display luminescence; (ii) some mechanism must exist for creating the triplet state in a state of spin alignment; (iii) this population imbalance, once created, must persist for a time sufficient to allow interaction between microwaves and the system.

The first condition, although by no means universally satisfied, is not a serious limitation, since many molecules, both organic and inorganic, display phosphorescence when in glass or solid solution at reduced temperatures (77 K or lower). The second of these conditions is generally readily satisfied, inasmuch as intersystem crossing processes not only exist with significant yields for many molecular systems, but also are subject to spin–orbit symmetry selection rules which cause them preferentially to populate specific sublevels in the triplet manifold.

Satisfaction of the third condition depends on the rates of the spin–lattice relaxation processes by which the spin sublevels interconvert; these rates are highly temperature dependent and also depend on the environment within which the molecular system is placed. In order to maintain a steady-state population imbalance of the triplet sublevel, the rates of sublevel interconversion must be slower than the rates at which the sublevels depopulate back to the singlet manifold. To reduce the spin–lattice relaxation rates to this level requires temperatures of the order of 5 K or lower; in some solid matrices, this spin–lattice 'freezing' (also termed 'spin alignment') is effectively complete at 4.2 K, whereas in other matrices it may not be attained even at 1.1 K. Whether or not spin–lattice processes can be 'frozen' at liquid helium temperatures may even depend on the solid phase of the crystals. For pyrazine in cyclohexane[61], for example, slow cooling of the solution to below its freezing point prior to immersion in liquid helium gives a spin-polarised system, but rapid cooling yields a system in which spin–lattice relaxation rates are too fast even at 1.4 K for polarisation to occur.

In principle, optical detection of microwave resonances should also be possible, even in the presence of fast spin–lattice relaxation, e.g. at 77 K, using a pulsed light source to create an instantaneous triplet sublevel population imbalance and observing the effect of microwaves before the populations have the opportunity to relax. Although such an effect has been observed by use of e.s.r. techniques[51], the experimental difficulties involved in doing optical observations of this type have so far prevented it from being observed in the optical mode at 77 K.

The basic instrumentation which is required to do experiments in optical detection of microwave resonances is as follows: (i) an excitation light source (for many experiments, a 75-watt mercury or xenon lamp is sufficient) equipped with lenses and filters sufficient to pump the sample radiatively to an electronic excited state from which the triplet state is efficiently populated; (ii) a variable-frequency microwave generator or sweeper in the 0.2–12 GHz range, together with the hardware required to conduct microwaves to the sample (connectors, coaxial cables, stainless-steel coaxial rod); (iii) a conventional quartz-tipped liquid helium Dewar for spectroscopic measurements, modified suitably to conduct microwaves to the sample; (iv) detection components of the type commonly used for time-varying emission spectroscopic measurements, such as a phototube or photomultiplier tube and electrometer or oscilloscope. As in most spectroscopic experiments, the level of sophistication of the apparatus can range from quite simple to extremely complex, depending on the specific requirements of the experiment in hand. Some of the possibilities for apparatus variations are described in the subsections which follow.

A schematic sketch of the typical experimental apparatus is shown in

Figure 4.4. The sample is inserted into a copper coil slow-wave helix structure which is attached to a stainless steel coaxial waveguide. This assembly is immersed in liquid helium in a double-jacketed, bubble-free, quartz window Dewar assembly. Exciting light from a u.v. light source suitable for excitation of the phosphorescence emission of the sample is focused on the sample through suitable filters which block unwanted light that would otherwise increase detection noise. The emitted light from the sample is collected by a lens system (optional), passed through a filter or monochromator, and detected by a phototube, whose output is amplified and displayed by use of suitable electronics.

The experiment then consists of changing the frequency output of the microwave generator, whereupon, at the frequency corresponding to a transition between magnetic sublevels of the sample, an intensity change in

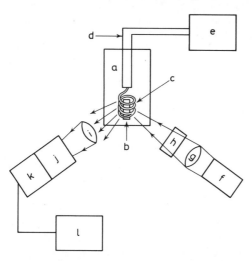

Figure 4.4 Schematic diagram of components for a typical experiment to do optical detection of microwave transitions; (a) liquid helium cryostat, (b) sample, (c) copper coil helix, (d) coaxial waveguide, (e) microwave generator, (f) light source, (g) quartz lens, (h) filters, (i) collecting lens, (j) monochromator (k) photomultiplier and (l) electrometer and recorder

the phosphorescence emission will be observed, provided that the microwave power and population difference between the levels are both sufficiently large. This basic scheme remains the same for the various modifications used in this field of modifications which represent ways of enhancing the observed signal, either by increasing the selectivity of the detection network or by increasing the population difference at the instant of passage through the microwave resonance.

In a few special cases the intensity change upon passage through a microwave transition may be observed with the naked eye. For pyrimidine in benzene, for example, the total emission intensity increases by c. 20% upon passage through the 0.923 GHz (2E) transition (this change is, however, extremely sensitive to the microwave sweep rate; see below), and the resulting flashes of light can be readily seen with the naked eye. Such large changes are unusual; intensity variations of the order of a few percent are the norm.

In the next three Sections, a description is given of the different applications of p.m.d.r. spectroscopy to the different research areas involved in the study of the magnetic, optical and non-radiative properties of the triplet state.

4.6 PHOSPHORESCENCE–MICROWAVE DOUBLE RESONANCE AND MAGNETIC STUDIES

4.6.1 Zero-field transitions

The first optical determination of z.f. transitions was by Schmidt and van der Waals[39] for quinoxaline in durene. For molecules with atoms having nuclear spin, transitions between hyperfine levels of different z.f. levels are also observed. The optical detection of z.f. transitions has also been reported for 2,3-dichloroquinoxaline[40], 8-chloroquinoline[62,63], quinoline[63], sym-1,2,4,5-tetrachlorobenzene[64] in durene, as well as its x-traps[49] and a few other systems.

Figure 4.5 shows the microwave spectra obtained when the $\tau_L \leftrightarrow \tau_M$ and $\tau_N \leftrightarrow \tau_L$ z.f. transitions of 2,3-dichloroquinoxaline are optically detected. The observed structure is similar to that found for quinoxaline in durene[39]. Both molecules have two nuclei each with spin $I = 1$. The structure is explained[41,42,63] as arising from the ^{14}N quadrupolar interaction and a second order ^{14}N hyperfine interaction. These experiments yield, for the first time, a measurement of excited state ^{14}N and $^{35,37}Cl$ coupling constants[41,42,63,65,66] for the excited triplet state. As has been observed in the e.s.r. spectra, transfer hyperfine interaction resulting from the transfer of a small but finite electron spin density from the excited triplet state of a guest to the nucleus of the host has also been optically detected[67] in z.f. as was observed in e.s.r.[68,69].

The detection of what is believed to be optically detected signals for the z.f. transition of triplet excitons has been recently reported[49]. By fitting the observed band shape to a theoretical expression based on a one-dimensional model[70], a minimum coherent length of 50–100 Å was obtained for the exciton in sym-1,2,4,5-tetrachlorobenzene at 3.2 K.

A very elegant experiment has been performed by van der Waals and his group recently[48] in which the phosphorescence modulation due to coherent coupling of the triplet spin system with a strong microwave field of resonance frequencies was demonstrated. The system was quinoxaline-d$_7$ in durene and the z.f. transition at 1.001 GHz was coupled to the microwave field. The experiment was performed in the following manner. The excitation was cut off after the steady state was accomplished; a few seconds later, the radiative level was empty, yet the other level was still densely populated. At this instant, powerful microwave pulses of 1.001 GHz frequency, but of different pulse widths, were applied in different experiments. The population of the radiative level, and thus the intensity of pulse of the phosphorescence produced, is shown to be a periodic function of the interaction time with the microwave field, i.e. a periodic function of the pulse width. In agreement with theory, the maximum intensity is found to occur when $\gamma H_1 t = \pi$ and the minimum occurs when $\gamma H_1 t = 2\pi$, where γ, H_1 and t are the gyromagnetic ratio, microwave magnetic field and the pulse width, respectively. This type of experiment was discussed in the same year by Harris[58].

The main advantages of the optical detection of z.f. transitions are: (i) The need to use single crystals and magnets is eliminated (ii) z.f. transitions for very short-lived triplet states may be detected as long as the

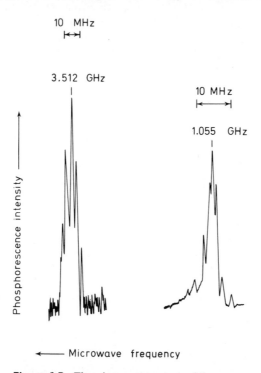

10 MHz

3.512 GHz

10 MHz

1.055 GHz

Phosphorescence intensity

⟵——— Microwave frequency

Figure 4.5   The microwave spectrum of the $\tau_L \leftarrow \tau_M$ (right) and $\tau_L \leftarrow \tau_N$ (left) zero-field transitions of 2,3-dichloroquinoxaline in durene, as monitored by detecting the changes of the 0–0 band intensity of the $^3B_2^{\pi,\pi^*} \rightarrow {}^1A_1$ phosphorescence at 1.9 K as the microwave frequency changes.
(From Tinti, D. S. *et al.*[40] by courtesy of North-Holland Publishing Company, Amsterdam)

cause of the short lifetime is an allowed radiative process. (iii) The cost of the experimental apparatus is not as expensive as the e.s.r. equipment. The disadvantages are: (i) A liquid helium apparatus is required, at least at the present time; (ii) the elimination of a direction indicator, e.g. the magnetic field in the e.s.r. experiments, renders the p.m.d.r. method (as described above) incapable of determining the polarisation and thus the assignment of the z.f. transitions, and (iii) for molecules with a single radiative z.f. level, only two of the three z.f. levels are optically detected. The transition involving the dark z.f. levels will not be detected if spin–lattice relaxation is absent.

Electron–electron double resonance techniques[49], described later, could be used to determine the third transition. Other methods that would slightly increase the coupling of the levels, e.g. heating or application of small magnetic fields, might also be useful in detecting the third transition.

An assignment of the z.f. transitions can be made by use of p.m.d.r. techniques by one of the following two methods. If the phosphorescence spectrum is analysed and its polarisation is determined for symmetric type molecules, the response of the different vibronic bands upon saturating the different z.f. transitions could give results from which the assignment could be made. On the other hand, the assignment could be made directly if the polarisation of the magnetic z.f. transitions is determined. P.M.D.R. techniques have been extended[46, 71] to make these measurements as discussed later.

4.6.2   Polarisation of z.f. transitions from p.m.d.r.[46, 71]

The polarisation of the z.f. transitions is determined in the absence of a laboratory magnetic field by using a single mixed crystal oriented in the microwave helix or cavity with one of its crystal axes parallel to the axial axis of the helix. This method makes use of the fact that the magnetic field of the microwaves is not isotropic inside the helix. An equation is derived which relates the change in the optical signal to the power of the microwave radiation in resonance with the z.f. transition whose polarisation is to be determined. At low microwave powers, the equation gives a straight line relationship between dI/dP and $1/P$ where I is the phosphorescence intensity from a certain z.f. level and P is the microwave power. From the slope to the intercept ratio, a quantity which is proportional to the microwave absorption cross-section along the crystal axis used is determined. Repeating the experiment with the crystal oriented differently, a quantity proportional to the absorption cross-section along the other axis is determined. From these two values, the polarisation ratio of the magnetic z.f. transition can be determined. From a knowledge of the crystal structures, these ratios can be translated to ratios in the molecular framework of the guest molecule. It should be pointed out that the polarisation ratios for the z.f. transitions of 2,3-dichloroquinoxaline in durene[71] agree with the oriented gas model, more so than the values obtained for the phosphorescence emission of the same crystal. This might be due to the fact that depolarisation effects are larger for high-energy oscillators in the optical region than for low-frequency ones in the microwave region.

This method was used to determine the polarisation of the z.f. transition of sym-1,2,4,5-tetrachlorobenzene[64] in durene. In this work, it was pointed out that an assignment of all the z.f. transitions as well as the relative order of the z.f. levels can be made from the lifetimes of the z.f. levels, the spectra and from the knowledge of the polarisation of one crucial z.f. transition.

4.6.3   Optical detection of multiple resonance magnetic transitions

In magnetic resonance, electron–nuclear double resonance experiments have been performed[72–74]. Using optical techniques, similar experiments can be performed on the triplet state in the absence of the laboratory magnetic field, but in the z.f. of the triplet state. Two types of these optical detection methods are described below.

4.6.3.1 Optical detection of electron–nuclear double resonance (e.n.d.o.r.) transitions in zero field[41, 42]

If a molecule contains an atom with a nucleus having nuclear spin $I > \frac{1}{2}$, e.g. N or Cl, the observed microwave spectrum will be more complex due to the electron–nuclear hyperfine interaction as well as nuclear quadrupole interactions. For each of the three electron spin directions, there are a number of different nuclear spin quantisation directions. For nitrogen, $I = 1$, thus for each electron spin level there are three different nuclear spin directions. These are indicated by τ_{11}, τ_{12} and τ_{13} for z.f. (electron spin), level 1; τ_{21}, τ_{22} and τ_{23} for z.f. level 2, etc. (see Figure 4.6). The intersystem crossing processes ($S_1 \leadsto T_1$ and $T_1 \rightarrow S_0$) are determined by the electron spin–

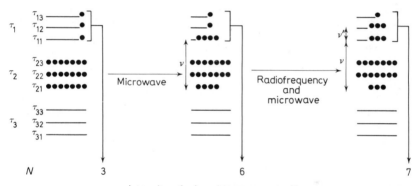

Intensity of phosphorescence $\propto N$

Figure 4.6 The effect of saturation of the electron spin transitions (with microwave radiation) and nuclear spin transitions (with radiofrequency radiation) on the population density of the emitting level (τ_1) of triplet molecules with a nucleus of spin $= 1$ in the triplet state. The effect of saturating the electron spin transition ($\tau_2 \rightarrow \tau_1$) on the phosphorescence intensity is shown by the double-headed arrow on the left hand side. The double-headed arrows on the right hand side demonstrate the principle involved in optical detection of e.n.d.o.r. The number of solid circles in each hyperfine level is proportional to its population density. This figure assumes the absence of spin–lattice relaxation processes, equal rate constants for the $\tau_1 \leadsto S_0$ and the $\tau_2 \leadsto S_0$ processes, and the absence of nuclear polarisation (i.e. the population of $\tau_{11} = \tau_{12} = \tau_{13}$ and that for $\tau_{21} = \tau_{22} = \tau_{23}$) (From El-Sayed, M. A.[59] by courtesy of the American Institute of Physics)

electron orbit interactions. If the rate of formation of the different z.f. states of the triplet state from S_1 is too fast compared with the nuclear spin reorientation time in the magnetic field of the two unpaired electrons, then nuclear polarisation will not be produced in the intersystem crossing process. Thus the steady-state population is expected to be equal for the different nuclear spin levels belonging to the same electron spin level at very low temperatures in zero field. If this is true, then the nuclear transitions in any z.f. level of the triplet state could not be detected optically by a direct method since the corresponding levels would have equal populations and their transitions would not have net radiowave absorption. One can make use of the difference in population of the different z.f. (electron spin) levels to create a net population difference in the hyperfine level. If one saturates two electron–nuclear spin levels continuously with microwave radiation, e.g. τ_{21} and τ_{11} levels in Figure 4.6, then the nuclear levels under saturation would have different populations from the other nuclear levels in the same z.f. manifold (see Figure 4.6). If a radiofrequency sweeper is then used, net absorption would occur, and a corresponding change in the phosphorescence intensity would be observed at frequencies corresponding to nuclear transitions between the nuclear spin levels under saturation with microwave and other nuclear spin levels in the same z.f. manifold, as shown in Figure 4.6. The sensitivity of the optical methods in detecting e.n.d.o.r. transitions can be predicted from the following expression[59].

$$\frac{I^{\nu,\nu'}}{I^{\nu}} = \frac{6k_1K_1 + 2k_1K_2 + k_2K_1}{3k_1K_1 + k_1K_2 + 2k_2K_1} \cdot \frac{k_1 + k_2}{2k_1 + k_2} \tag{4.14}$$

where I^{ν} is the phosphorescence intensity of the emission from z.f. level τ_1 in Figure 4.6 when a pure electron spin transition between z.f. level 1 and 2 is saturated and $I^{\nu,\nu'}$ is the intensity of the emission from z.f. level 1 when, in addition, a nuclear transition in z.f. level 1 is saturated; k and K were defined previously (see equation (4.1)). A number of interesting limits has been discussed in the same reference[59].

It is thus obvious that the energies required to change the direction of the nuclear spin in the field of the electron spin can be determined optically in zero field. This is an optically detected n.m.r.-type experiment whereby the laboratory field is replaced by the field of the two unpaired spins (z.f.) of the triplet state in the molecular framework. The e.n.d.o.r. frequencies are important in determining hyperfine and quadrupole parameters in the excited triplet state[41, 42, 63, 65, 66].

4.6.3.2. Optical detection of electron–electron double resonance (e.e.d.o.r.) transitions in zero field[44]

For molecules whose emission originates from only one z.f. level, e.g. symmetrical aromatic hydrocarbons and their N-heterocyclics, optical methods could only determine the two z.f. transitions involving the emitting level at very low temperatures. In order to determine the third transition, an electron–electron double resonance experiment has to be performed. Let us assume the three electron spin (z.f.) levels are τ_1, τ_2 and τ_3. If τ_1 is the only emitting level, then the saturation of the $\tau_2 \leftrightarrow \tau_3$ transition at low temperature would induce no change in the population of the emitting level τ_1. However, if the $\tau_1 \leftrightarrow \tau_2$ transition is continuously saturated with microwave radiation ν_1 and another microwave sweeper is being swept, the population of the emitting τ_1 level changes, and a corresponding change in the phosphorescence intensity from I^{ν_1} to I^{ν_1,ν_2} is observed when the frequency of the variable frequency microwave v_2 is equal to $(E_2 - E_3)/\hbar$. The following equation[59] shows the sensitivity of the e.e.d.o.r. method

$$\frac{I^{v_1,v_2}}{I^{v_1}} = \frac{K_1 + K_2 + K_3}{K_1 + K_2} \cdot \frac{k_1 + k_2}{k_1 + k_2 + k_3} \tag{4.15}$$

It should be pointed out that the e.e.d.o.r. method is also useful in determining z.f. transition energy between two levels, e.g. $\tau_1 \leftrightarrow \tau_2$ whose steady-state population is the same in the absence of microwave irradiation. If one disturbs this equality by saturating the $\tau_2 \leftrightarrow \tau_3$ or the $\tau_1 \leftrightarrow \tau_3$ transition and then by scanning the $\tau_1 \leftrightarrow \tau_2$ energy region, the energy for the latter transition could be optically detected.

4.6.4   Optical detection in a magnetic field; microwave–optical double resonance (m.o.d.r.)

Optical detection of the Zeeman transitions of the triplet state preceded the optical detection in z.f. (see Section 4.1). Since these former experiments resemble those in ionic crystals, researchers in this field called this technique m.o.d.r. (microwave–optical double resonance). The assignment of the z.f. transitions as well as the relative order of the z.f. levels could be obtained from m.o.d.r. techniques as well as from p.m.d.r. The first reported m.o.d.r. experiment was that by Sharnoff[33] in which the $\Delta m = 2$ transition of the $C_{10}D_8$ triplet state in a biphenyl host was observed by use of amplitude modulation of the microwave power. A few months later, Kwiram reported the optical detection of the $\Delta m = \pm 1$ for phenanthrene in biphenyl[34]. The experiments were performed at a fixed microwave frequency (8.88 GHz) but the magnetic field was slowly swept through the resonance position. The output signal from a photomultiplier detector was stored in a computer of average transients (CAT); one hundred sweeps were required. Four months later the optically detected spectrum of quinoxaline in durene was reported[35]. These workers compared two techniques: in one method, the field was modulated at 12 Hz and phase-sensitive techniques were used for detection; in the other, the microwave was amplitude modulated and phase-sensitive detection was again employed.

These m.o.d.r. techniques have the same advantages as p.m.d.r. methods as they are able to determine z.f. transition of short-lived triplet states. They have the added advantage of being able to modulate either the field or the microwave and thus are able to obtain better signal-to-noise ratios. Furthermore, it is not necessary that the spin–lattice relaxation processes be absent for successful optical detection with the m.o.d.r. methods. In these cases, however, the temperature has to be sufficiently low (e.g. 4.2 K) so that a detectable Boltzmann population difference can be maintained. P.M.D.R. is more convenient than m.o.d.r. for a determination of the z.f. energies since single crystals of known structures are not required for the former technique. A detailed examination of the different factors affecting the sensitivity of m.o.d.r. has been given by Sharnoff[75]. An equally extensive examination of the different experimental techniques used in p.m.d.r. has also recently been made[60].

M.O.D.R. techniques have been used to obtain the magnetic properties of new short-lived systems, e.g. pyrazine[76, 77], benzophenone[78, 79], the oxidation product of durene (carbonyl compound)[80] and cyclopentanone[75].

In addition to the optical detection and assignment of z.f. transitions, m.o.d.r. has also been used by Veeman and van der Waals to show level anticrossing and cross-relaxation effects between the Zeeman levels of the lowest triplet state of benzophenone[47]. If the magnetic field is aligned exactly along one of the magnetic principal axes of a triplet molecule, the field mixes the function corresponding to the other two orthogonal magnetic axes to give the $m = \pm 1$ states. The state which corresponds to the magnetic axis along the field is that correlated with the $m = 0$ high field state. At fields when the $m = 0$ and $m = +1$ (or $m = -1$) states have equal energies, or have energy separation smaller than the radiative line width, a coherent interference takes place. Since the two states have different symmetry properties, their levels cross one another as the magnetic field changes its value. If the magnetic field is not exactly aligned with one of the molecular magnetic principal axes, the $m = 0$ and $m = +1$ (or $m = -1$) would no longer be orthogonal to one another, but will be mixed. Each state would have a small part of the other state. This would result in a repulsion between their energy levels and the phenomenon of level anticrossing, rather than crossing, would be observed. This is found to be the case for the benzophenone[47] crystal due to the difficulty of exact alignment of the magnetic field with the principal magnetic axes. Furthermore, at high exciting light intensities, the phenomenon of cross-relaxation between the electron spin directions of the Zeeman states of different adjacent equivalent triplet molecules has been demonstrated[47]. This possibility was discussed previously by Gorter[81].

M.O.D.R. has not progressed at the same rate as p.m.d.r. The number of contributed papers concerning p.m.d.r. during the past few years is more than five times those concerning m.o.d.r. One important contributing factor is the convenience of using powder, rather than single crystals, in the p.m.d.r. method. This eliminates the difficulties of aligning the single crystal in the magnetic field. Another equally important contributing factor is the fact that the measurements performed with p.m.d.r. are characteristic of the molecule without the external perturbation of the field. A third factor is a result of the convenience in using sweepers and helices rather than magnets and tuned cavities. These factors made p.m.d.r. methods attractive to researchers not only in the magnetic field, but also in other fields, e.g. optical spectroscopy as well as the field of non-radiative relaxation. These will be discussed in the following Sections.

4.7 PHOSPHORESCENCE–MICROWAVE DOUBLE RESONANCE IN OPTICAL SPECTROSCOPY

The first experiments performed on optical detection of the Zeeman transitions were actually focused on determining the z.f. origin of the total intensity of the phosphorescence emission[33–35]. The previous workers did not use a spectrometer as a part of their detection system. Fortunately, most of the phosphorescence intensity of the systems studied by m.o.d.r. originates from a single z.f. level. No optical spectroscopy has been carried out yet with the m.o.d.r. methods in spite of the very important information that can be obtained from it. This is probably due to the fact that researchers in m.o.d.r. came to it by way of e.s.r. It is certain that such studies will soon be made. The importance of p.m.d.r. methods in phosphorescence spectroscopy was first realised[40] in 1969 and its use has since been extended[52, 62, 64, 82–84]. In the next four Sections, the different applications of p.m.d.r. to optical spectroscopy will be briefly discussed.

4.7.1   Amplitude-modulated p.m.d.r. and z.f. origin of emission bands

It has been shown that p.m.d.r. is useful in identifying the z.f. origin(s) of the different vibronic bands in the phosphorescence spectrum[40, 52, 53, 82]. If a vibronic band originates from a single z.f. level and if the z.f. transitions are known and assigned, then the response of the intensity of this band to the microwave saturation of the different z.f. transitions could determine its z.f. origin. On the other hand, the band could be emitted from two z.f. levels. If the molecule has reasonable symmetry (not too high and not too low), the radiation emitted from each z.f. level would have different polarisation characteristics. In these systems, microwave saturation of the z.f. transitions connecting the two radiative z.f. levels would have little effect on the intensity of the band but could have large effects on the polarisation characteristics. This is due to the fact that microwave saturation has the effect of transferring the excess molecules from one level to the other. If the radiative rate constant of the two levels are comparable, the excess molecules emit radiation at the same rate but with different polarisation. It is thus convenient to conclude the z.f. origin of the vibronic bands from the following simple rules:

(i) For bands of single origin, saturation of two z.f. transitions should change its intensity, but not its polarisation. Saturation of the third transition or the one connecting the levels that do not contribute to the intensity of the band, should not have any effect on the intensity or the polarisation of the band.

(ii) For bands of mixed origin, with two z.f. levels emitting, in molecules of reasonable symmetry, saturation of all three z.f. transitions could change the intensity; the one with the least effect should cause large changes in the polarisation of the band. The latter transition connects the z.f. origins of the band.

If the symmetry increases or decreases drastically, the spin–lattice relaxation process is fast, or the degree of spin alignment is not very large, the above rules would not be as definite. A convenient experimental method by which these experiments are performed is the use of phase-sensitive detection. The microwave frequency that is capable of saturating a certain z.f. transition is modulated, and the phosphorescence intensity change is detected at the same modulation frequency while the optical spectrum is slowly scanned. The spectrum obtained is termed[53] the amplitude-modulated phosphorescence–microwave double resonance spectrum (a.m.p.m.d.r.) (the second spectrum in Figure 4.7). It is better to take this spectrum when the solute is in a single crystal of the host and an analyser, parallel to one of the crystal

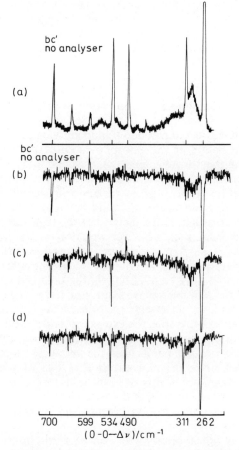

bc'
no analyser

(a)

bc'
no analyser

(b)

(c)

(d)

700   599 534 490   311 262

$(0-0-\Delta\nu)/cm^{-1}$

Figure 4.7 Part of the phosphorescence spectrum of 2,3-dichloroquinoxaline in durene at 1.6 K; (a) conventional spectrum, (b) amplitude-modulated phosphorescence–microwave double resonance (a.m.p.m.d.r.) spectrum, saturating the τ_L–τ_M transitions at 1.055 GHz. The polarised a.m.p.m.d.r. spectra, (c) and (d) show the phosphorescence resolved into the sub-spectra that originate from the τ_L and τ_M z.f. levels, respectively

(From El-Sayed, M. A. *et al.*[53] by courtesy of North-Holland Publishing Company, Amsterdam)

faces, is used with its transmission maximum along one of the crystal axes. The spectrum is also recorded with the analyser axis rotated by 90 degrees. The spectrum[53] obtained is termed the polarised (a.m.p.m.d.r.) spectrum (see Figure 4.7). In this spectrum, bands that originate from one single level should show only a change in intensity upon changing the direction of the polariser, while bands of mixed origin could show additional change in phase (or phase angle).

It should be pointed out that once the z.f. origin of the different bands, in particular the true and false origins of the spectrum, is determined, a complete description could be given for the spin–orbit perturbations that give the lowest triplet state its radiative properties. A very extensive work is by Tinti and El-Sayed[52] in which different perturbations, e.g. heating, applying a magnetic field, and saturation of the z.f. transitions with microwave radiation are used to determine the property of the individual z.f. levels. The effect of these perturbations, not only on the phosphorescence spectrum, but also on the observed decays and polarisations, has been examined. Limits on the importance of the different spin–orbit interactions are obtained. This spectroscopic work represents the type of experiments that can be done and the kind of information that can be obtained from p.m.d.r. and other methods.

4.7.2 Amplitude-modulated p.m.d.r. and spectral analysis

One very powerful utilisation of the a.m.p.m.d.r. studies is to differentiate between bands that belong to the molecule of interest from those belonging to impurities or emission from different sites. In many instances, one obtains bands that are separated from each other by lattice-vibration frequencies. However, they could very well be due to emission from different sites of the guest molecule under examination in the lattice, as shown in Figure 4.8. These emissions complicate the analysis of the emission, in particular in the region of the spectrum away from the 0—0 band (true origin). The fact that sites (or impurities) do have different z.f. energies that could be resolved in the microwave region makes a.m.p.m.d.r. spectroscopy an attractive technique that could be used to obtain the spectrum of each site (separately). It could also be used to obtain the spectrum of the molecules under study without any impurity bands. Furthermore, progressions originating from different z.f. origins could be obtained in different spectra or in the same spectrum with a different sign for the intensity changes observed when different z.f. transitions are saturated. This would identify false origins, assist in assigning the symmetry of the observed vibrations and greatly assist in making the spectral analysis.

4.7.3 Amplitude-modulated p.m.d.r. and assignment of triplet-state symmetry

The spacial symmetry of the emitting state can be uniquely determined for symmetrical molecules from a knowledge of the z.f. origin of the 0–0 band of the phosphorescence emission and its spin–orbit origin as determined from the polarisation of the emission. On the other hand, if the z.f. origin and the symmetry of the emitting state is known, the spin–orbit interaction responsible for the emission could be identified.

In spite of the extensive optical research that has been carried out on the spectroscopy of the phosphorescence of benzene, no definitive and direct

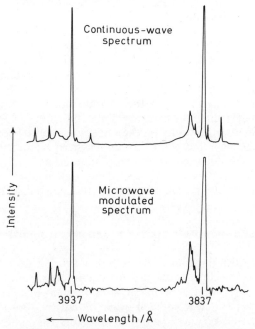

Figure 4.8 The use of a.m.p.m.d.r. spectroscopy in differentiating between site or impurity bands and emission bands of the molecules of interest. Top spectrum is the conventional phosphorescence spectra of pyrazine in benzene at 1.6 K. There are few bands close in energy to each strong vibronic band. These could either be sites, impurities or phonon bands. The spectrum at bottom is an a.m.p.m.d.r. spectrum obtained while modulating the 10.1 GHz z.f. transition at 1.6 K. The disappearance of the high-energy bands, both in this spectrum as well as in another spectrum obtained when modulating the other z.f. transitions, strongly supports the assignment of these bands to sites. The low-intensity band at the low-energy side of each of the strong bands belongs to the same band system to which the strong vibronic bands belong (From Tinti, D. S. and El-Sayed, M. A.[83] by courtesy of Academic Press Inc., New York)

proof of the assignment of the emitting state could be given[17b] from the properties of the observed emission. The reason for this is that the studies were made at 77 K, at which temperature the structure of benzene is D_{6h}. In this high symmetry, the emission is symmetry- as well as spin-forbidden. The 0–0 band is thus not observed. The vibronically induced part of the spectrum, which constitutes the total emission intensity, can be successfully described by spin–orbit schemes equally well for the $^3B_{1u}$, as well as for the $^3B_{2u}$ state. C_6H_6 in C_6D_6 at 1.6 K is found by e.s.r.[85] to be of D_{2h} symmetry in the triplet state. The emission is thus symmetry-allowed and the 0–0 band is allowed by internal direct spin–orbit perturbation. The determination of the z.f. origin of this band could identify not only the special symmetry of the state but also the spin–orbit scheme responsible for the radiation. The emitting state is found[84] to correlate with the $^3B_{1u}$ state in D_{6h}, in agreement with previous identifications. Furthermore, in agreement with the observed polarisation of the 0–0 band of condensed aromatic hydrocarbons, e.g. naphthalene and phenanthrene, the direct spin–orbit interaction between the $T(\pi,\pi^*)$ state and the σ,π^* singlet state is found to be responsible for the 0–0 band intensity of benzene as well[84]. More recently, the symmetry as well as the relative order of the z.f. levels has been determined[64] for sym-1,2,4,5-tetrachlorobenzene. This is done from a knowledge of the spin–orbit coupling routes (both direct and vibronically induced), the polarisation of the magnetic z.f. transitions and the lifetimes of the individual z.f. levels.

4.8 PHOSPHORESCENCE–MICROWAVE DOUBLE RESONANCE AND THE NON-RADIATIVE PROPERTIES OF THE TRIPLET STATE

4.8.1 Intramolecular process

The first discussion of the mechanism of the intersystem crossing process in N-heterocyclic molecules was made in 1961[86, 87]. In this work, it was suggested that the strong phosphorescence properties and weak fluorescence characteristics of these and aromatic carbonyls is due to the strong spin–orbit interaction between n,π^* and π,π^* singlet and triplet states. This would suggest that if the lowest excited singlet state is of the n,π^* type, the most probable intersystem crossing is to a triplet state of the π,π^* type, and vice versa.

The first experimental confirmation of the above proposal was made from the studies involving the detection of the state of spin alignment[38, 88] from the optical decay at very low temperatures. The results showed that the strong radiative level is populated most quickly by the intersystem crossing process. From a knowledge of the symmetry of the spin radiative level, it was easily concluded that, in quinoxaline, the most favoured intersystem crossing process is $S(n,\pi^*) \rightsquigarrow T(\pi,\pi^*)$. The first quantitative measure of the relative intersystem crossing probability to the different z.f. levels, and thus the relative importance of the different mechanisms, was made[89] in 1968. This was done by breaking down the decay curve of pyrazine at 1.4 K. The three z.f. levels of the lowest triplet state of this molecule are shown[90] to be radiative. At very low temperatures, when the spin–lattice relaxation is assumed to be absent, the steady-state initial intensity ratios of the different decay components of the decay curve are given by:

$$I^0_i/I^0_j = k^r_i\, n_i/k^r_j\, n_j \approx k_i\, n_i/k_j\, n_j = K_i/K_j \qquad (4.16)$$

In this expression, the observed decay constants are assumed to be radiative, i.e., the quantum yield of the phosphorescence is assumed to be unity. Later other expressions, using p.m.d.r. methods, are given from which the ratio of the radiative constants may be determined[45].

The important equations used in the different methods of determining the ratios of the intersystem crossing rates are given in Section 4.4.3 by equations (4.5), (4.7)–(4.9) and (4.11)–(4.13). These equations are to be used in determining the intersystem crossing ratios by the different techniques, when the radiation monitored is emitted from a single z.f. level. Some of these equations, and thus methods, have not as yet been demonstrated, but they certainly will be useful in the future. A detailed description of the different experimental techniques used at the present time, together with equations derived for molecules whose emission results from two z.f. levels are given in a recent review of the subject[91]. In this Section, we will only give a brief description of these p.m.d.r. methods. In all but the pulsed excitation method, the following assumptions are made: (i) The spin–lattice relaxation is absent and (ii) the microwave radiation either saturates or inverts the population of the z.f. levels being coupled. The pulsed-excitation method makes the second assumption, as well as the assumption that pumping the different z.f. levels occurs from the same excited singlet state, presumably the lowest one, in a time shorter than the decay and spin–lattice relaxation times.

4.8.1.1   Microwave-induced delayed phosphorescence (m.i.d.p.) method[43]

In addition to its usefulness in determining the K_i/K_j ratios, this technique is powerful in resolving the values of the decay constants of the different z.f. levels, even if they are comparable in magnitude. The technique has been applied to obtain K_i/K_j for the case when one z.f. level is strongly radiative and the other two levels are long lived. The principle involved depends on the fact that if level i has a very short radiative lifetime as compared to j, then after a time $t \gtrsim 5/k_i$, the population of level i decays to an almost zero value, while that of level j is still appreciable. If a microwave pulse of frequency ν_{ij} and of sufficient power to saturate the $\tau_i \leftrightarrow \tau_j$ transition is now applied, $\frac{1}{2} n_j(t)$ will now be in level τ_i which then gives a pulse of phosphorescence signal whose leading edge, $h(t)$, is proportional to $n_j(t)k_i^r$. Starting with the same steady-state population for each level, prior to cutting off the excitation, these pulses can be applied with different delay times. The time dependence of $h(t)$ reflects the time dependence of $n_j(t)$. A plot of $\ln(h(t))$ v. time should give a straight line with a slope, k_j. The decay of any one pulse in this case is determined by k_i. In the special case for systems having one strongly radiative level, i, the intensity at $t = 0$ is given by:

$$I_0 = C n_i^0 k_i^r \tag{4.17}$$

At any time $\qquad h(t) = \frac{1}{2} C n_j(t) k_i^r = \frac{1}{2} C n_j^0 e^{-k_j t} \cdot k_i^r \tag{4.18}$

$\therefore \quad I_0/h(t) = 2 n_i^0/(n_j^0 e^{-k_j t}) = \dfrac{2 K_i/k_i}{(K_j/k_j)e^{-k_j t}} = 2(K_i/K_j)(k_j/k_i) e^{+k_j t} \tag{4.19}$

Thus, from the initial intensity and the leading edge of the microwave-induced phosphorescence pulse resulting from the microwave pulse given at a delay time t after cutting off the excitation, the ratio K_i/K_j can be determined.

Equations have recently been derived[91] which extends the m.i.d.p. method to systems whose monitored emission originates from two z.f. levels.

4.8.1.2 Continuous-wave method

In some systems, a detectable state of spin alignment can exist in the steady-state mode of excitation and decay. The application of microwave radiation with resonance frequencies could change the observed phosphorescence intensity from a single z.f. level[59] or the relative intensity from any two z.f. levels[45, 91]. For the case of an emission from a single z.f. level, equation (4.5) has been used previously[59, 52] to determine the intersystem-crossing rate ratios. The continuous wave (c.w.) techniques seem to be reliable. However, the fact that in a number of systems the degree of spin alignment is not large under stead-state conditions might limit its usefulness. A better signal-to-noise ratio could be obtained if the mode of excitation and decay remains under steady conditions but the microwave is swept rapidly to equalise or to invert the population of the levels. In this case, equation (4.7) is used if the microwave saturates the z.f. transition being swept, and equation (4.11) if the microwave radiation leads to a complete population inversion. This technique has not as yet been demonstrated.

4.8.1.3 Pulsed-excitation method[50]

In this method, the system studied is pumped and irradiated at microwave frequencies at a faster rate than its decay rate. In this case, the ratio of the intersystem crossing rates is measured by the ratio of the population of the z.f. field. The latter can be measured from the intensity ratio before and after the fast microwave sweep, as given in equations (4.9) or (4.13) depending on whether the microwave radiation saturates or completely inverts the population of the levels. These equations assume that the pumping of the z.f. level occurs from the same excited singlet state, presumably the lowest excited singlet state. This technique does not require the spin–lattice relaxation to be frozen, as long as the pulsed excitation and sweep of the microwaves occur in a time short compared with the spin–lattice relaxation times. This can only occur if (i) the intersystem crossing process is faster than the spin–lattice relaxation processes, (ii) a very short excitation pulse is used, and (iii) fast microwave sweep or very short microwave pulse are employed. This, in principal, could be used at temperatures higher than liquid He temperatures. In order to increase the microwave sweep rate and still be able to saturate or invert the z.f. population, high microwave powers will be required.

4.8.1.4   Important results

In addition to the results obtained by the optical methods in z.f., a number of important results concerning the relative intersystem crossing processes in a number of aromatic hydrocarbons have been obtained by use of e.s.r. methods[92] without selective excitation of the guest. It is thus expected that the ratio of the pumping of the z.f. levels obtained in these experiments is a result of a combination of host–guest energy transfer as well as intra-molecular intersystem crossing processes following direct absorption of the guest molecules.

The important results obtained so far concerning the relative intersystem crossing processes can be summarised as follows:

(i) In N-heterocyclic molecules, the z.f. level of the lowest triplet state which is most favoured is the one which could be involved in an intersystem-crossing process involving direct spin–orbit coupling between singlet and triplet states of different electronic type, e.g., $S(n,\pi^*) \rightsquigarrow T(\pi,\pi^*)$ or $S(\pi,\pi^*) \rightsquigarrow T(n,\pi^*)$. This is in agreement with previous proposals concerning the inter-system crossing process in these systems[86, 87].

(ii) In aromatic hydrocarbons[92, 93], the z.f. level favoured in the inter-system crossing process is the one which is coupled to the lowest singlet state by spin–orbit coupling in a non-planar configuration. One cannot conclude at the moment whether the non-planarity is a result of Herzberg–Teller type interaction or simple non-planar equilibrium configuration for some of the excited states involved in the intersystem crossing process.

It has been pointed out recently[94] that the above two types of results obtained for aromatic hydrocarbons and N-heterocyclic molecules are consistent with the following conclusion: The strongly radiative z.f. level is also the most favoured in the intersystem crossing process. This conclusion is not hard to explain as it merely re-states the fact that the z.f. level which interacts strongly with the singlet manifold via spin–orbit interaction would be favoured in both the non-radiative and radiative processes involving a change in the spin quantum number. In aromatic hydrocarbons, strong spin–orbit coupling involves the lowest triplet π,π^* state and the high energy σ,π^* (or π,σ^*) singlet states. On the other hand, in N-heterocyclic molecules the coupling is between the π,π^* and n,π^* states. It is thus clear that in N-heterocyclic molecules, the triplet manifold is mixed with the lowest singlet state by spin–orbit interaction and there is no need for vibronic or geometrical changes to cross over from the lowest singlet state to the triplet manifold. In aromatic hydrocarbons, the triplet states are mixed with singlet σ,π^* states, whereas the lowest singlet state is of the π,π^* type. In order for the σ,π^* and π,π^* singlet states to mix, a non-totally symmetric vibration or geometrical distortion that would eliminate the plane of symmetry is required. This is in agreement with the experimental conclusions.

4.8.2   Intermolecular triplet–triplet energy transfer

The state of spin alignment in the lowest triplet state can be prepared[59] by a number of other methods in addition to its preparation by singlet–singlet absorption, e.g. by ground state singlet → triplet absorption, excitation by electron scattering and donor–accepter energy transfer. In these cases, the relative pumping rates are no longer related to relative intersystem rates. The equations derived in Section 4.4.3 are still applicable if K_i and K_j are rates of pumping. For excitation using the $S_0 \to T_{\tau i}$ and $S_0 \to T_{\tau j}$, the K values are the radiative rate constants (or absorption cross-sections) for these two transitions. For pumping using triplet–triplet energy transfer, the pumping relative rates depend on the relative orientation between the magnetic axes of the donor and accepter as well as the relative population of the z.f. levels of donor molecules. The equations governing these processes are given elsewhere[59].

An interesting experiment has been carried out which showed[95] that (i) spin alignment can be produced from triplet–triplet energy transfer and (ii) if the triplet–triplet energy transfer is due to an exchange interaction, the probability of the transfer between the z.f. level i of the donor and the j level of the accepter is proportional to the square of the projection of the magnetic axis corresponding to one level onto that for the other. The latter observation showed that the spin direction is conserved in the triplet–triplet energy transfer process. Furthermore, it confirmed the exchange-type mechanism previously proposed[96] for the transfer. Similar experiments have been carried out more recently by other workers[79, 97, 98] using conventional e.s.r. or m.o.d.r. methods.

Acknowledgements

The author wishes to thank the effort and stimulation of his collaborators for the past few years, in particular, Drs Chen, Gwaiz, Hall, Kalman, Owens and Tinti. The author wishes to thank Professor Moomaw for reading the manuscript. The financial support of the U.S. Atomic Energy Commission is gratefully acknowledged. This is contribution No. 2903 from the Chemistry Department, University of California, Los Angeles.

References

1. Wiedemann, E. and Schmidt, G. C. (1895). *Amer. Phys.*, **56**, 201
2. Sklar, A. (1937). *J. Chem. Phys.*, **5**, 599
3. Jablónski, A. (1935). *Z. Phys.*, **94**, 38
4. Lewis, G., Lipkin, D. and Magel, T. (1961). *J. Amer. Chem. Soc.*, **63**, 3005
5. Terenin, A. (1943). *Acta Physico Chim. USSR*, **18**, 210; (1944). *Zhur. Fiz. Kkim.*, **18**, 1
6. Lewis, G. and Kasha, M. (1944). *J. Amer. Chem. Soc.*, **66**, 2100
7. Lewis, G. and Calvin, M. (1945). *J. Amer. Chem. Soc.*, **67**, 1232
8. Lewis, G., Calvin, M. and Kasha, M. (1949). *J. Chem. Phys.*, **17**, 804
9. Weissman, S. and Lipkin, D. (1942). *J. Amer. Chem. Soc.*, **64**, 1916
10. McClure, D. (1949). *J. Chem. Phys.*, **17**, 905
11. Hutchison, C. A., Jr. and Mangum, B. W. (1958). *J. Chem. Phys.*, **29**, 952
12. Hutchison, C. A., Jr. and Mangum, B. W. (1960). *J. Chem. Phys.*, **34**, 908
13. Kasha, M. (1950). *Discuss. Faraday Soc.*, **9**, 14
14. McClure, D. (1952). *J. Chem. Phys.*, **20**, 682
15. Hameka, H. and Oosterhoff, L. (1958). *Mol. Phys.*, **1**, 358
16. Mizushima, M. and Koide, S. (1952). *J. Chem. Phys.*, **20**, 765

17a Lower, S. K. and El-Sayed, M. A. (1966). *Chem. Rev.,* **66,** 199

17b McGlynn, S. P., Azumi, T. and Kinoshita, M. (1969). *Molecular Spectroscopy of the Triplet State,* (New York: Prentice-Hall)

18a El-Sayed, M. A. and Pavlopoulos, T. (1963). *J. Chem. Phys.,* **39,** 1899

18b Pavlopoulos, T. and El-Sayed, M. A. (1964). *J. Chem. Phys.,* **41,** 1082

19a Clarke, R., Hochstrasser, R. M. and Marzzacco, C. J. (1969). *J. Chem. Phys.,* **51,** 5015

19b Chaudhuri, N. K. and El-Sayed, M. A. (1965). *J. Chem. Phys.,* **43,** 1423

20. Yager, W. A., Wasserman, E. and Cramer, R. M. R. (1962). *J. Chem. Phys.,* **37,** 1148

21. Wasserman, E., Snyder, L. C. and Yager, W. A. (1964). *J. Chem. Phys.,* **41,** 1763

22. van der Waals, J. H. and de Groot, M. S. (1959). *Mol. Phys.,* **2,** 333

23. de Groot, M. S. and van der Waals, J. H. (1960). *Mol Phys.,* **3,** 190

24. Kottis, P. and Lefebure, R. (1964). *J. Chem. Phys.,* **41,** 3660

25. Lhotse, J. M., Haug, A. and Ptak, M. (1966). *J. Chem. Phys.,* **44,** 648, 654

26. El-Sayed, M. A. and Siegel, S. (1966). *J. Chem. Phys.,* **44,** 1416

27. Fermi, E. and Rasetti, F. (1925). *Nature (London),* **115,** 764

28. Breit, G. and Ellet, A. (1925). *Phys. Rev.,* **25,** 888

29. Varsangi, F., Wood, D. C. and Schawlow, A. L. (1959). *Phys. Rev. Lett.,* **3,** 544

30. Geschwind, S., Collins, R. and Schawlow, A. L. (1959). *Phys. Rev. Lett.,* **3,** 545

31. Brossel, J., Geschwind, S. and Schawlow, A. (1959). *Phys. Rev. Lett.,* **3,** 549

32. For a review see: Bernhein, R. A. (1965). *Optical Pumping: An Introduction.* (New York: W. A. Benjamin, Inc.)

33. Sharnoff, M. (1967). *J. Chem. Phys.,* **46,** 3263

34. Kwiram, A. L. (1967). *Chem. Phys. Lett.,* **1,** 272

35. Schmidt, J., Hesselmann, I. A., de Groot, M. S. and van der Waals, J. (1967). *Chem. Phys. Lett.,* **1,** 434

36. Schwoerer, M. and Wolf, H. C. (1966). *Proc. of the XIVth Collogue Ampére,* Ed. by Blinc, R. p. 87 (Amsterdam: North Holland)

37. Hornig, A. W. and Hyde, J. S. (1963). *Mol. Phys.,* **6,** 33

38. de Groot, M. S., Hesselmann, I. A. and van der Waals, J. H. (1967). *Mol. Phys.,* **12,** 259

39. Schmidt, J. and van der Waals, J. H. (1968). *Chem. Phys. Lett.,* **2,** 640

40. Tinti, D. S., El-Sayed, M. A., Maki, A. H. and Harris, C. B. (1969). *Chem. Phys. Lett.,* **3,** 343

41. Chan, I. Y., Schmidt, J. and van der Waals, J. H. (1969). *Chem. Phys. Lett.,* **4,** 269

42. Harris, C. B., Tinti, D. S., El-Sayed, M. A. and Maki, A. H. (1969). *Chem. Phys. Lett.,* **4,** 409

43. Antheunis, D. A., Schmidt, J. and van der Waals, J. H. (1970). *Chem Phys. Lett.,* **6,** 255

44. Kuan, T. S., Tinti, D. S. and El-Sayed, M. A. (1970). *Chem. Phys. Lett.,* **4,** 507

45. El-Sayed, M. A. (1970). *J. Chem. Phys.,* **52,** 6438

46. El-Sayed, M. A. and Kalman, O. F. (1970). *J. Chem. Phys.,* **52,** 4903

47. Veeman, W. S. and van der Waals, J. H. (1970). *Chem. Phys. Lett.,* **7,** 65

48. Schmidt, J., Van Dorp, W. G. and van der Waals, J. H. (1971). *Chem. Phys. Lett.,* **8,** 345

49. Francis, A. H. and Harris, C. B. (1971). *Chem. Phys. Lett.,* **9,** 188

50. El-Sayed, M. A. and Olmsted, J. (1971). *Chem. Phys. Lett.,* in the press

51. Levenon and Weissman, S. (1971). *J. Amer. Chem. Soc.,* **93,** 4309

52. Tinti, D. S. and El-Sayed, M. A. (1971). *J. Chem. Phys.,* **54,** 2529

53. El-Sayed, M. A., Owens, D. V. and Tinti, D. S. (1970). *Chem. Phys. Lett.,* **6,** 395

54. El-Sayed, M. A., Tinti, D. S. and Owens, D. V. (1969). *Chem. Phys. Lett.,* **3,** 339

55. Azumi, T., O'Donnell, C. M. and McGlynn, S. P. (1966). *J. Chem. Phys.,* **45,** 2735

56. Hall, L., Armstrong, A., Moomaw, W. and El-Sayed, M. A. (1968). *J. Chem. Phys.,* **48,** 1395

57. El-Sayed, M. A., Moomaw, W. R. and Tinti, D. S. (1969). *J. Chem. Phys.,* **50,** 1888

57a Hall, L. H., Owens, D. V. and El-Sayed, M. A. (1971). *Mol. Phys.,* **20,** 1025

58a Harris, C. S. (1970). *Proc. of 5th Molecular Crystal Symposium,* Philadelphia, Pennsylvania

58b Harris, C. S. (1971). *J. Chem. Phys.,* **54,** 972

59. El-Sayed, M. A. (1971). *J. Chem. Phys.,* **54,** 680

60. Olmsted, J. and El-Sayed, M. A. (1972). *Experimental Methods on Phosphorescence–Microwave Double Resonance,* to be published in *The Creation and Detection of the Excited State,* (New York: Marcel Dekker)

61. Hall, L. and El-Sayed, M. A., unpublished results.

62. Owens, D., El-Sayed, M. A. and Ziegler, S. (1970). *J. Chem. Phys.,* **52,** 4315

63. Buckley, M. J. and Harris, C. B. (1970). *Chem. Phys. Lett.*, **5**, 205
64. Chen, C. R. and El-Sayed, M. A. (1971). *Chem. Phys. Lett.*, **10**, 307
65. Buckley, M. J., Harris, C. B. and·Maki, A. H. (1970). *Chem. Phys. Lett.*, **4**, 591
66. Schmidt, J. and van der Waals, J. H. (1969). *Chem. Phys. Lett.*, **3**, 546
67. Fayer, M. D., Harris, C. B. and Yuen, D. A. (1970). *J. Chem. Phys.*, **53**, 4719
68. Hutchison, C. A., Jr. and Pearson, G. A. (1967). *J. Chem. Phys.*, **47**, 520; and Hutchison, C. A. and Kohler, B. E. (1969). *J. Chem. Phys.*, **51**, 3327
69. McCalley, R. C. and Kwiram, A. L. (1970). *Phys. Rev. Lett.*, **24**, 1279
70. Francis, A. H. and Harris, C. B. (1971). *Chem. Phys. Lett.*, **9**, 181
71. Kalman, O. F. and El-Sayed, M. A. (1971). *J. Chem. Phys.*, **54**, 4414
72. For basic references see: Abragam, A. (1961). *The Principles of Nuclear Magnetism*, (Oxford: The Clarendon Press)
73. Pople, J. A., Schneider, W. G. and Bernstein, H. J. (1959). *High Resolution Nuclear Magnetic Resonance*, (New York: McGraw-Hill Book Co.)
74. Slichter, C. P. (1963). *Principles of Magnetic Resonance* (New York: Harper and Row Publishers, Inc.)
75. Sharnoff, M. (1969). *Mol. Crystallogr.*, **9**, 265
76. Sharnoff, M. (1968). *Chem. Phys. Lett.*, **2**, 498
77. Cheng, L. J. and Kwiram, A. L. (1969). *Chem. Phys. Lett.*, **4**, 457
78. Sharnoff, M. (1969). *J. Chem. Phys.*, **51**, 451
79. Chan, I. Y. and Schmidt, J. (1969). *Symposium on Magneto Optical Effects*, Faraday Soc. No. 3
80. Sharnoff, M. (1969). *Mol. Crystallogr.*, **5**, 297
81. Gorter, C. J. (1969). in *Polarisation, Matiére et rayonnement*, p. 259. (Paris: Societe Francais de Physique)
82. Tinti, D. S. and El-Sayed, M. A. (1969). Presentation at the *International Conference on Molecular Luminescence*, Newark, Delaware, August 1969
83. Tinti, D. S. and El-Sayed, M. A. (1971). *Organic Scintillators and Liquid Scintillation Counting*, p. 563, (New York: Academic Press)
84. Gwaiz, A. A., El-Sayed, M. A. and Tinti, D. S. (1971). *Chem. Phys. Lett.*, **9**, 454
85. de Groot, M. S., Hesselmann, I. A. and van der Waals, J. H. (1967). *Mol. Phys.*, **13**, 583
86. El-Sayed, M. A. (1962). *J. Chem. Phys.*, **36**, 573
87. El-Sayed, M. A. (1963). *J. Chem. Phys.*, **38**, 2834
88. van der Waals, J. H. and de Groot, M. S. (1967). *The Triplet State*, Ed. by Zahlan, A. B., p. 101, (London: Cambridge University Press)
89. El-Sayed, M. A. (1968). in *Proceedings of the International Conference on Molecular Luminescence*. (Loyola University, Chicago, Illinois). Ed. by Lim, E., p. 715. (New York: W. Benjamin Inc.)
90. El-Sayed, M. A., Moomaw, W. and Tinti, D. (1969). *J. Chem. Phys.*, **50**, 1888
91. El-Sayed, M. A. (1972). *Advances in Electronic Excitation and Relaxation*, Ed. by Lim, E., Vol. 1, in the press, (New York: Academic Press)
92. Sixl, H. and Schwoerer, M. (1970). *Z. Naturforsch.*, **25a**, 1383
93. El-Sayed, M. A. and Chen, C. R. (1971). *Chem. Phys. Lett.*, **10**, 313
94. El-Sayed, M. A. (1971). *J. Chem. Phys.*, to be published
95. El-Sayed, M. A., Tinti, D. S. and Yee, E. M. (1969). *J. Chem. Phys.*, **51**, 5721
96. Dexter, D. L. (1953). *J. Chem. Phys.*, **21**, 836
97. Clarke, R. H. (1970). *Chem. Phys. Lett.*, **6**, 413
98. Sharnoff, M. and Iturbe, E. B. (1971). *Phys. Rev. Lett.*, **27**, 576
99. El-Sayed, M. A. (1971). *Accounts Chem. Res.*, **4**, 23

5
Band Contour Analysis

J. C. D. BRAND
University of Western Ontario, London

5.1   INTRODUCTION

The history of band contour analysis can be traced to the early decades of
the century. In 1913, Burmeister[1] obtained rough values of the rotational
constant B from the separation of PR-maxima in the infrared bands of CO and
HCl. The treatment was extended to parallel bands of symmetric rotors by
Gerhard and Dennison in 1933 [2]. Asymmetric rotors are considerably more
complex and were discussed in a number of papers, culminating in the classic
work of Badger and Zumwalt[3] which laid the foundations for the assignment
of infrared fundamentals from vapour-phase spectra. Coriolis effects were
first taken into account in connection with the envelopes of spherical rotor
bands[4].

   These early developments were concerned with features rather than detail
in the rotational envelope and were more or less restricted to vibration–
rotation bands by the extra computational problems that arise when the
moment of inertia changes are other than small. Attempts to account for
incompletely resolved rotational structure using punched-card methods were
almost overwhelmed, even for vibrational bands of small molecules, by the
weight of computation[5]. Rapid evolution has occurred only in the past
decade, corresponding to the availability of fast digital computers. Nearly all
the successful applications to electronic bands have taken place within this
period.
   In the electronic and vibrational bands of small molecules the intensity is
partitioned among about one thousand lines spread over a frequency range
of a few hundred cm^{-1}. A spectrum of this type may be described as rotationally resolved since, with the possible exception of crowded regions near the
band centre, the separation of neighbouring lines is, on average, greater than
the line-width. The density of lines increases rapidly with molecular size,
partly because the levels are more densely packed in both upper and lower
states of the transition, but also because the number of thermally populated
ground-state levels rises with decreasing spacing. Each contributes approximately in proportion to the moment of inertia, so that the overall density of
lines in the band increases with the fourth power of the molecular dimension.
A typical large molecule such as benzene or naphthalene then has 20 000–
50 000 individual transitions compressed into a region covering a few tens of
cm^{-1}; the average density of lines – of the order 1000 per cm^{-1} – is such that
transitions cannot possibly be seen separately. In the instrumentation of
electronic spectroscopy, conventional grating spectrometers have a limiting
resolution of 0.03–0.05 cm^{-1}, while ultraviolet–visible interferometer spectrometers as well as high-resolution infrared spectrometers have a wavenumber
resolution in the range 0.01–0.02 cm^{-1}. Electronic spectra of interest are
normally recorded under conditions that the line-width is primarily Doppler-
determined (that is, the width is 0.03 cm^{-1} or more), thus even the grating
spectrometer is capable of resolving as many separate peaks as may exist in
the band envelope. The higher resolution of the interferometer spectrometer
can be expected to give more detailed information on the profile of individual
features in the fine structure though no developments in this direction have
yet been reported. In vibrational spectra the much smaller Doppler width
(c. 0.002 cm^{-1}) means that considerations of instrumental resolving power
usually dominate the band contour.
   The objectives of band contour analysis are of course the same as those
of rotational analysis of a resolved spectrum, namely, the determination of
electronic or vibrational symmetry and the measure of molecular constants.
Although it may give useful corroborative evidence, vibrational analysis of
an electronic band system is seldom sufficient by itself to establish the excited-
state symmetry. To the present time, contour analysis has tended to concentrate on rotationally unresolved spectra but many examples are known
of molecules where partially resolved rotational structure is certainly
approachable by the contour method; possibly a puritan work ethic has
dictated that such cases are reserved for traditional methods. The existence
of rotational 'fine structure' in electronic bands of large molecules has been

recognised for about three decades; the first example – benzene – was recorded in 1942 [6]. By the end of the 1950s, similar observations were available for the isomeric xylenes[7] and very detailed, regular structure had been observed in the π^*-n bands of several aza derivatives of benzene[8–10]. The spectrum of naphthalene vapour, though showing only one or two main features in the rotational structure, provided an early example of the determination of vapour state electronic symmetry from a band profile[11]. What follows is an account of developments spurred by these pioneer observations, especially those of the past five years when computational problems have been largely overcome.

5.2 THEORY OF MOLECULAR ROTATION

5.2.1 Rotational energy levels of rigid molecules

The well-known theory of molecular rotation[12, 13a] is here outlined briefly as an introduction to the approximations valid for large molecules.

The *rigid* rotor Hamiltonian may be written:

$$(1/hc)H_{rot} = AP_a^2 + BP_b^2 + CP_c^2 \tag{5.1}$$

in which the P's are angular momentum operators and the rotational constants A, B, C are so chosen that $A \geqslant B \geqslant C$. A symmetric rotor corresponds to one of the limits $A = B$ (oblate rotor) or $B = C$ (prolate rotor), whilst the spherical rotor has $A = B = C$. The energy matrix for the symmetric (and spherical) rotor is diagonal, corresponding to the familiar closed-form expressions,

$$F(J, K_c) = BJ(J+1) - (C-B)K_c^2 \quad \text{(oblate)} \tag{5.2}$$

and

$$F(J, K_a) = BJ(J+1) + (A-B)K_a^2 \quad \text{(prolate)} \tag{5.3}$$

where J and K_c or K_a are, respectively, the quantum numbers for total angular momentum and its component on the molecular figure axis. In the absence of external magnetic or electric fields, space quantisation (quantum number, M) contributes to the degeneracy but not to the energy of a given level. The eigenfunctions of H_{rot}, except for a normalisation factor, are the Wigner functions $D^J(\theta\phi\chi)$ [14] which for the special case of $M = 0$ reduce to the ordinary spherical harmonics.

For the asymmetric rotor ($A \neq B \neq C$), closed-form solutions like those in equations (5.2) and (5.3) are not available. The usual approach utilises a basis of symmetric rotor wave functions, when the matrix elements of H_{rot} are diagonal in J but have non-zero off-diagonal elements in K of the form $K, K \pm 2$. Evidently, these off-diagonal elements mix levels of the same parity in K so that there is factorisation into two sub-matrices E (even K) and O (odd K), and a further twofold factorisation can be achieved by forming symmetric ($+$) and antisymmetric ($-$) combinations of the functions $|JkM\rangle$ ($K = |k|$). No more simplification is possible; thus, for each value of J, the problem of computing asymmetric rotor eigenvalues reduces to that of diagonalising four sub-matrices E^+, E^-, O^+ and O^- of dimension $\sim\frac{1}{2}J$, obtained by factorisation of a matrix of rank and order $2J+1$. The twofold K-degeneracy of the symmetric rotor is lifted in the asymmetric rotor ('asymmetry splitting'), the magnitude of the splitting being greatest for low K and high J. Although K is not a good quantum number for the asymmetric rotor it is useful as an index discriminating between different sub-levels of given J. The conventional notation $J_{K_aK_c}$ reflects the need for two such indices in order to distinguish the split levels of an asymmetric rotor by means of the quantum numbers in the symmetric rotor limits.

A number of analytic methods have been developed for obtaining approximate asymmetric rotor eigenvalues without entering into the matrix-diagonalisation procedure described above. Each of these methods has a range of validity described by the quantum numbers J and K_a, K_c and by the asymmetry $\kappa = (2B-A-C)/(A-C)$. The availability of large, fast computers has rendered these methods unnecessary for extended or accurate calculations, but for qualitative purposes and for preliminary analysis of band contours they are sometimes extremely useful. If one understands the structure of an asymmetric rotor band the process of assigning quantum numbers, the most difficult part, is simplified considerably. The remainder of this Section outlines two levels of approximation which are of practical value in organising the early stages of analysis.

The first method utilises the property that the highest and lowest asymmetric rotor eigenvalues for a given J are determined essentially by the constants A and C. This property was first applied to rotational analysis by Cross, Allen and their collaborators[13c, 15]. The zero-order approximations are,

$$F(J_{0J}) \approx F(J_{1J}) \approx CJ^2 \tag{5.4}$$

and

$$F(J_{J0}) \approx F(J_{J1}) \approx AJ^2 \tag{5.5}$$

Take for example the low-K_c Q-branch transitions of a C-type electronic or vibrational band of a near-oblate rotor. The transitions are $J_{1J} \leftarrow J_{0J}$ (r-form) and $J_{0J} \leftarrow J_{1J}$ (p-form): at sufficiently high values of J both sub-branches obey the expression,

$$v \approx v_0 + (C'-C'')J^2 \tag{5.6}$$

In this approximation, the p- and r-form transitions coincide at each value of J. The intrinsically high intensity in this branch is reinforced by this coincidence so that the branch may stand out from an unresolved background. If the lines are individually resolved the second difference $d^2v/dJ^2 = 2(C'-C'')$ can be used to estimate the C-constant change in the transition.

Another approximation[16] takes advantage of the similarity between the energy matrix for the asymmetric rotor and the matrix arising from the Mathieu equation. A diagram of Mathieu function eigenvalues[17] shows that a general asymmetric rotor has a region of K_a degeneracy *and* a region of K_c degeneracy, the limits of each being determined by the degree of asymmetry. A near-prolate rotor, for instance, has a considerable area of K_a degeneracy (the 'symmetric rotor' limit) but also some K_c degeneracy in the high-J, low-K_a region [see equation (5.4)]. Series solutions to the Mathieu equation converge rapidly under conditions of K_a or K_c degeneracy or near-degeneracy, though unfortunately the solution must be written in slightly different form at each limit. When K_c degeneracy occurs the series solution, first due to Gora[16], commences[18]:

$$F(J_{K_aK_c}) = CJ(J+1) - \tfrac{1}{4}m_2(A+B-2C) + \tfrac{1}{2}m_1(2J+1)[(B-C)(A-C)]^{\frac{1}{2}}$$
$$\{1 - (m_2+1)(1-\delta^2)/32\delta J(J+1)\} + \dots \quad (5.7)$$

where

$$m_n = (m+1)^n + m' \quad (5.8)$$

$$m = J - K_c \quad (5.9)$$

and $\delta = \tfrac{1}{2}(\kappa+1)$. This formula is valid when K_a or $(1+K_a) \ll [\delta J(J+1)]^{\frac{1}{2}}$ according to the symmetry of the wavefunction, a condition which ensures K_c degeneracy. A glance at Figure 5.1 should make clear the relationship of the

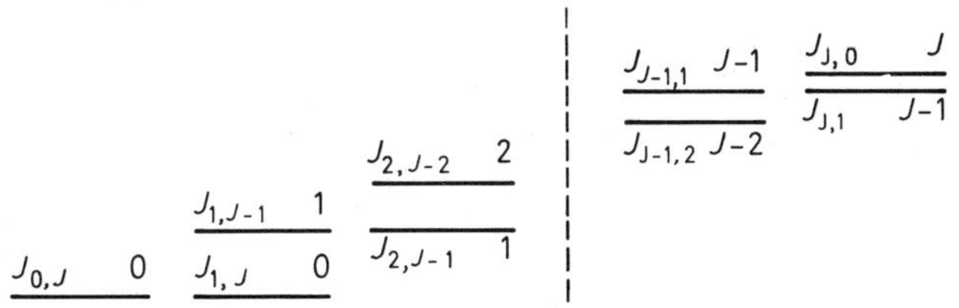

Figure 5.1 Energy levels of an asymmetric rotor for given J. The levels are indexed on the left by $J_{K_aK_c}$ and on the right by $m = J - K_c$; the m index runs $0, 1, 2, \dots, J-1$ or J. Gora's energy expression, equation (5.7), is valid through the range in which levels of the same m are degenerate (K_c degeneracy)

index m to K_c degeneracy. In the limit that $\kappa \to +1$, equation (5.7) reduces after some manipulation to the energy expression for a near-oblate rotor,

$$F(J, K_c) = \bar{B}J(J+1) - (C-\bar{B})K_c^2 \quad (5.10)$$

with $\bar{B} = \tfrac{1}{2}(B+C)$. The corresponding formula for K_a degenerate levels was developed by Brown[19]: it can be constructed from equation (5.7) by exchanging A and C, with m redefined as $J - K_a$ and δ as $-\tfrac{1}{2}(\kappa-1)$. As $\kappa \to -1$ this expression simplifies to the ordinary formula for a near-prolate rotor,

$$F(J, K_a) = \bar{B}J(J+1) + (A-\bar{B})K_a^2 \quad (5.11)$$

The useful range of equation (5.7) and of the analogous K_a equation is of course much greater than that of equation (5.10) or (5.11).

For the special case of a planar rigid molecule $[(B-C)(A-C)]^{\frac{1}{2}} = C$. Under conditions that terms in $[J(J+1)]^{-1}$ can be neglected, Gora's formula, equation (5.7), then simplifies to,

$$F(J_{K_aK_c}) = CJ(J+1) + \tfrac{1}{2}C(2J+1)(2m+1) - \tfrac{1}{4}(2m^2+2m+1)(A+B-2C) \quad (5.12)$$

where the index m covers both components of K_c degeneracy. Consider again the Q-branch of a C-type band, with the added restriction to a planar molecule so that equation (5.12) is applicable. The branches calculated previously correspond to $m = 0$; for them equation (5.12) gives

$$F(J_{K_aK_c}) = C(J+1)^2 - \tfrac{1}{4}(A+B) \tag{5.13}$$

so that the expression for the branch is,

$$v = \{v_0 - \tfrac{1}{4}\Delta(A+B)\} + \Delta C(J+1)^2 \tag{5.14}$$

a superior approximation to that in equation (5.6). The second difference, however, has the same significance in both equations.

5.2.2   Calculation of intensity

Unfortunately there are no short cuts in intensity calculations. The absorption intensity of each rovibronic line in a band is given by the expression,

$$\int_{line} \alpha(v)\mathrm{d}v = \frac{8\pi^3 vN}{3hcQ} \exp\left(-\frac{F''hc}{kT}\right) S^{ev'r'}_{ev''r''} \tag{5.15}$$

where $\alpha(v)$ is the absorbance at the transition frequency v, N is the overall density of molecules in the initial state, Q is the partition function and $S^{ev'r'}_{ev''r''}$ the line strength. For practical purposes v can be treated as constant across an infrared or vibronic band. The factor $S^{ev'r'}_{ev''r''}$ represents the square of the transition moment relative to space-fixed axes, summed over all transitions contributing to the line,

$$S^{ev'r'}_{ev''r''} = \Sigma_{m''s''}\Sigma_{m's'}\,3\,|\,\langle ev''r''\,|\,\mu_\alpha\Phi_{\alpha f}\,|\,ev'r'\rangle\,|^2 \tag{5.16}$$

Here, $\Phi_{\alpha f}$ is the direction cosine between the molecule-fixed axis α and a space-fixed axis f, μ_α is the electric dipole moment operator, and the factor 3 encompasses the sum over space-fixed axes (equivalent in absence of external fields). The initial summations cover the spatial quantum number m and the nuclear spin s. Some simplification results if the transition moment can be written as a simple product of vibronic and rotational matrix elements, since the expression for the line strength may then be factorised,

$$S^{ev'r'}_{ev''r''} = S^{ev'}_{ev''}S^{r'}_{r''} = |\,\langle ev''\,|\,\mu_\alpha\,|\,ev'\rangle\,|^2\,\Sigma_{m''s''}\Sigma_{m's'}3\,|\,\langle r''\,|\,\Phi_{\alpha f}\,|\,r'\rangle\,|^2 \tag{5.17}$$

and the vibronic factor $\langle ev''\,|\,\mu_\alpha\,|\,ev'\rangle$ treated as a constant for a given band. This factorisation is always possible for transition moments between basis functions.

The rotational matrix elements $\langle r''\,|\,\Phi_{\alpha f}\,|\,r'\rangle$ between symmetric rotor functions can be written as a product of two factors, one depending on J,m and the direction assigned to the space-fixed axis f, the other on J,k and the molecule-fixed axis α,

$$\langle r''\,|\,\Phi_{\alpha f}\,|\,r'\rangle = \langle J''m''\,||\,\Phi_f\,||\,J'm'\rangle\langle J''k''\,||\,\Phi_\alpha\,||\,J'k'\rangle \tag{5.18}$$

where the double bars indicate 'reduced', not true matrix elements. Except for the effects due to external magnetic or electric fields, perturbations to the symmetric rotor (including asymmetry) are diagonal in J and m, so that the first factor in equation (5.18) can be squared and summed over m regardless of whether perturbations are present or not. This factor,

$$\Sigma_{m''}\Sigma_{m'}3\,|\,\langle J''m''\,||\,\Phi_f\,||\,J'm'\rangle\,|^2 \tag{5.19}$$

is given in column 2 of Table 5.1. Columns 3–5 of the table contain the matrix elements of the second factor in equation (5.18) in a form suited to asymmetric rotor calculations[20]. Pairwise products of these expressions must be multiplied by the nuclear statistical-weight factor in order to obtain the $S^r_{r''}$. In symmetric rotor calculations it is usual to consider the operator $\Phi_x \pm i\Phi_y$ instead of the individual Φ_x and Φ_y[21].

Asymmetric rotor wave functions are linear combinations of symmetric rotor functions with coefficients given by the eigenvectors of the Hamiltonian

Table 5.1 Factors in the line-strength formulae for transitions between $|Jk\rangle$ wave functions

J'	$3\Sigma_m\|\langle Jm\|\|\Phi Z\|\|J'm\rangle\|^2$	$\langle Jk\|\|\Phi_z\|\|J'k\rangle$	$\langle Jk\|\|\Phi_x\|\|J'k\pm 1\rangle$	$\langle Jk\|\|\Phi_y\|\|J'k\pm 1\rangle$
$J+1$	$(J+1)^{-1}$	$+[(J+1)^2-k^2]^{\frac{1}{2}}$	$\mp\frac{1}{2}[(J\pm k+1)(J\pm k+2)]^{\frac{1}{2}}$	$+\frac{1}{2}i[(J\pm k+1)(J\pm k+2)]^{\frac{1}{2}}$
J	$(2J+1)[J(J+1)]^{-1}$	$+k$	$+\frac{1}{2}[(J\mp k)(J\pm k+1)]^{\frac{1}{2}}$	$\mp\frac{1}{2}i[(J\mp k)(J\pm k+1)]^{\frac{1}{2}}$
$J-1$	J^{-1}	$+[J^2-k^2]^{\frac{1}{2}}$	$\pm\frac{1}{2}[(J\mp k)(J\mp k-1)]^{\frac{1}{2}}$	$-\frac{1}{2}i[(J\mp k)(J\mp k-1)]^{\frac{1}{2}}$

matrix, thus the rotational line strengths $S^r_{r''}$ are given by the square of a linear combination of transition moments in the basis functions. Each term in this linear combination may be taken from Table 5.1. It is most important that individual terms in this combination appear with the correct sign, for in squaring the transition moments to obtain the line strength positive and negative cross terms add and subtract from the positive square terms. To obtain correct final intensities it is therefore essential to use consistent sign and phase conventions for the rotational basis functions.

The method just outlined can be applied to line strength calculations in the presence of any type of perturbation diagonal in J and m.

5.2.3 Non-rigid rotors

A rigid-rotor treatment disregards the effects of centrifugal distortion, finite vibrational amplitude and incomplete separation of vibration and rotation. Centrifugal distortion is of secondary importance in contour calculations and is frequently omitted altogether. It is obviously not practicable to include centrifugal distortion in a calculation unless the measured contour shows considerable detail, effectively a restriction to relatively small molecules. Owing to the vibrational motion, the rotational constants determined by analysis are average quantities whose mild variation from one vibrational state to another is largely determined by the anharmonic (cubic, quartic, ...) potential constants in the force field. These effects are of considerable interest for very small molecules but their interpretation looks at present to be unmanageably complex for larger systems. Coriolis coupling has a major effect on band profiles, often providing the motive for contour analysis of infrared bands. In electronic spectroscopy the tendency has been to concentrate on 0–0 band contours where Coriolis effects are minimal, but this simple state of affairs is not likely to extend to bands other than the 0–0 band.

Centrifugal distortion involves the addition of matrix elements of the operators $P^4_{\alpha\beta\gamma\delta}$ to the rotational energy matrix[13b]. For contour purposes, where small extra constants are an embarrassment, it may be sufficient to retain diagonal elements only (symmetric-rotor approximation), or even to zero all the centrifugal constants except one. Electronic band contour analyses have tended to concentrate on large cyclic molecules where, judged from cases where the ground-state centrifugal constants are known from microwave work, centrifugal distortion does not significantly affect the fit. (Excited-state centrifugal constants are of course unmeasured in these cases, so that in forming this conclusion the force field is presumed to be not greatly different from that of the ground state.) Coriolis coupling is another matter altogether. Very mild Coriolis perturbations can be absorbed into the rotational constants which then assume effective values. This situation typically arises in the 0–0 band of planar molecules, where the inertial defect,

$$\Delta = I_c - I_b - I_a \qquad\qquad (5.20)$$

is dominated by residual effects of Coriolis coupling. In the zero-point vibrational state of planar molecules Δ is normally a small positive number of marginal importance in contour calculations. In higher vibrational states Δ may change sharply from this value, reflecting generalised changes of the rotational constants with vibrational quantum number, an effect in no way specialised to planar molecules. The dependence of rotational constants upon vibrational state is in part due to Coriolis interactions and in part to vibrational anharmonicity. The theory of such effects is fully worked out in small molecules[24, 25], but its application to large structures[26, 27] involves serious practical difficulties.

Much larger effects occur in degenerate vibrational states of symmetric rotors and in the coupling of near-degenerate states of symmetric and asymmetric rotors. z-Axis coupling affecting degenerate symmetric rotor states is well-known. In this case the effective Hamiltonian is diagonal unless the l-type doubling components of different k are degenerate or nearly degenerate, when matrix elements of the form $(\Delta l = \pm 2, \Delta k = \pm 2)$, $(\Delta l = \pm 2, \Delta k = \mp 2)$ and $(\Delta l = \pm 2, \Delta k = \mp 1)$ must be included in the matrix[28]. An oblate symmetric rotor having $C \approx \frac{1}{2}B$ (e.g. benzene) meets resonance conditions of this type when $\zeta_z = +1, -1$ and $-\frac{1}{2}$, respectively. xy-Axis coupling in symmetric rotors is also diagonal in J but introduces off-diagonal elements $\Delta k = \Delta l = \pm 1$ into the energy matrix (second-order coupling); the eigenvalues of the problem must therefore always be obtained by matrix diagonalisation and intensities calculated by means of the eigenvectors, in qualitatively the same manner described for the rigid asymmetric rotor[21]. Programs devised to compute the envelope of coupled bands of asymmetric rotors are available for the case when coupling occurs about the approximate figure axis[29]. So far as can be foreseen the analysis of coupled bands arising from xy-axis coupling in symmetric rotors, and for asymmetric rotors in general, is likely to be restricted to vibrational spectra. The reason for this lies in the different selection rules operating in vibrational as compared with electronic spectra. Infrared selection rules tend to allow both transitions in a coupled band, so that the contour can be fitted to the two components of the coupling. But in electronic spectra the transition to one component is normally forbidden and thus, unless the coupling is very strong, one-half the information needed to complete the analysis is missing or extremely weak in the spectrum.

A similar problem, specialised to molecules with substituents (e.g. CH_3) attached by a formal single bond to a 'rigid' frame, is that of internal rotation. In cases where internal rotation is hindered by a large barrier, as in the molecule of phenol, the problem reduces to that of the ordinary rigid asymmetric rotor; but, in the limit that the barrier vanishes, levels of the torsional vibration collapse to those of free internal rotation and create an extremely high density of motional levels. Matrix elements for the problem of a symmetrical group attached to a rigid asymmetric frame have been fully worked out[30-32], but the number of eigenvalues and intensities to be computed is so large that contour calculations of this type are uneconomic. Such calculations as exist refer to the symmetric rotor approximation (see Table 5.4) and their reliability might be questioned.

5.2.4 Effect of electron spin

The inclusion of electron spin appears to be practicable though no contour calculations of this type have yet been reported.

With few exceptions stable molecules have singlet ground states; thus the most important group of transitions showing effects due to electron spin are triplet–singlet transitions. In non-linear molecules where orbital electronic angular momentum is quenched, the coupling scheme in the triplet state can be expected to be close to the limit of Hund's case (b) in which the energies associated with spin–rotation and spin–spin coupling vanish. In this limit the energy levels are determined by the ordinary rotational motion of the molecular framework but, owing to the spin, each level is threefold degenerate. The effective rigid-rotor Hamiltonian then becomes[33],

$$(1/hc)H_{rot} = AN_a^2 + BN_b^2 + CN_c^2 \qquad (5.21)$$

where the operator $N = J - S$ relates to the pure rotation of the molecule in absence of any source of internal angular momentum other than electron spin. The symmetric rotor basis functions are the functions $| NJSk \rangle$ of the 'coupled' representation[14] and in this basis the energy matrix for an asymmetric rotor with pure case (b) coupling has the same form as that for the rigid, spin-free rotor. However, because the quantum numbers N, J, S obey the same relation $N = J - S$ as the operators, each N corresponds to three coincident sub-levels $J = N$ and $N \pm 1$; and the selection rule $\Delta J = 0, \pm 1$ is replaced by $\Delta N = 0, \pm 1, \pm 2$. The $\Delta N = \pm 2$ branches are characteristic of case (b) coupling and are not observed unless the coupling scheme is close to this limit.

The case (b) limit of vanishing magnetic interactions is also that of vanishing spin–orbit coupling, a condition under which the transition probability of a triplet–singlet combination also vanishes. A contour calculation must therefore take some account of spin–orbit effects. The effective Hamiltonian for an asymmetric rotor in a triplet state[34-36] makes use of a Van Vleck transformation whereby the spin–orbit interaction between separated singlet and triplet states is replaced by interaction matrix elements within the triplet manifold. To a good approximation the complete energy matrix for near-case (b) coupling factorises into three sub-matrices, one for each spin component $N = J - 1, J, J + 1$). Coupling cases other than near-case (b) have also been discussed[36, 37].

Rotational-line strength factors are available for case (b) and other coupling schemes[33, 36, 38]. Unless there are restrictions arising from the symmetries of the triplet state and components of the spin–orbit coupling operator, a triplet–singlet transition 'borrows' its intensity from singlet–singlet transitions of x-, y- or z-polarisation; thus a maximum of three independent transition moments appear in the intensity expressions.

5.3   CALCULATION OF BAND CONTOURS

In outline, the procedure is simple. A computer program simulates a band profile by calculating frequencies and intensities for those rotational lines which contribute to the absorption, and the constants for the combining states are then varied systematically until the calculated and observed profiles match one another.

Two criteria can be applied. First, the relative frequency of calculated and measured features in the band should fit within a certain tolerance. This involves considerations not unlike those which arise in conventional analyses of rotationally-resolved bands and is an important part of the overall fitting procedure for bands with considerable fine structure. The second criterion concerns the relative intensity of different features in the envelope. For various reasons this poses more complicated questions, part experimental and part theoretical. The experimental difficulty evolves from the fact that high-resolution electronic spectra are for the most part recorded photographically; in order to convert the band envelope, usually available as a microdensito-meter trace, to a linear intensity scale the photographic emulsion must there-fore be calibrated. This tedious and somewhat inaccurate procedure is usually ignored in favour of a representation of the calculated profile on a log or log-squared intensity scale which roughly matches the microdensito-meter output. The problem does not arise with photometric recording but to the present this method has been largely restricted to the infrared region. Double-beam recording may distort a profile if a small frequency mis-match occurs in the sample and reference beams, especially in a region where the absorbance changes rapidly with frequency. More subtle questions enter in the theoretical intensity calculation. The line shape function $\alpha(v)$, equation (5.15), is expected to be Gaussian when the profile is wholly determined by Doppler broadening, or Lorentzian when collision-broadening dominates; but a blend of the two can also arise because a Lorentzian function (pro-portional to v^{-2}) still contributes to the wings of a line when the Lorentzian half-width is small. Since the Doppler width is proportional to frequency, electronic spectral lines tend to be Doppler broadened whereas infrared lines are more likely to be collision broadened and the line-shape function should be chosen accordingly. The difficulties associated with hybrid profiles are not resolved but it should be stressed that the experimental conditions for achieving a pure Doppler profile are not easily realised; when the ratio of Doppler to Lorentzian half-width is two, the Doppler broadening contributes less than 10% to the effective width of strong lines because of the greater Lorentzian contribution in the wings[39]. The extra parameters involved in hybrid line profiles are an unwelcome complication and calculations of this type have not been attempted for large molecules.

A contour program calculates frequencies and intensities to preset limits of quantum number and line strength. The individual lines are then broadened according to the chosen line-shape functions and their intensity partitioned among a number of frequency steps ('boxes'). When all the lines have been so distributed, the total intensity in each box is summed and plotted at the frequency corresponding to the box centre. If the frequency step and line width are much smaller than the instrumental resolution, the intensity distribution is scanned with a slit function in preparation for the final output from an XY-plotter. A calculation based on the $\alpha(v)$ of equation (5.15) gives an output in absorbance units. Figure 5.2 is an example of the fit obtained for

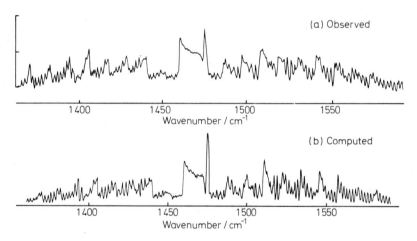

Figure 5.2 (a) Infrared absorption and (b) simulated contour of the $v_2(a_1)$, $v_5(e)$ xy-axis coupled fundamentals of CH_3F, 1350–1600 cm^{-1}
(From di Lauro, C. and Mills, I. M.[21], by courtesy of Academic Press)

observed and simulated spectra in recent work. The usual philosophy in these calculations is that the molecular constants are improved by trial and error, guided where possible by the asymptotic energy expression in Section 5.2.1.

About five asymmetric-rotor programs appear to be in use at the present time. Their chief attributes are approximately as noted in Table 5.2, but all of them are in process of evolution and none of these programs has been published in detail. The pioneer program by Parkin[40] relies upon interpolation of frequency and intensity over preset intervals of J, partly because of the small core size of then-available computers and partly in an attempt to reduce computation time. It would now be considered that interpolation is not

Table 5.2

Program	Method	Line-shape function	Comments
(a) Parkin[40]	All eigenvalues for $J \leqslant 10$ calculated: thereafter 10 'spot' values of frequency and intensity, with inter-mediate values obtained by interpolation.	Gaussian	Operable with modest CPU capacity. Frequency and/or intensity interpolation is not always reliable.
(b) Christoffersen[41]	Complete asymmetric rotor calculation of frequency and intensity. $J_{max} \sim 120$.	Triangular	Time-saving feature allows truncation of asymmetric rotor matrices. Additional lines calculated in near-symmetric rotor approxi-mation.
(c) Pierce[42]	Complete asymmetric rotor calculation of frequency and intensity; full pro-vision for asymmetric rotor centrifugal distor-tion. $J_{max} \sim 120$.	Gaussian. Slit function also provided	Compact and extremely fast
(d) Ueda and Shimanouchi[43, 44]	Complete asymmetric rotor calculation of frequency and intensity.	Lorentzian. Triangular slit function provided	Special consideration given to calculation of intensity.
(e) Birss and Ramsay[18]	Complete asymmetric rotor calculation including centrifugal distortion.	Choice of tri-angular, Gaussian or Lorentzian	Provision for very fast repetitive calculation

necessarily faster than carefully designed diagonalisation procedures. Pro-gram (c) is designed for efficient, high-speed execution. Program (e) has the feature that the eigenvectors of an initial matrix diagonalisation are used to compute energy derivatives which are retained on file and employed to generate contours for a revised set of constants, without repeating the time-consuming matrix diagonalisation or intensity calculation. The Ueda–Shimanouchi program has recently been extended to include effects of Coriolis interaction in coupled asymmetric-rotor bands[29].

Specialised symmetric-rotor programs designed to include the effects of l-type resonance in degenerate states[28] and second-order xy-axis Coriolis coupling[21] have been mentioned previously (Section 5.2.3). Calculation of symmetric-rotor envelopes with c- (or a-) axis coupling is relatively straight-forward and found an early, decisive application in the electronic spectrum of benzene[45]. Finally, there are programs devised mainly for astrophysical calculations, centring upon diatomic molecules. The program of this type by Arnold, Whiting and Lyle[46] is of interest in that it assumes a Voigt line-shape function (intermediate between the Gaussian and Lorentzian shapes) for intensity calculations.

It is often helpful to illustrate the formation of features in a contour by means of a Fortrat diagram, or series of such diagrams, in which the track of contributing sub-branches is plotted for a range of values of J and K. A program which combines these frequency plots with a qualitative impression of the cumulated intensity has recently been described by Kirby[47].

5.4 APPLICATIONS

5.4.1 Scope of the contour method

No contour analysis can achieve much unless fine structure is present in the rotational envelope. Whether a fine structure is observed, and in what detail, depends on the individual line-widths and the density of transitions. Small symmetrical molecules have a low density of transitions which, if they approximate to the rigid symmetric-rotor formulae, are also stacked in a regular way so as to produce a detailed microstructure in the band envelope. For large asymmetric molecules the most that can be hoped for is that systematic coincidences produce piles of lines, collectively of sufficient intensity that they protrude from an unresolved background. Even so, a microstructure formed in this way is of limited value unless at least some of its features can be assigned by quantum numbers. The reason is that the microstructure depends strongly on the changes in molecular constants accompanying the transition, so that an unguided grid search is tedious and possibly uneconomic. Instead, the assigned microstructural features are used to obtain approximate values for some molecular constants and the analysis then refined by matching the simulated and observed envelopes. The detailed steps in this process vary a great deal from one example to another and some typical case-histories are discussed later in this Section. No contour calculation is completely self-contained, in the sense that a minimum initial input consists of a set of principal rotational constants for the ground state of the transition. If these constants are not available from microwave measurements they must be computed from x-ray or electron-diffraction data or, if necessary, from a model geometry.

It is interesting to notice that contour analyses of infrared bands on the one hand and of ultraviolet–visible bands on the other have followed radically different paths. Infrared analyses have been oriented towards the partially resolved bands of small molecules whose contours are rich in detail, with the object of determining the vibration–rotation constants α or the parameters describing Coriolis or other perturbations in the rotational structure. A classification by magnitude of the molecular parameters in vibration–rotation spectra is given in the following table, after Cartwright and Mills[28]:

Molecular constant		Magnitude
ω_r	vibration frequency	ω
$x_{rr'}, g_{tt'}$	anharmonic constants	$\kappa^2\omega$
$A, B(C); A\zeta_t (C_{\zeta_t})$	rotational constants	$\kappa^2\omega$
$\alpha_r^A, \alpha_r^B, \alpha_r^{A\zeta_t} (\approx \zeta_t\alpha_r^A)$	vibrational–rotation constants	$\kappa^4\omega$
$q_t^{(+)}, q_t^{(-)}, \gamma_t$	l-doubling constants	$\kappa^4\omega$
D_J, D_{JK}, D_K	centrifugal constants	$\kappa^6\omega$

Here, κ is a Born–Oppenheimer expansion parameter $(m/M)^{\frac{1}{4}} \sim \frac{1}{10}$, where m and M denote electron and nuclear masses respectively, and the magnitude follows from the consideration that, for example, a constant $\alpha \sim B^2/\omega$. A typical infrared contour analysis is directed at the constants of magnitude $\kappa^4\omega$. Thus, contour-based analyses of the $v_1 v_5$ fundamental bands of CH_2O [44, 48], where direct vibrational microwave measurements are impracticable, have recently completed the determination of the set of α rotational constants for this molecule. Much attention has also been given to the analysis of perturbed bands of symmetric rotors. One example is represented by the coupled $v_2(a)$, $v_5(e)$ bands of methyl fluoride[21] shown in Figure 5.2, where the Coriolis interaction is too strong for satisfactory analysis by conventional methods; another by the $v_{11}(e')$ fundamental of cyclopropane in which the Q-branch (see Figure 5.3) is split by l-type resonance[28].

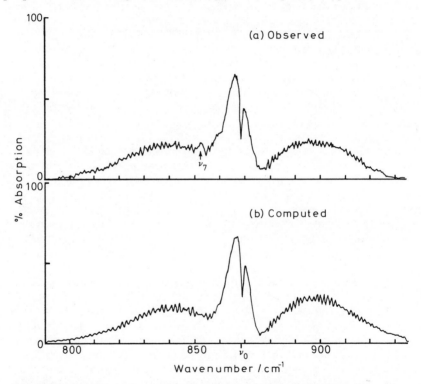

Figure 5.3 (a) Observed and (b) computed contour of the $v_{11}(e')$ fundamental band of cyclopropane. The weak $v_7(a_2'')$ fundamental is not included in the calculation (From Cartwright, G. J. and Mills, I. M.[28], by courtesy of Academic Press)

All this contrasts sharply with electronic band analyses which are usually undertaken for much larger molecules with less-structured bands so that, apart from the band polarisation, one does not expect to obtain more than the principal constants A, B, and C (magnitude $\kappa^2\omega$). This different attitude is enforced by the size of chromophoric groups, many of which have more than the critical number of four massive nuclei which marks an upper bound to the

Table 5.3 Contour analyses of infrared bands

No.	Molecule	Point group	Vibrational mode/cm⁻¹	Parameters	Comment	Reference
1	O_3 (ozone)	C_{2v}	$\nu_2(a_1)$ 700.9	$\alpha_2^A, \alpha_2^B, \alpha_2^C$.	Contour used to select unblended lines	68
2	ClO_2 (chlorine dioxide)	C_{2v}	$\nu_3(b_2)$	$\alpha_3^A, \alpha_3^B, \alpha_3^C$.	Illustrative of determination of α values from low-resolution envelope	69
3	CH_2O (formaldehyde)	C_{2v}	$\nu_1(a_1)$ 2782.4 $\nu_5(b_2)$ 2843.2	$\alpha_1^A, \alpha_1^B, \alpha_1^C$. $\alpha_5^A, \alpha_5^B, \alpha_5^C$.	Contour matched to partially resolved rotational structure. Possible weak c-axis Coriolis coupling is discussed	48
			$\nu_2(a_1)$ 1746.1 $\nu_3(a_1)$ 1500.1	$\alpha_2^A, \alpha_2^B, \alpha_2^C$. $\alpha_3^A, \alpha_3^B, \alpha_3^C$.	x,y-Axis Coriolis interactions ν_3, ν_4 and ν_3, ν_6. ν_2 analysis confirms values of $\alpha_2^A, \alpha_2^B, \alpha_2^C$ available from vibrational microwave measurements	44
			$\nu_4(b_1)$ 1167.2 $\nu_6(b_2)$ 1249.1	$\alpha_4^A, \alpha_4^B, \alpha_4^C$. $\alpha_6^A, \alpha_6^B, \alpha_6^C$. $\zeta_{64}^a(\approx 2A\,\zeta_{64}^a), \eta_{64}^{bc},$ $(\partial\mu_z/\partial Q_4)/(\partial\mu_b/\partial Q_6)$.	Analysis of first-order Coriolis interaction. Upper state of ν_4, ν_6 coupled band is also analysed as the initial state of transitions in the electronic $^1A_2 \leftarrow {}^1A_1$ system of CHDO	29, 70
4	CH_3F (methyl fluoride)	C_{3v}	$\nu_2(a_1)$ 1460.0 $\nu_5(e)$ 1468.3 $\nu_3(a_1)$ 1048.6 $\nu_6(e)$ 1182.4	$\lvert \zeta_{2,\,5a} \rvert, \zeta_{55}^z,$ $(\partial\mu_x/\partial Q_{5a})/(\partial\mu_z/\partial Q_2)$. $\alpha_3^A, \alpha_3^B, \lvert \zeta_{3,\,6a} \rvert,$ $\alpha_6^A, \alpha_6^B, \zeta_{6}^z$.	Analysis of simultaneous first- and second-order Coriolis interaction	21
5	CD_3Cl (chloroform)	C_{3v}	$\nu_2(a_1)$ 1029.9 $\nu_5(e)$ 1059.7	$\lvert \zeta_{2,\,5a} \rvert, \zeta_{55}^z,$ $(\partial\mu_x/\partial Q_{5a})/(\partial\mu_z/\partial Q_2)$.	Analysis of simultaneous first- and second-order Coriolis interaction	21
6	C_3H_4 (allene)	D_{2d}	$\nu_9(e)$ 999 $\nu_{10}(e)$ 841	$(\partial\mu_x/\partial Q_9)/(\partial\mu_z/\partial Q_{10})$.	Intensity perturbation in ν_9, ν_{10} Coriolis coupled bands	21
7	C_3H_6 (cyclopropane)	D_{3h}	$\nu_9(e')$ 1438.5 $\nu_{10}(e')$ 1028.5 $\nu_{11}(e')$ 868.4	$\alpha_9^B, \alpha_9^C, \zeta_9, q_9^{(+)}$. $\alpha_{10}^B, \alpha_{10}^C, \zeta_{10}, q_{10}^{(+)}$. $\alpha_{11}^B, \alpha_{11}^C, \zeta_{11}, q_{11}^{(+)}$.	Analysis of l-type resonance. $q^{(+)}$ is the coefficient appearing in matrix elements $\Delta k = \Delta l = \pm 2$	28, 71, 72
8	CH_3CN (acetonitrile)	C_{3v}	$\nu_3(a_1)$ 1390 $\nu_6(e)$ 1448.0 $(\nu_7 + \nu_8)$ 1410.2	$\lvert \zeta_{36}^a \rvert$. $\alpha_6^A, \alpha_6^B, \zeta_{6}^z,$ $\alpha_{78}^A, \alpha_{78}^B, \zeta_{eff}^{zz}$ $[\approx -(\zeta_7^z + \zeta_8^z)], k_{678},$ $M(\nu_7 + \nu_8)/M(\nu_6)$.	Analysis of $\nu_3, \nu_6, \nu_7 + \nu_8$ Fermi–Coriolis coupled bands	73
9	CH_3C_2H (propyne)	C_{3v}	$\nu_4(a_1)$ 1390.6 $\nu_7(e)$ 1450.9 $(\nu_8 + \nu_{10})$ 1371.1 $(\nu_5 + \nu_9)$ 1549	$M(\nu_4)/M(\nu_7)$, $\alpha_4^A, \zeta_7^z, k_{7810}$, $\alpha_{810}^A, \zeta_{eff}^{zz} [\approx -(\zeta_8^z + \zeta_{10}^z)],$ $k_{579}, M(\nu_8 + \nu_{10})/M(\nu_7)$.	Analysis of $\nu_4, \nu_7, \nu_8 + \nu_{10}$ Fermi–Coriolis coupled bands. Additional localised perturbation by $(\nu_5 + \nu_9)$	74

observation of rotationally resolved spectra. The further question of whether molecular size imposes an upper bound to the observation of microstructured contours is an interesting one which will be developed more fully in later Sections. In principle no such limit exists but the conditions governing the aggregation of lines in sufficient numbers to stand out from the unresolved background probably become more and more exacting as the line density increases. Eventually the problem of volatility may be decisive.

5.4.2   Development of fine structure in electronic bands

The features observed in an electronic band contour range from major peaks of 10^3–10^4 individual lines to microstructural elements just distinguishable from the unresolved background. The minimum requirement for observable microstructure is the coincidence within Doppler width of at least several strong rotational transitions, so that the envelope is a sensitive record of coincidences of this kind. A regular microstructure forms for one or other of a small number of specific reasons — for example, $\Delta \bar{B} \sim 0$ in type A bands of prolate molecules or type C bands of oblate molecules — and thus the assignment of microstructure is a key step in an analysis.

There are essentially two ways to account for systematic coincidences or near-coincidences of lines. The first, familiar from small-molecule spectra, is associated with frequency reversals or 'turns' in the sub-branch structure. How much intensity accumulates in this way depends on the individual line strengths, the curvature of the Fortrat diagram and the number of contributing sub-bands. Major peaks in a band envelope are almost always the result of turning sub-branches. McHugh and Ross[49] point out that the formation of major peaks will not be observed unless the percentage change of the controlling rotational constant falls within quite narrow limits; if the turn occurs too quickly the number of contributing lines is small, and if it occurs too late the individual intensities are low. The second reason for coincidences arises from a regular superposition of lines belonging to different sub-bands. One version of this mechanism, well-known in infrared bands of near-symmetric rotors, is mentioned at the end of the previous paragraph; but more versatile mechanisms which occur especially for planar asymmetric rotors evolve from the asymptotic energy expression in equations (5.7) and (5.12). The theory was developed by McHugh, Ramsay and Ross[18] and by Brown[19]. By introducing a number n,

$$n = J + m \tag{5.22}$$

(where $m = J - K_c$; see Figure 5.1) the energy expression (equation (5.12)) for a planar or near-planar rotor becomes,

$$F(J_{K_a K_c}) = (n+1)^2 C - S \tag{5.23}$$

in which

$$S = \tfrac{1}{4}(2m^2 + 2m + 1)(A + B) \tag{5.24}$$

The equations predict near-coincidences between lines in sub-branches for which $\Delta J = \Delta K_c$, and hence $\Delta m = 0$. Consider first an A-type band: the intense branches which meet the specification $\Delta J = \Delta K_c$ are the ^{qr}R- and

^{qp}P- sub-branches for which the selection rule in n becomes,

$$\Delta n = \Delta J + \Delta m = \Delta J \qquad (5.25)$$

so that the branches obey the expressions,

$$v(R) = v_0 - \Delta S + (2n+3)C' + (n+1)^2 \Delta C \qquad (5.26)$$

$$v(P) = v_0 - \Delta S - (2n+1)C'' + n^2 \Delta C \qquad (5.27)$$

Imagine for the moment that $\Delta S = 0$. The remaining terms on the right side of equations (5.26) and (5.27) predict multiple coincidences because for given $n = J + m$, levels of different J and m combine to give the same n. These coincidences are:

$$^{qr}R(J,m),\ ^{qr}R(J-1, m+1),\ ^{qr}R(J-2, m+2), \ldots \qquad (5.28)$$

and

$$^{qp}P(J,m),\ ^{qp}P(J-1, m+1),\ ^{qp}P(J-2, m+2), \ldots \qquad (5.29)$$

Such coincidences persist through the range of m values for which equation (5.23) is valid, essentially the region of K_c degeneracy. If $\Delta S \neq 0$ the exact coincidences break down and the number of lines contributing to the cluster depends on the relative values of ΔS and the Doppler line width. A glance at Figure 5.1 shows that the parity of K_c is the same for all transitions which make up the clusters (equations (5.28) and (5.29)); therefore if the c-axis is a twofold or higher symmetry axis the clusters possess whatever intensity alternation is implied by the nuclear statistics.

Similar cluster formation occurs in ^{rr}R-, ^{pr}R-, ^{rp}P- and ^{pp}P- sub-branches of a B-type band, and in ^{rq}Q- and ^{pq}Q- sub-branches of a C-type band. The P- and R- branches of the B-type band obey equations (5.26) and (5.27); the expression for the C-type Q-branches has been given earlier (equation (5.14)). If ΔC is a negative quantity – as often happens in an electronic transition – the P-branch clusters in an A- or B-type band are red-degraded, whilst the R-branch clusters turn at some frequency higher than the band origin v_0. The location of the turn relative to the band origin is obtained by differentiation of equation (5.26) and is given by,

$$v(R)_{turn} = v_0 - \Delta S + C'\{1 - C'/\Delta C\} \qquad (5.30)$$

This turn provides the dominant feature of A- and B-type band contours of many cyclic molecules.

Equations (5.23)–(5.27) are applicable to near-planar asymmetric molecules, prolate or oblate, in the domain of K_c degeneracy. Allen's approximation for this domain, equation (5.4), predicts that the line-positions in P- and R-branches will conform to the expression,

$$v(R,P) = v_0 \pm (2J \pm 1)C' + \Delta CJ^2 \qquad (5.31)$$

If ΔC is small, as frequently happens, the rough formula (5.31) is almost equivalent to equations (5.26) and (5.27). The main drawback to Allen's approximation is that it affords no means of estimating the likelihood of cluster formation (this depends upon ΔS), but one should emphasise that Allen's treatment is not specialised to planar molecules and in this sense has the wider range of validity.

The asymptotic energy expression in the domain of K_a degeneracy does not lead to an equivalent set of conditions for clustering, essentially because the condition $I_c = I_a + I_b$ for a planar rigid molecule is unsymmetrical in I_a and I_c[19]. It does, however, explain another type of microstructure observed in transitions of prolate molecules when $\Delta \overline{B} \approx \Delta B + \Delta C$ is small, a condition satisfied in the first singlet transitions of a number of simple derivatives of benzene. The zero-order expression (5.11) predicts that Q-branches in a B-type band degrade monotonically to low frequency if $\Delta \overline{B}$ is negative, or to high frequency for positive $\Delta \overline{B}$. The asymptotic formula for K_a degenerate levels, written according to the prescription immediately following equation (5.10) disregarding terms in $[J(J+1)]^{-1}$, gives after reorganisation an improved energy expression,

$$F(J_{K_aK_c}) = \overline{B}J(J+1) + (A-\overline{B})K_a^2 + (2J+1)(2J-2K_a+1) \times (B-C)^2/16(A-B)$$
(5.32)

The last term in equation (5.32) is necessarily positive, thus the interval between energy levels of given K_a increases with J more rapidly than expected from the near-prolate formula (5.11); and the increments depend on $(B-C)^2$ and so are sensitive to a change in asymmetry. If $A-\overline{B}$ is small a different rate of increase in the initial and final states of the transition may cause the Q-branches to turn. A microstructure formed in this way can be fitted to a polynomial in K_a with a second difference of approximately $2\Delta(A-\overline{B})$ and shows whatever intensity alternation is implied by the statistics of nuclei exchanged by the operation C_2^a. This structure cuts off quite sharply in higher sub-bands as the cancellation of lines overtakes the turn. It should be emphasised that the limits of equation (5.32) are those of K_a degeneracy: for high values of J or low values of K_a the asymmetry splitting is appreciable and the Q-branch rapidly divides into separated, possibly head-forming, sub-branches which however cannot be analysed by the asymptotic formulae.

Molecular size is immaterial to the mechanisms described in this Section. In spite of this, molecular size does have an effect on the visibility of microstructure, apparently because an increasing density of rotational lines must be matched by an increased number of lines within the cluster or turn if it is to stand out from unresolved background. One notices that $2J-K_c$ clusters are easily seen in spectra of medium-sized molecules, such as phenol and benzonitrile (Nos 6 and 8, Table 5.5) even though the $m=0$ values of $\Delta S = \frac{1}{4}\Delta(A+B)$ are relatively large, -0.006 and -0.004 cm^{-1}, and therefore unfavourable to clustering; whilst similar clusters are reported[49] as absent from the simulated contour of naphthalene where the value of $\frac{1}{4}\Delta(A+B)$ $\approx -0.001\,8$ cm^{-1} is more favourable to their formation. In the azulene spectrum, where conditions are very favourable ($\Delta A = 0$), the clusters reappear. It may be premature, however, to draw conclusions until information on the line widths, or radiationless transition probabilities, is available.

5.4.3   Examples

5.4.3.1   Benzene

Because benzene is a symmetric rotor the rotational analysis of bands in its 2600 Å system is not typical of large molecules generally, having more in common with the infrared band analyses discussed in Section 5.4.1. All theories agree that the observed bands are $E_{1u} \leftrightarrow A_{1g}$ vibronic transitions; therefore, since a planar rigid oblate rotor has $B = 2C$ and the B'' value is available from the rotational Raman spectrum[59], the two parameters of the electronic bands are B' (or ΔB) and the effective Coriolis coupling coefficient ζ_{eff} for the E_{1u} vibronic state. The expression for ζ_{eff} varies from one vibronic state to another; for the 6_0^1 vibronic transition it is, $\zeta_{eff} = -\zeta_6^z$. Callomon

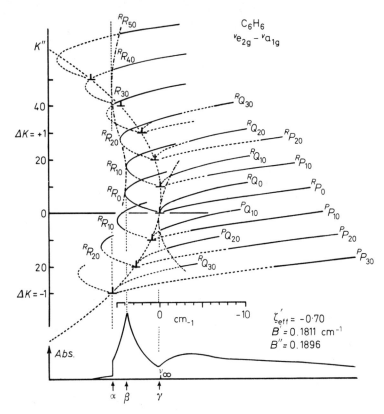

Figure 5.4 Structure of the 6_0^1 band of C_6H_6. The J-type Fortrat curves are drawn every 10th K-sub-band; vertical K scale three times the J scale. Note that $\zeta'_{eff} = -0.70$ is not the final value adopted
(From Callomon, J. H. *et al.*[45], by courtesy of the Royal Society)

et al.[45] show in detail how the varied profiles observed in different vibronic bands depend upon ζ_{eff}.

The structure of the principal vibronic origin, 6_0^1, is shown in Figure 5.4. The main features of the envelope are the cliff α, the principal maximum β and the intensity minimum γ. (The band has microstructural features not shown in the profile which appear to have no simple explanation and were not used in the analysis.) Individual sub-bands are illustrated by Fortrat parabolas for $K_c = 0, 10, 20, \ldots 50$ in the r-form and $K_c = 0, 10, \ldots 30$ in the

Table 5.4 Non-aromatic cyclic and acyclic compounds

No.	Molecule	Point group	Electronic transition	T_0/cm⁻¹	Vibronic band	Band type	Rotational constants/cm⁻¹ Initial state	Change	Final state	Comments	Reference
1	C_2D_6 (ethane)	D_{3d}	$\tilde{A}\,E_u \leftarrow \tilde{X}\,A_{1g}$	71 099	0–0	⊥	A 1.341 6 B 0.459 73 ζ_e	$-0.011\,6$ $0.003\,9$	1.330 0.463 6 $+0.32$	Single main peak with very minimal structure. Line-width (FWHM) ~3 cm⁻¹. Corresponding band of C_2H_6 is diffuse, the estimate line-width being >30 cm⁻¹.	75
2	H_2CO (formaldehyde)	$C_{2v}(C_s)$*	$\tilde{A}\,A_2 \leftarrow \tilde{X}\,A_1$	28 188.0	0+124 (4_0^1) 0+542 (4_0^2)	B A				Contour calculated using constants available from previous rotational analyses.	40
	HCDO	$C_s(C_1)$*	$\tilde{A}\,A_2 \leftarrow \tilde{X}\,A_1$	28 243	0−1050 ($4_1^1,\,6_0^1$)		$\tfrac{1}{2}(A_4'' + A_6'')$ ζ_{46}''	6.585 9 0.44		Rotationally resolved contour of the transition from the coupled v_4'', v_6'' lower state.	70
3	$CSCl_2$ (thiophosgene)	$C_{2v}(C_s)$*	$\tilde{A}\,A_2 \leftarrow \tilde{X}\,A_1$	18 716	0+725 ($2_0^1 3_0^1 4_0^1$)	B	A (0.116 0) B (0.111 9) C (0.057 0) $\Delta\dagger$ (0)	$-0.002\,0$ $-0.002\,0$ $-0.000\,64$ -1.82	0.114 0 0.109 9 0.056 3	Single main peak with red-degraded J-structure. Franck-Condon structure gives $A = 0.119$, $B = 0.106$, $C = 0.056\,5$ cm⁻¹, in poor agreement with rotational analysis. $r(CS) = 1.73$, $r(CCl) = 1.75$ Å, $<ClCCl = 112$ degrees, out-of-plane $< = 27$ degrees.	76
4	trans-C_2H_3CHO (trans-propenal)	C_s	$\tilde{A}\,A'' \leftarrow \tilde{X}\,A'$	25 860	0–0	C	A 1.579 5 B 0.155 4 C 0.141 5	$+0.097$ $-0.007\,4$ $-0.005\,5$	1.677 0.148 0 0.136 0	Single main peak with well-defined K-structure. Line-width parameter (FWHM) 2 cm⁻¹ from contour analysis. 1 cm⁻¹ from electric field-induced spectrum.	77, 78
	cis-C_2H_3CHO (cis-propenal)	C_s	$\tilde{A}\,A'' \leftarrow \tilde{X}\,A'$	24 638	0+333 0–0	A–B C	Contour calculated with 0–0 band constants. A (0.739 8) B (0.215 7) C (0.167 0)	$-0.014\,8$ $-0.004\,3$ $-0.003\,4$	0.725 0 0.211 4 0.163 6	Three main peaks with no observable fine structure. Line-width parameter, 2 cm⁻¹. Ground state of cis-propenal 700±40 cm⁻¹ above that of trans-propenal. $\mu_a' = 1.75 \pm 0.2$ D.	77

5	CF_2N_2 (difluorodiazirine)	C_{2v}	$\tilde{A} B_2 \leftarrow \tilde{X} A_1$	28 389	0–0	B	A B C	0.296 95 0.219 27 0.152 93	+0.002 42 −0.002 50 −0.002 67	0.299 37 0.207 77 0.150 26	Extremely detailed envelope with different regions dominated by high-K and low-K R-branches, and by low-K P-branches. $\Delta rNN = +0.060$, $\Delta rFF = -0.034$ Å. $\mu' = 1.5 \pm 0.20$ (μ'' assumed zero).	42
6	C_5H_8O (cyclopentanone)	$C_{2v}(C_2)$*	$\tilde{A} A_2 \leftarrow \tilde{X} A_1$	30 514	Various [$\Delta\nu'_{23}$ odd, $\Delta\nu'_{18}$ odd]	C	A B C	(0.220 6) (0.111 7) (0.080 3)	−0.006 4 +0.000 3 +0.000 8	0.214 2 0.112 0 0.081 1	Broad peaks with no fine structure. Type A and B contours are also presented. $r(CO) \sim 1.35$, out-of-plane <32 degrees.	79
7	$C_6H_4O_2$ (1,4-benzoquinone)	D_{2h}	$\tilde{A} B_{1g} \leftarrow \tilde{X} A_g$	20 041	0+946 0+140	A B	A B C	(0.175 4) (0.056 3) (0.042 6) Contour calculated with 0+946 band constants.	+0.008 3 −0.001 30 −0.000 20	0.183 7 0.055 0 0.042 4	Single main peak with blue-degraded K-structure.	80
8	FCHO (formyl fluoride)	$C_s(C_1)$*	$\tilde{A} A'' \leftarrow \tilde{X} A'$	37 488	0+2793 ($2_0^0 6_0^1$)	C	A B C	3.040 6 0.392 3 0.346 8	−0.874 3 −0.001 9 −0.018 6	2.166 3 0.390 4 0.328 2	Rotational structure is partially-resolved. Franck–Condon structure assuming 35 degrees out-of-plane angle gives $A' = 2.16$, $B' = 0.393$, $C' = 0.336$ cm^{-1} in fair agreement with rotational analysis.	81, 82

*See footnote † in Table 5.5   †Δ in a.m.u. Å²

Table 5.5 Benzene and its mono- and di-substituted derivatives

No.	Molecule	Point group	Electronic transition	T_0/cm⁻¹	Vibronic band	Band type	Rotational constants/cm⁻¹ Initial state	Change	Final state	Comments	Reference
1	C_6H_6 (benzene)	D_{6h}	$\tilde{A}\,B_{2u} \leftarrow \tilde{X}\,A_{1g}$	38 086.1	0+522 (6₀¹)	⊥	$\tilde{B}_6$ 0.189 6 0	−0.008 6	0.181 0 +0.60	Excited state structure: $r(CC)$ = 1.435, $r(CH)$ = 1.07 Å. Change of profile in other vibronic bands qualitatively accounted for by changes in $\bar{\zeta}_{eff}$.	45
	−d_6			38 289.1	0+498		$\tilde{B}_6$ 0.156 8 0	−0.006 4	0.150 4 +0.43		
2	C_6H_5Br (bromobenzene)	C_{2v}	$\tilde{A}\,B_2 \leftarrow \tilde{X}\,A_1$	36 991.5	0–0	B	A 0.189 05 B 0.033 19 C 0.028 23	−0.009 6 −0.000 02 −0.000 23	0.179 5 0.033 17 0.028 00	Two main peaks with fine structure of ᴾQ-sub-branches. Intensity alternation in fine structure.	83
					0+525(6b₀¹)	A	Contour calculated using 0–0 band constants.				
3	C_6H_5Cl (chlorobenzene)	C_{2v}	$\tilde{A}\,B_2 \leftarrow \tilde{X}\,A_1$	37 048.2	0–0	B	A 0.189 23 B 0.052 59 C 0.041 151	−0.009 5 −0.000 3 −0.000 64	0.179 7 0.052 3 0.040 51	Three main peaks with sub-structure of ᴾQ turns and $2J − K_c$ clusters.	84, 85
4	C_6H_5F (fluorobenzene)	C_{2v}	$\tilde{A}\,B_2 \leftarrow \tilde{X}\,A_1$	37 813.9	0–615(6b₁⁰) 0–0	A B	Contour uses 0–0 band constants. A 0.188 92 B 0.085 75 C 0.058 97	−0.011 2 −0.001 0 −0.001 59	0.177 7 0.084 7 0.057 38	Three main peaks with fine structure contain·ng ᴾQ turns and $2J − K_c$ clusters. $|\Delta\mu|$ = 0.30±0.07 D.	86, 87
5	$C_6H_5NH_2$ (aniline)	$C_{2v}(C_s)$†	$\tilde{A}\,B_2 \leftarrow \tilde{X}\,A_1$	34 029.1	0⁺–0⁺	B	A 0.187 38 B 0.086 52 C 0.059 28 −0.41	−0.011 2 +0.001 36 −0.000 59	0.176 2 0.087 88 0.058 69 −0.27	Three main features with red-degraded fine structure of ᴾQ turns. $|\Delta\mu|$ = 0.85±0.15 D.	88,89
					0–0⁻ (0+293)	B	A 0.187 32 B 0.086 47 C 0.059 27 −0.51	−0.011 3 +0.001 31 −0.000 62	0.176 0 0.087 78 0.058 65 −0.40	Intensity alternation in ᴾQ-sub-branches opposite to that in 0⁺–0⁺ band.	
6	C_6H_5OH (phenol)	$C_{2v}(C_s)$†	$\tilde{A}\,B_2 \leftarrow \tilde{X}\,A_1$	36 348.8	0–0	B	A 0.188 48 B 0.087 37 C 0.059 70	−0.011 4 +0.000 15 −0.001 15	0.177 1 0.087 52 0.058 55	Two main peaks with fine-structure of ᴾQ turns and $2J − K_c$ clusters. $|\Delta\mu|$ = 0.2±0.2 D. Independent analyses are in satisfactory agreement.	41, 89, 90
					0–0	B	A 0.188 48 B 0.087 37 C 0.059 70	−0.011 2 +0.000 14 −0.011 1	0.177 3 0.087 51 0.058 59		
	C_6H_5OD (phenol-d_1)				0+519 (6₀¹)	A	Contour calculated using 0–0 band constants.				
7	$C_6H_5CH_3$ (toluene)	$C_{2v}(C_s)$†	$\tilde{A}\,B_2 \leftarrow \tilde{X}\,A_1$		0–0	B				Incomplete results indicate similarity to p-fluorotoluene (No. 16).	91

No.	Molecule	Point group	Transition	Band origin	Band	Type	Axis				Remarks	Ref.
8	C_6H_5CN (benzonitrile)	C_{2v}	$\tilde{A}\,B_2 \leftarrow \tilde{X}\,A_1$	36 512.9	0–0	B	A	0.188 63	−0.006 1	0.182 5	Contour has two main peaks with red-degraded microstructure due to $2J - K_c$ clusters.	92, 93
							B	0.051 58	−0.001 2	0.050 4		
							C	0.040 50	−0.001 03	0.039 47		
	–4d			36 540.2	0–0	B	A	0.188 53	−0.006 2	0.182 3		
							B	0.049 91	−0.001 2	0.048 7		
							C	0.039 46	−0.001 01	0.038 45		
	$-^{15}N$			36 513.3	0–0	B	A	0.188 60	−0.006 1	0.182 5	Single main peak ($^R R$) without observable microstructure. Corresponding bands of benzonitrile and benzonitrile-4d are doubled by an unidentified resonance.	
							B	0.050 09	−0.001 2	0.048 9		
							C	0.039 57	−0.000 99	0.385 8		
					$0+514$ ($6b_0^1$)	A					Contour calculated using 0–0 band constants.	
9	C_6H_5CCH (ethynylbenzene)	C_{2v}	$\tilde{A}\,B_2 \leftarrow \tilde{X}\,A_1$	35 877.3	0–0	B	A	0.188 77	−0.007 1	0.181 7	Analysis based on separation and relative intensity of main peaks. Microstructure present (not analysed) is apparently due to 'accidental' rather than systematic coincidences.	94
							B	0.051 01	−0.000 92	0.050 09		
							C	0.040 15	−0.000 89	0.039 26		
					$0+492$ ($6b_0^1$)	A					Contour calculated using 0–0 band constants. Single main peak of $^R R$ sub-branches.	
10	$C_6H_5CHCH_2$ (styrene)	C_s	$\tilde{A}\,A' \leftarrow \tilde{X}\,A'$	34 758.8	0–0	A	A	(0.169 4)	−0.006 15	0.163 2	Single main peak with considerable fine structure ($^R R$ contribution analysed). No type B component detected in this band. Excited state appears planar: the possibility that the ethylenic group is twisted is rejected. $|\Delta\mu| = 0.13$ D.	52, 95
							B	(0.053 8)	−0.000 62	0.053 2		
							C	(0.040 8)	−0.000 72	0.040 1		
11	$1:4\text{-}C_6H_4Cl_2$ (p-dichlorobenzene)	D_{2h}	$\tilde{A}\,B_{2u} \leftarrow \tilde{X}\,A_g$	35 740.5	0–0	B	A	(0.188 7)	−0.010 5	0.178 2	Single main peak with red-degraded K-structure formed from a composite of $^P Q$ and $^P P$-sub-branches.	96
							B	(0.022 36)	+0.000 17	0.022 53		
							C	(0.019 99)	+0.000 1	0.020 00		
					$0+538$	A					Contour uses 0–0 band constants.	
12	$1:4\text{-}ClC_6H_4F$ (p-chlorofluorobenzene)	C_{2v}	$\tilde{A}\,B_2 \leftarrow \tilde{X}\,A_1$	36 275.1	0–0	B	A	(0.188 4)	−0.011 2	0.177 2	Single main peak with red-degraded fine structure of $^P Q$-sub-branches showing well-defined intensity alternation. $\mu' = 1.35$ (or $0.70)\pm0.1$ D.	97, 98
							B	(0.031 90)	+0.000 25	0.032 15		
							C	(0.027 28)	−0.000 07	0.027 21		

Table 5.5 (continued)

No.	Molecule	Point group	Electronic transition	T_0/cm^{-1}	Vibronic band	Band type	Rotational constants/cm^{-1}			Comments	Reference
								Initial state	Change	Final state		
13	1:4-C$_6$H$_4$F$_2$ (p-difluorobenzene)	D_{2h}	$\tilde{A}\,B_{2u} \leftarrow \tilde{X}\,A_g$	36 837.5	0–0	B	A	(0.189 6)	−0.011 8	0.177 8	Two main peaks with red-degraded fine structure showing intensity alternation. Fine structure assigned to ^{R}Q-sub-branches. Force field calculation for excited state supports a structure in which $\Delta r(\mathrm{CF}) = -0.04$ Å. Independent analyses are in good agreement.	99, 100
							B	(0.047 8)	+0.000 22	0.048 0		
							C	(0.038 2)	−0.000 37	0.037 8		
				36 837.9	0–0	B	A	(0.188 1)	−0.011 9	0.176 2		
							B	(0.047 6)	+0.000 23	0.047 9		
							C	(0.038 0)	−0.000 36	0.037 6		
14	1:4-FC$_6$H$_4$NH$_2$ (p-fluoroaniline)	$C_{2v}(C_s)$†	$\tilde{A}\,B_2 \leftarrow \tilde{X}A_1$	32 653.1	0$^+$–0$^+$	B	A	0.186 6	−0.011 1	0.175 5	Broad main peaks with distinguishable ^{pp}Q substructure and red-degraded K-structure of ^{P}Q turns. Contour is insensitive to in range 0–0.3 amu Å²: $\|\Delta\mu\| = 0.82\pm0.09$ D.	101, 102
							B	0.048 34	+0.001 0	0.049 33		
							C	0.038 43	+0.000 8	0.038 51		
15	1,4-FC$_6$H$_4$OH (p-fluorophenol)	$C_{2v}(C_s)$†	$\tilde{A}\,B_2 \leftarrow \tilde{X}\,A_1$	35 115.9	0–0	B	A	(0.187 7)	−0.012 5	0.175 2	Single main peak with red-degraded K-structure of ^{P}P turns ($K_a = 12$–32). $\|\Delta\mu\| = 0.4\pm0.1$ D.	101, 102
							B	(0.048 6)	+0.000 7	0.049 3		
							C	(0.038 6)	−0.000 1	0.038 5		
16	1,4-FC$_6$H$_4$CH$_3$ (p-fluorotoluene)	$C_{2v}(C_s)$†	$\tilde{A}\,B_2 \leftarrow \tilde{X}\,A_1$	36 859.6	0–0	B	A_f	0.190 22‡	−0.011 2	0.179 0	Contour calculated in symmetric rotor approximation assuming free internal rotation. Transition moment is located in the frame ($\pi^*\pi$ transition). Fine structure in the band envelope is not accounted for. Results for p-cresol and p-chlorotoluene are reported briefly.	91
							A_t	5.270‡	+0.08	5.35		
							B	0.047 71	+0.000 1	0.047 8		
							C	0.038 15	−0.000 42	0.037 7		

*Δ in amu Å²

†The point symmetry of these molecules is, at most, C_s. A 'molecular symmetry' (non-rigid) classification leads to a group of order 8 for aniline and p-fluoraniline (isomorphic with D_{2h}), 12 for toluene and p-fluorotoluene (isomorphic with D_{3h}) and its derivatives (isomorphic with C_{2v}). C_{2v} is a sub-group of D_{2h} and D_{3h} and forms a convenient common basis for classification.

‡Subscripts f and t denote frame (phenylene group) and top (methyl group) respectively.

p-form sub-bands. With the condition $B = 2C$, the expression for sub-band origins in an $E \leftarrow A$ transition of an oblate symmetric rotor is,

$$\nu_0^{sub} = \nu_0 - \tfrac{1}{2}B'(1+2\zeta'_{eff}) \mp B'(1+\zeta'_{eff})K_c - \tfrac{1}{2}\Delta BK_c^2 \qquad (5.33)$$

and hence the ν_0^{sub} fall on a parabola for which the value of K_c at the turn occurring in the R-form sub-bands is given by,

$$K_c^{turn} = -B'(1+\zeta'_{eff})/\Delta B \qquad (5.34)$$

Values of ΔB and ζ'_{eff} which fit the observed contour are $\Delta B = -0.008\,6\ \mathrm{cm}^{-1}$ ($B' = 0.181\,0\ \mathrm{cm}^{-1}$) and $\zeta'_{eff} = -0.60$, thus $K_c^{turn} \approx 8$. The corresponding turn of ^rR-heads form the main feature β. Cancellation of lines in the ^rR-sub-branches creates a profile (dot-dash curve) which causes the intensity to drop sharply at the cliff α; beyond this point there are no more ^rR lines, though some weak ^pR-sub-branches ($\Delta J = -\Delta K_c$) make a small contribution to intensity. The minimum γ value marks the high-frequency limit of intensity due to the red-degraded Q- and P-branches and so coincides closely with the band origin. Final values for the parameters, given in Table 5.5 were obtained by matching the features α-β-γ in the observed and simulated contour. Although the fit did not extend to the microstructure, so that the error limits ($\Delta B = -0.008\,6 \pm 0.000\,5\ \mathrm{cm}^{-1}$) are quite high, the results for C_6H_6 and C_6D_6 are sufficient to determine the excited state geometry, $r(CH) = 1.07$ and $r(CC) = 1.43_5$ Å, a unique achievement among large molecules.

Unfortunately neither rotational nor vibrational analysis provides *experimental* proof that the 2600 Å bands of benzene belong to a $^1B_{2u} \leftarrow {}^1A_{1g}$ electronic transition; both sets of observations are equally consistent with a $^1B_{1u}$ final state. The most comprehensive evidence for the B_{2u} assignment comes from the correlation with the first singlet–singlet band systems of benzene derivatives. It was first pointed out by Dunn[51] that if the substituent perturbation is mild the lowest singlet states of C_{2v} derivatives of benzene are B_2 electronic states if the benzene state is B_{2u}, or A_1 states if the latter is B_{1u}. With one possible exception, the first excited states of all simple benzene derivatives are found to be B_2 electronic states (Table 5.5). The apparent exception is styrene[52] which gives rise to hybrid bands with more type A character than expected on simple grounds.

5.4.3.2 Naphthalene and azulene

These two molecules have small rotational constants and consequently a very high density of rotational lines contributing to the band contour. Analyses are available for a vibronic origin in the first singlet–singlet band system of naphthalene (3100 Å), and for the electronic origin of the second singlet–singlet system of azulene (3480 Å). The first singlet system of azulene, responsible for the blue colour, is too diffuse for the contour method.

The electronically allowed sub-system of naphthalene bands in the 3100 Å region is made up of bands with a single main feature, easily distinguished from the stronger vibronic sub-systems of bands showing two main features separated by about $3\ \mathrm{cm}^{-1}$ [11]. A densitometer trace of the $0+438\ \mathrm{cm}^{-1}$ vibronic origin is reproduced in Figure 5.5(a). This band is historically

interesting as the subject of the first attempt to deduce polarisation from the
unresolved rotational envelope of an electronic band[11], and also of the first
calculation by modern methods of a simulated contour for a large asymmetric molecule. In the contour calculation Innes *et al.*[53] adopted two sets of
rotational constants; one set corresponding to the ground state x-ray
structure[54], the other calculated from an excited-state structure estimated

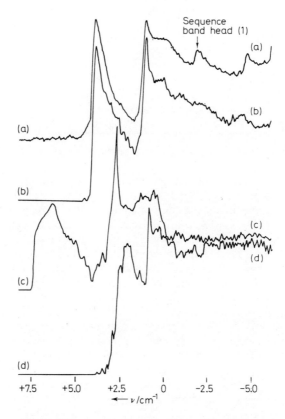

Figure 5.5 (a) Densitometer trace of the $32\,458\,\mathrm{cm}^{-1}$
band of naphthalene vapour. The lower curves are simulated
B-type contours: (b) $\Delta A = -0.002\,6$, $\Delta B = -0.000\,9$,
$\Delta C = -0.000\,7\,\mathrm{cm}^{-1}$; (c) $\Delta A = -0.001\,3$, $\Delta B = -0.000\,6$,
$\Delta C = -0.000\,4\,\mathrm{cm}^{-1}$; (d) $\Delta A = -0.003\,9$, $\Delta B = -0.001\,3$,
$\Delta C = -0.001\,1\,\mathrm{cm}^{-1}$
(From Innes, K. K. *et al.*[53], by courtesy of Academic Press)

from semi-empirical theoretical considerations[55]. For b-axis polarisation
these constants produce the contour in Figure 5.5(b), in remarkably good
agreement with the observed envelope. The remaining curves show the effect
of changes of the order 1% in the upper-state rotational constants, a shift
which destroys the agreement completely. These calculations were not the
first experimental demonstration of polarisation in the $3100\,\text{Å}$ bands of
naphthalene (this was accomplished earlier by McClure[56] in low-temperature

crystal studies) but their success left no one in doubt of the potential of the method for asymmetric molecules.

In dissecting the contour one notes that the planarity condition reduces the number of parameters in the problem to two, which may be chosen as ΔA and ΔC. One of the main peaks must mark the turn of $2J - K_c$ clusters in the R-branches of low K_a. The expression for the $''R$- and ^{pr}R-sub-branches of low K_a is given by equation (5.26) and the relative location of their turn by equation (5.30). Curve (b) corresponds to the constants $C'' = 0.029\ 7$, $\Delta C = -0.000\ 7$ cm^{-1}, thus the estimated position of the turn in the simulated contour is $v_0 + 1.2$ cm^{-1} which matches the position of the main peak at lower frequency. The high-frequency peak belongs to the high $K_a\ 'R$-branches in which, for a type B band, the most intense lines $J \approx K_a$ connect levels whose energy is given by equation (5.5) and so turn at the position ($A'' = 0.104\ 4$, $\Delta A = -0.002\ 6$ cm^{-1}),

$$v_{\text{turn}} = v_0 + A'(1 - A'/\Delta A) = v_0 + 4.1 \text{ cm}^{-1} \qquad (5.35)$$

This estimate also agrees quite well with the final contour (b). If preliminary values for the excited state constants had not been available, one would have a single measurement of frequency — the separation of main peaks — and two unknown constants, ΔA and ΔC; the analysis then depends on matching relative intensities in the observed and simulated profiles using compatible values of ΔA and ΔC in each calculation of the contour. Curves (c) and (d) of Figure 5.5 show that the intensity fit is a sensitive criterion.

It is interesting that the type B assignment for the $0 + 438$ cm^{-1} vibronic band was correctly deduced several years earlier by Craig et al.[11] whose work marked the first attempt to obtain electronic polarisation from band contour considerations. Limited computer facilities restricted the calculation to $J \leqslant 40$, and the contour was plotted in terms of the number of lines per frequency interval against frequency; intensity as such was not calculated. An envelope with two main peaks was shown to be consistent with type B structure. However, the calculation would not now be considered correct in detail because the simulated contour refers to very small changes in the rotational constants so that the peaks have the same composition as the P- and R-maxima in an infrared type B band. The 0–0 band of the 3100 Å system of naphthalene has a single main peak and was similarly identified by Craig et al. as type A[11]. A complete contour calculation is not available, but it is almost certain that the main peak represents the turn of low $K_a\ ^{qr}R$-branches obeying equation (5.26), therefore turning at the frequency given by equation (5.30).

A profile of the azulene band at 28 757 cm^{-1}, given in Figure 5.6, shows a single main peak in the high-frequency wing and a broad maximum merging into well-developed regular fine structure in the low-frequency wing. Because no intensity alternation is observed in the fine structure, its formation is probably not due to successive K_a sub-bands where the relative statistical weights are 7:9 (a ratio which, however, might not be easily distinguished from unity). The analysis by McHugh et al.[18] proves that the band is A-type, with the fine structure attributed to the low $K_a\ ^{qp}P$-clusters of equation (5.27) and the major peak to the turns of low $K_a\ ^{qr}R$-clusters at the position indicated by equation (5.30). A fit of the fine structure 'lines' to the quadratic

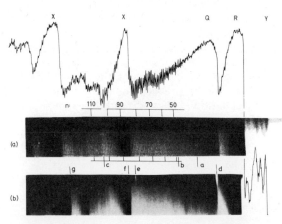

Figure 5.6 Absorption (0–0 band) of (a) azulene (28 757 cm^{-1}) and (b) azulene-d$_8$ (28 865 cm^{-1}) The densitometer trace refers to the azulene absorption; X denotes interference by sequence bands. Iron arc calibration lines are identified by the lower case letters a–q
(From McHugh, A. J. *et al.*[18], by courtesy of C.S.I.R.O.)

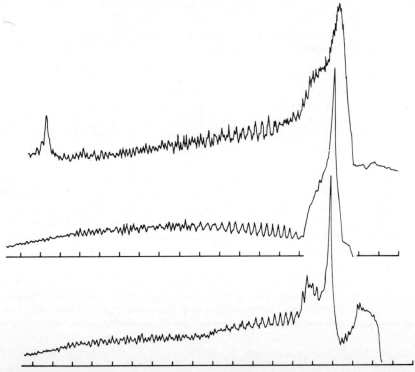

Figure 5.7 Densitometer trace of the 0–0 band of indole at 35 232 cm^{-1}. The middle and lower curves are simulated contours for type *A* and type *B* selection rules respectively. Indicators are separated by 1 cm^{-1}
(From Mani, A. and Lombardi, J. R.[59], by courtesy of Academic Press)

expression (5.27) gives a value for ΔC, a well-determined parameter, and for the n-numbers with a probable uncertainty of ± 1. The parameter ΔA must be determined from the overall fit of the simulated contour: the value adopted, $\Delta A = 0.000\,00 \pm 0.000\,05$ cm^{-1}, favours development of the $2J - K_c$ clusters for which the ideal condition is $\Delta(A + B) = 0$. Constants for the upper state (Table 5.7) are arrived at by combining ΔA and ΔC with the ground state constants determined from microwave data[57]. The band origin essentially coincides with the intensity minimum, a condition reminiscent of the benzene profile; in both cases, the rise in intensity towards lower frequency is explained by the onset of Q- and P-branches. Analysis of the corresponding band of azulene-d$_8$ follows similar lines.

The case histories of naphthalene and azulene underline the interdependence of rotational and vibrational analyses, a feature which is by no means peculiar to large molecules. Rotational analysis gives information on the vibronic symmetries of the combining states; vibrational analysis allows this to be translated into formation on electronic symmetry. A wrong assignment for the 0–0 band in the $^1B_2 \leftarrow {}^1A_1$ electronic band system of phenylacetylene has recently been corrected on the basis of rotational analysis[58].

5.4.3.3 Indole: structure of a hybrid band

If the orientation of principal axes is not fixed in the molecule by symmetry the contour calculation must superimpose envelopes of different polarisation. At present, all such calculations[59–62] refer to π^*–π electronic transitions of planar molecules and therefore to hybrids of a- and b-axis polarisation.

Mani and Lombardi's analysis[59] of an indole band at 35 232 cm^{-1} is typical of the steps involved. The molecule has fused benzene and pyrrole rings; the mass distribution is such that the b-axis of inertia is essentially parallel to the bond shared by both rings. A microdensitometer trace of the absorption (Figure 5.7) shows a single main peak with red-shaded microstructure of two types with distinctively different intervals of 0.3–0.4 cm^{-1} for one type and c. 0.2 cm^{-1} for the other. By analogy with azulene the single main peak and the 0.2 cm^{-1} fine structure must be assigned to ^{qr}R- and ^{qp}P-sub-branches of low K_a, indicating a dominant A-type component. It should be emphasised that the low K_a ^{rr}R-, ^{pr}R, ^{rp}P- and ^{pp}P-sub-branches of a B-type component coincide with the corresponding A-type ^{qr}R- and ^{qp}P-sub-branches, so that the preference for a mainly A-type hybrid structure is based on the *absence* of a second main feature in the envelope observed. The second difference in the series of ^{rp}P- and ^{pp}P-clusters is $2\Delta C$.

The second type of fine structure is generated by turns in the ^{qp}Q-sub-branches for $K_a = 21 - 36$. If $\Delta \bar{B}$ is small and negative these sub-bands are initially red-degraded, but under proper conditions their direction is reversed by the asymmetry shift and the cumulation of lines at the turn stands out from the background. The second difference in this structure yields a preliminary value for $\Delta(A - \bar{B})$, thus two parameters of the band, ΔC and $\Delta(A - \bar{B})$, can be estimated from the fine-structure analysis. The third parameter, the direction of the transition moment is arrived at by matching the observed contour by a blend of the simulated A- and B-type profiles

Table 5.6 Heterocyclic compounds with one ring

No.	Molecule	Point group	Electronic transition	T_0/cm^{-1}	Vibronic band	Band type	Rotational constants/cm^{-1}			Comments	Reference
							Initial state	Change	Final state		
1	C$_6$H$_5$NO (pyridine-N-oxide)	C_{2v}	$\tilde{A}\,^1B_2 \leftarrow \tilde{X}\,^1A_1$	29 295.4	0–0	B	A 0.196 79 B 0.093 22 C 0.063 25	−0.009 7 +0.000 21 −0.000 93	0.187 1 0.093 43 0.062 32	Two main peaks with red-degraded K-structure. Profile closely resembles that of the isovalent molecule phenol. The electronic transition is π^*–π.	103
2	C$_5$H$_4$N-NH$_2$ (2-amino-pyridine)	$C_s(C_1)$	$\tilde{A}\,A' \leftarrow \tilde{X}\,A'$	33 471.1	0–0	A–B	A 0.1923 B 0.0901 C 0.0614	−0.0120 +0.0014 −0.000 65	0.1803 0.0915 0.0608	Three main peaks with detailed fine structure. Hybrid band composition type $A{:}B \approx 55{:}45$. Change of constants similar to that observed for aniline. The electronic transition is π^*–π.	62
3	1,4-C$_4$H$_4$N$_2$ (1,4-diazine)	D_{2h}	$\tilde{B}(?)\,B_{3u} \leftarrow \tilde{X}\,A_g$	30 875.8	0 + 383 (5^1_0)	A	A 0.2128 B 0.1975 C 0.10244	+0.001 −0.0035 −0.000 83	0.2140 0.1940 0.101 61	^qP-Branches have regular fine structure. The ^qR-branches however are less regular and unexpectedly intense, possibly indicating an intensity perturbation caused by rotational coupling with a higher B_{2u} state near 38 000 cm^{-1}.	104
	1,4-C$_4$D$_4$N$_2$ (1,4-diazine)	D_{2h}	$\tilde{R}_2\,B_{2u} \leftarrow \tilde{X}\,A_g$ (Rydberg)	55 288	0–0	A				Two main peaks with no observable fine structure. Possible values of A' and B' are discussed.	105
4	1,3-C$_4$H$_4$N$_2$ (1,3-diazine)	C_{2v}	$\tilde{A}\,B_1 \leftarrow \tilde{X}\,A_1$	31 072.3	0–0	C		ΔC −0.000 75		Analysis of Q-branch structure $(2J - K_a$ clusters).	106
5	1,2-C$_4$H$_4$N$_2$ (1,2-diazine)	C_{2v}	$\tilde{A}\,B_1 \leftarrow \tilde{X}\,A_1$	26 648.8	0 + 373 $(6a^1_0)$	C	A 0.2082 B 0.1988 C 0.1017	−0.0012 −0.0018 −0.0005	0.2070 0.1970 0.1012	Band shows exceptionally detailed fine structure. Small changes in rotational constants cause the contour to resemble closely that of an infrared C-type band. Rotational analyses of other bands are discussed.	107
6	s-C$_2$H$_4$N$_4$ (s-tetrazine)	D_{2h}	$\tilde{A}\,B_{3u} \leftarrow \tilde{X}\,A_g$	18 128.0	0–0	C		ΔC +0.000 589		Analysis of Q-branch $(2J - K_a$ clusters) for $J = 39$–92.	65, 108

Table 5.7 Polycyclic aromatic and heterocyclic compounds

No.	Molecule	Point group	Electronic transition	T_0/cm^{-1}	Vibronic band	Band type	Rotational constants/cm^{-1}			Comments	Reference
							Initial state	Change	Final state		
1	$C_{10}H_8$ (naphthalene)	D_{2h}	$\tilde{A}\,B_{2u} \leftarrow \tilde{X}\,A_g$	32 020	$0+438$ ($B_{1u} \leftarrow A_g$)	B	A (0.104 4) B (0.041 5) C (0.029 7)	0.002 6 −0.000 9 −0.000 7	0.101 8 0.040 6 0.029 0	Main peaks resulting from ^rR (high K_a) and ^rR-^{pr}R (low K_a) turns. No fine structure observed. Type A and C contours calculated with same constants. See Section 5.4.3.2.	11, 53
2	$C_{10}H_8$ (azulene)	C_{2v}	$\tilde{B}\,A_1 \leftarrow \tilde{X}\,A_1$	28 756.8	0-0	A	A 0.094 80 B 0.041 86 C 0.029 04	0.000 00 −0.000 84 −0.000 40	0.094 80 0.041 02 0.029 65	Main peak results from ^qR turns at low K_a. Detailed $2J-K$ microstructure in ^{qP} branches. See Section 5.4.3.2.	18
	$-d_8$			28 865.3	0-0	A	A (0.081 4) B (0.037 3) C (0.025 6)	0.000 0 −0.000 7 −0.000 3	0.081 4 0.036 6 0.025 3		
3	C_9H_8 (indene)	C_s	$\tilde{A}\,A' \leftarrow \tilde{X}\,A'$	34 723.0	0-0	A-B	A 0.125 6 B 0.052 6 C 0.037 3 A 3.2	−0.003 49 −0.000 41 −0.000 64	0.122 1 0.052 1 0.036 7 1.7	Single main peak with detailed fine structure of $2J-K_c$ clusters. Hybrid band composition is type $A:B \approx 88:12$. Transition moment inclined at $\pm20\pm5°$ relative to a inertial axis.	60
4	C_8H_7N (indole)	C_s	$\tilde{A}\,A' \leftarrow \tilde{X}\,A'$	35 232.1	0-0	A-B	A 0.126 6 B 0.053 8 C 0.037 8	−0.005 15 −0.000 76 −0.000 84	0.121 5 0.053 1 0.036 9	Single main peak with detailed sub-band structure and $2J-K_c$ clusters. Hybrid band composition is type $A:B \approx 80:20$. Transition moment inclined at $\pm26\pm7°$ to a inertial axis.	59
5	C_8H_6O (benzofuran)	C_s	$\tilde{A}\,A' \leftarrow \tilde{X}\,A'$	35 920.0	0-0	A-B	A 0.127 4 B 0.052 9 C 0.037 4	−0.005 89 −0.000 65 −0.000 84	0.121 5 0.052 3 0.036 6	Single main peak with K-structure and $2J-K_c$ clusters. Hybrid band composition is type $A:B \approx 90:10$. Transition moment inclined at $\pm18\pm5°$ relative to a inertial axis.	61
6	C_8H_6S (thionaphthene)	C_s	$\tilde{A}\,A' \leftarrow \tilde{X}\,A'$	34 055.7	0-0	A-B	A 0.104 3 B 0.042 8 C 0.030 3	−0.002 80 −0.000 45 −0.000 46	0.101 5 0.042 3 0.029 9	Single main peak with well-developed K-structure and $2J-K_c$ clusters. Hybrid band composition is type $A:B \approx 10:90$.	61
7	$C_6H_4N_2S$ (2,1,3-benzothiadiazole)	C_{2v}	$\tilde{A}\,B_2 \leftarrow \tilde{X}\,A_1$	30 410.5	0-0	B	A 0.130 1 B 0.041 8 C 0.031 7	+0.000 8 −0.001 3 −0.000 8	0.130 9 0.040 5 0.030 9	Single main peak with K-structure in high-frequency and $2J-K_c$ clusters on low-frequency wing.	109
					$0+527$	A	Contour computed using 0-0 band constants.				

shown in the lower portion of Figure 5.7. A step-out present to high frequency of the main peak in the observed envelope, not adequately reproduced in the pure A-type contour, requires some fairly small addition of type B character. By careful examination of this and other regions in the band, Mani and Lombardi conclude that B-type character probably accounts for 20% of the band. The transition moment direction relative to the a-axis (perpendicular to the bond shared by both rings) is then calculated to be $\pm 27 \pm 7$ degrees, where the first $\pm$ sign indicates the indeterminacy of the *sense* of the angle.

5.4.4   Error limits in contour analysis

Error limits in the rotational constants determined from contours vary with the assignable structure present in the observed envelope. Under favourable circumstances the change in a rotational constant is good to two significant figures with uncertainty entering in the third figure. Very favourable circumstances are required before all three constants can be determined to this standard which for planar or nearly planar molecules represents a level at which the inertial defect becomes significant; a much more common result is that one of the constants is appreciably better determined than the others. Changes in the constants are normally an order of magnitude smaller than the constants themselves. When ground-state constants are available from microwave data, the *best* constants for a higher electronic state are therefore reliable through four significant figures, with uncertainty in the fifth; otherwise, uncertainty may enter in the fourth, possibly in the third, figure.

A test of error estimates is feasible where the same molecule has been studied independently in different laboratories using different instrumentation, different programs, and possibly a different approach to the analysis (for example, the independent constants for a planar molecule could be chosen as A and $\bar{B}$, or A and C). Two comparisons of this sort are available. For phenol (Table 5.5, No. 6) the analyses have a common base in the ground-state microwave measurements of Forest and Dailey[63], so that the final state constants are directly comparable. Differences occur in the fourth significant figure, the discrepancy for A' and C' being of the order of 1 part in 1000. The p-difluorobenzene analyses (Table 5.5, No. 13) are based on model geometries for the ground state, therefore on different input constants, but the agreement as to changes in constants implies the same relative uncertainty noted for phenol. In each case the analysis is a two-parameter problem for a molecule with well-developed fine structure, hence the indicated uncertainties are probably less than average for contour determinations as a whole.

5.4.5   Geometry of excited states

Except for molecules of high symmetry, the outlook for unaided structure determination based on contour analysis is not promising; the rotational constants are not sufficiently well-determined for isotopic methods to be generally useful. This is not surprising when one recalls that for unsymmetrical molecules the rotational constants are required to standards approaching microwave accuracy in order to undertake a complete structure determination from isotopic data.

Benzonitrile provides a cautionary example. The 0–0 band of the $^1B_2 \leftarrow {}^1A_1$ transition has been analysed for the normal molecule, 4-d- and ^{15}N-isotopes (see Table 5.5, No. 8) in order to establish the n-numbers in a series of $2J - K_c$ clusters, but the data could equally be used to calculate the overall 4-H–N distance in the $^1B_2$ state using Kraitchman's equation,

$$a_s^2 = \Delta I_b (M + \Delta m)/M \Delta m$$

If the uncertainty in the B' rotational constants is taken to be $\pm 0.000\,02$ cm^{-1} the end-to-end internuclear distance in the $^1B_2$ state is calculated to be 6.51 ± 0.04 Å. A complete r_s-structure for the ground state determined by Bak and his colleagues[64], gives the distance as 6.456 ± 0.001 Å. Evidently the sense of the change, $+0.05 \pm 0.04$ Å, is marginally significant. Moreover, the error estimate for B' used in this illustration is extremely optimistic – a more realistic value is $\pm 0.000\,1$ cm^{-1}, five times as great – in that it ignores uncertainties which proliferate from the neglect of centrifugal distortion and the assumption of zero inertial defect. Isotopic substitution has little to offer in cases of this type. The opposite extreme is represented by benzene where, granted the regular hexagonal structure indicated by the vibrational analysis, one isotopic substitution is sufficient for a structure determination of the $^1B_{2u}$ state. The work of Merer and Innes[65] on the structure of s-tetrazine, in the $^1B_{3u}$ state, though not by contour methods, probably indicates the useful limits of isotopic substitution at the present time. Another generation of instruments with the opportunity for more detailed contour fitting may ultimately alter its competitive position.

A blend of rotational and vibrational analysis offers interesting possibilities but except for small or symmetrical molecules the method has not worked well in the past. One difficulty is that Franck–Condon analyses require good normal coordinates, at least for the ground state, which are not usually available for large molecules; another difficulty is that the structure has a sign ambiguity for each totally symmetrical coordinate; and a third is that the method assumes factorisation of the vibronic wave functions into independent electronic and vibrational parts. The Franck–Condon and contour methods are in excellent agreement as to the structure of $^1B_{2u}$ benzene, where high symmetry favours the normal-coordinate calculation[45]; but agreement for the $^1A_2$ state of $CSCl_2$ (Table 5.4, No. 3) is poor, a discouraging result for a relatively small molecule.

A more hopeful approach uses the experimental constants as a guide to theoretical considerations. Theoretical calculations of excited-state structures are increasing in number and sophistication, but this development lies outside the scope of this Chapter. An expressly empirical approach has been used by Cvitas, Hollas and Kirby[66] and by Bist and Verma[67] who have shown that the rotational constants for derivatives of benzene can be reproduced by a simple model in which CC and CH distances are transferred from the $^1B_{2u}$ state of benzene, and changes accommodated by a small contraction of the ring-to-substituent bond and some angular distortion of the ring.

5.4.6   Examples and appraisal

The rapid development of contour analysis in its modern form can be judged from the summary in Tables 5.3–5.7. Rather more than 30 determinations have been reported, particularly since 1968 which marks a division between the exploratory phase of the technique and its period of more routine development. The fact that the tables include rotational analyses by the contour method only gives an unbalanced representation where earlier analyses are available, based on frequency measurements. A review of azabenzene spectra[110], the class most affected, contains much supplementary information.

A survey of the tables shows a heavy commitment to aromatic and hetero-cyclic molecules, with a concentration approaching overkill for benzene derivatives. With few exceptions the method has been applied to first singlet states only. The direction of the transition moment is determined absolutely if it coincides with a symmetry axis or is perpendicular to a symmetry plane, otherwise its orientation in the principal axis frames can be obtained to c. ± 10 degrees though with an uncertainty in sign. Rotational constants and moments of inertia are determined with an accuracy of the order 1 part per 1000, which is adequate for a structure determination for small or symmetrical molecules with few independent geometrical parameters. The precision, however, is too low for routine structure determination of large unsymmetrical molecules. Dipole moment changes can be calculated from the effect of a Stark field on the contour.

To the present there has been no contour analysis of a triplet–singlet transition and relatively few analyses of valence shell–singlet transitions higher than the first, or of Rydberg transitions. Triplet–singlet transitions involve more parameters than singlet transitions and therefore greater difficulty in matching the contour but there seems to be no reason to consider them impracticable; many first transitions to triplet states are known to have structured bands. The prospects for higher singlet transitions are less clear. In general these transitions are diffuse, sometimes unstructured, presumably because the natural lifetime is shortened by photochemical or photophysical processes[111]. The range of line widths from mild broadening to total loss of definition in a rotational contour is not large, approximately 0.05–5 cm^{-1}, corresponding to lifetimes of 0.1 ns to 1 ps. First singlet states usually have lifetimes > 0.1 ns so that natural line width has not normally been a necessary parameter, although lifetime broadening is observed in the first transitions of ethane and propenal, and many other first systems show diffuseness in higher bands even when the electronic origin has narrow lines. How much information can be obtained for higher electronic states remains to be seen, but interesting possibilities exist for forming a connection with the time-resolved spectroscopy of short-lived states.

References

1. Burmeister, W. (1913). *Ber. Deutsch. Phys. Ges.,* **15,** 575
2. Gerhard, S. L. and Dennison, D. M. (1933). *Phys. Rev.,* **43,** 197
3. Badger, R. M. and Zumwalt, L. R. (1938). *J. Chem. Phys.,* **6,** 711
4. Edgell, W. F. and Moynihan, R. E. (1957). *J. Chem. Phys.,* **27,** 155

5. King, G. W., Cross, P. C. and Thomas, G. B. (1946). *J. Chem. Phys.*, **14**, 35
6. Turkevich, A. and Fred, M. (1942). *Rev. Mod. Phys.*, **14**, 246
7. Cooper, C. D. and Sponer, H. (1952). *J. Chem. Phys.*, **20**, 1248
8. Innes, K. K., Merritt, J. A., Tincher, W. C. and Tilford, S. C. (1960). *Nature (London)*, **187**, 500
9. Merritt, J. A. and Innes, K. K. (1960). *Spectrochim. Acta*, **16**, 945
10. Mason, S. F. (1959). *J. Chem. Soc.*, 1269
11. Craig, D. P., Hollas, J. M., Redies, M. F. and Wait, S. C. (1961). *Phil. Trans. Roy. Soc. (London)*, *A*, **253**, 543
12. Pauling, L. and Wilson, E. B. (1935). *Introduction to Quantum Mechanics*, Sec. 36 (New York: McGraw-Hill)
13. Allen H. C. and Cross, P. C. (1963). *Molecular Vib-Rotors*, (a) Sec. 2j, (b) Appendix 7, (c) Sec. 8c (New York: Wiley)
14. Rose, M. E. (1957). *Elementary Theory of Angular Momentum*, Ch., 4 (New York: Wiley)
15. Allen, H. C. (1961). *Phil. Trans. Roy. Soc. (London)*, *A*, **253**, 335
16. Gora, E. K. (1965). *J. Mol. Spectrosc.*, **16**, 378
17. Abramowitz, M. and Stegun, I. A. (1964). *Handbook of Methematical Functions*, Sec. 20 (Washington: U.S. Dept. of Commerce).
18. McHugh, A. J., Ramsay, D. A. and Ross, I. G. (1968). *Austral. J. Chem.*, **21**, 2835
19. Brown, J. M. (1969). *J. Mol. Spectrosc.*, **31**, 118
20. Townes, C. H. and Schawlow, A. L. (1955). *Microwave Spectroscopy*, Sec. 4–2 (New York: McGraw-Hill).
21. di Lauro, C. and Mills, I. M. (1966). *J. Mol. Spectrosc.*, **21**, 386
22. Herzberg, G. (1966). *Electronic Spectra and Electronic Structure of Polyatomic Molecules*, Sec. II-3b. (Princeton, N.J.: Van Nostrand)
23. Watson, J. K. G. (1966). *J. Chem. Phys.*, **45**, 1360 (and earlier references)
24. Oka, T. and Morino, Y. (1961). *J. Mol. Spectrosc.*, **6**, 472
25. Herschbach, D. R. and Laurie, V. W. (1964). *J. Chem. Phys.*, **40**, 3142
26. McHugh, A. J. and Ross, I. G. (1969). *Austral. J. Chem.*, **22**, 1
27. Brown, J. M. (1971). *J. Mol. Spectrosc.*, **37**, 179
28. Cartwright, G. J. and Mills, I. M. (1970). *J. Mol. Spectrosc.*, **34**, 415
29. Nakagawa, T. and Morino, Y. (1971). *J. Mol. Spectrosc.*, **38**, 84
30. Lin, C. C. and Swalen, J. D. (1959). *Rev. Mod. Phys.* **31**, 841
31. Hecht, K. T. and Dennison, D. M. (1957). *J. Chem. Phys.* **26**, 31
32. Sheppard, N. and Woodman, C. M. (1969). *Proc. Roy. Soc. (London)*, *A*, **313**, 149
33. Hougen, J. T. (1964). *Can. J. Phys.*, **42**, 433
34. Van Vleck, J. H. (1951). *Rev. Mod. Phys.*, **23**, 213
35. Raynes, W. T. (1964). *J. Chem. Phys.*, **41**, 3020
36. di Lauro, C. (1970). *J. Mol. Spectrosc.*, **35**, 461
37. Creutzberg, F. and Hougen, J. T. (1967). *Can. J. Phys.*, **45**, 1363
38. Creutzberg, F. and Hougen, J. T. (1971). *J. Mol. Spectrosc.*, **38**, 257
39. Benedict, W. S., Herman, R., Moore, G. E. and Silverman, S. (1956). *Can. J. Phys.*, **34**, 830, 850
40. Parkin, J. E. (1965). *J. Mol. Spectrosc.*, **15**, 483
41. Christoffersen, J., Hollas, J. M. and Kirby, G. H. (1968). *Proc. Roy. Soc. (London)*, *A*, **307**, 97
42. See Lombardi, J. R., Klemperer, W., Robin, M. B., Basch, H. and Kuebler, N. A. (1969). *J. Chem. Phys.*, **51**, 33
43. Ueda, T. and Shimanouchi, T. (1968). *J. Mol. Spectrosc.*, **28**, 350
44. Nakagawa, T., Kashiwagi, H., Kurihara, H. and Morino, Y. (1969). *J. Mol. Spectrosc.*, **31**, 436
45. Callomon, J. H., Dunn, T. M. and Mills, I. M. (1966). *Phil. Trans. Roy. Soc. (London)*, *A*, **259**, 499
46. Arnold, J. O., Whiting, E. E. and Lyle, G. C. (1969). *J. Quant. Spectrosc. Radiat. Transfer*, **9**, 775
47. Kirby, G. H. (1970). *Mol. Phys.*, **18**, 371
48. Yamada, K., Nakagawa, T., Kuchitsu, K. and Morino, Y. (1971). *J. Mol. Spectrosc.*, **38**, 70
49. McHugh, A. J. and Ross, I. G. (1970). *Spectrochim. Acta* **26A**, 441

50. Stoicheff, B. P. (1954). *Can. J. Phys.,* **32,** 339
51. Dunn, T. M. (1966). *Studies on Chemical Structure and Reactivity,* Ed. by Ridd, J. H., p. 103 (Methuen: London)
52. Hartford, A. and Lombardi, J. R. (1970). *J. Mol. Spectrosc.,* **35,** 413
53. Innes, K. K., Parkin, J. E. Hollas, J. M., Ervin, D. and Ross, I. G. (1965). *J. Mol. Spectrosc.,* **16,** 406
54. Cruickshank, D. W. J. (1957). *Acta Crystallogr.,* **10,** 504
55. McCoy, E. F. and Ross, I. G. (1962). *Austral. J. Chem.,* **15,** 573
56. McClure, D. S. (1956). *J. Chem. Phys.,* **24,** 1
57. Tobler, H. J., Bauder, A. and Gunthard, H. (1965). *J. Mol. Spectrosc.,* **18,** 239
58. King, G. W. and So, S. P. (1969). *J. Mol. Spectrosc.,* **33,** 377
59. Mani, A. and Lombardi, J. R. (1969). *J. Mol. Spectrosc.,* **31,** 308
60. Hartford, A. and Lombardi, J. R. (1970). *J. Mol. Spectrosc.,* **34,** 257
61. Hartford, A., Muirhead, A. R. and Lombardi, J. R. (1970). *J. Mol. Spectrosc.,* **35,** 199
62. Hollas, J. M., Kirby, G. H. and Wright, R. A. (1970). *Mol. Phys.,***18,** 327
63. Forest, H. and Dailey, B. P. (1966). *J. Chem. Phys.,* **45,** 1736
64. Bak, B., Christensen, D., Dixon, W. B., Hansen-Nygaard, L. and Rastrup-Andersen, J. (1962). *J. Chem. Phys.,* **37,** 2027
65. Merer, A. J. and Innes, K. K. (1968). *Proc. Roy. Soc. (London),* A, **302,** 271
66. Cvitas, T., Hollas, J. M. and Kirby, G. H. (1970). *Mol. Phys.,* **19,** 305
67. Bist, H. D. and Verma, A. L. (1970). *Chem. Phys. Lett.,* **4,** 577
68. Tanaka, T. and Morino, Y. (1970). *J. Mol. Spectrosc.,* **33,** 552
69. Richardson, A. W. (1970). *J. Mol. Spectrosc.,* **35,** 43
70, Sethuraman, V., Job, V. A. and Innes, K. K. (1970). *J. Mol. Spectrosc.,* **33,** 189
71 Duncan, J. L. and Ellis, D. (1968). *J. Mol. Spectrosc.,* **28,** 540
72. Masri, F. N. and Blass, W. E. (1971). *J. Mol. Spectrosc.,* **39,** 21
73. Duncan, J. L., Ellis, D. and Wright, I. J. (1971). *Mol. Phys.,* **20,** 673
74. Duncan, J. L., Wright, I. J. and Ellis, D. (1971). *J. Mol. Spectrosc.,* **37,** 394
75. Pearson, E. F. and Innes, K. K. (1969). *J. Mol. Spectrosc.,* **30,** 232
76. Lombardi, J. R. (1970). *J. Chem. Phys.,* **52,** 6126
77. Alves, A. C. P., Christoffersen, J. and Hollas, J. M. (1971). *Mol. Phys.,* **20,** 625
78. Haner, D. A. and Dows, D. A. (1970). *J. Mol. Spectrosc.,* **34,** 296
79. Howard-Lock, H. E. and King, G. W. (1970). *J. Mol. Spectrosc.,* **36,** 53
80. Christoffersen, J. and Hollas, J. M. (1969). *Mol. Phys.,* **17,** 655
81. Parkin, J. E. and Innes, K. K. (1965). *J. Mol. Spectrosc.,* **16,** 93
82. Fischer, G. (1969). *J. Mol. Spectrosc.,* **29,** 37
83. Cvitas, T. (1970). *Mol. Phys.,* **19,** 297 .
84. Bist, H. D., Sarin, V. N., Ojha, A. and Jain, Y. S. (1970). *Spectrochim. Acta,* **26A,** 841
85. Cvitas, T. and Hollas, J. M. (1970). *Mol. Phys.,* **18,** 101
86. Kirby, G. H. (1970). *Mol. Phys.,* **19,** 289
87. Huang, K-T. and Lombardi, J. R. (1970). *J. Chem. Phys.,* **52,** 5613
88. Christoffersen, J., Hollas, J. M. and Kirby, G. H. (1969). *Mol. Phys.,* **16,** 441
89. Lombardi, J. R. (1969). *J. Chem. Phys.,* **50,** 3780
90. Bist, H. D., Brand, J. C. D. and Williams, D. R. (1967). *J. Mol. Spectrosc.,* **24,** 413
91. Cvitas, T. and Hollas, J. M. (1971). *Mol. Phys.,* **20,** 645
92. Brand, J. C. D. and Knight, P. D. (1970). *J. Mol. Spectrosc.,* **36,** 328
93. Knight, P. D. *Thesis,* Vanderbilt University, 1970
94. King., G. W. and So. S. P. (1971). *J. Mol. Spectrosc.,* **37,** 543
95. Parker, H. and Lombardi, J. R. (1971). *J. Chem. Phys.,* **54,** 5095
96. Cvitas, T. and Hollas, J. M. (1970). *Mol. Phys.,* **18,** 801
97. Cvitas, T. and Hollas, J. M. (1970). *Mol. Phys.,* **18,** 261
98. Wu, C. Y. and Lombardi, J. R. (1971). *J. Chem. Phys.,* **54,** 3659
99. Udagawa, Y., Ito, M. and Nagakura, S. (1970). *J. Mol. Spectrosc.,* **36,** 541
100. Cvitas, T. and Hollas, J. M. (1970). *Mol. Phys.,* **18,** 793
101. Christoffersen, J., Hollas, J. M. and Kirby, G. H. (1970). *Mol. Phys.,* **18,** 451
102. Huang, K-T. and Lombardi, J. R. (1969). *J. Chem. Phys.,* **51,** 1228
103. Brand, J. C. D. and Tang, K-T. (1971). *J. Mol. Spectrosc.,* **39,** 171
104. Innes, K. K. and Parkin, J. E. (1966). *J. Mol. Spectrosc.,* **21,** 66
105. Parkin, J. E. and Innes, K. K. (1965). *J. Mol. Spectrosc.,* **15,** 407

106. Innes, K. K., McSwiney, H. D., Simmons, J. D. and Tilford, S. G. (1969). *J. Mol. Spectrosc.*, **31**, 76
107. Innes, K. K., Tincher, W. C. and Pearson, E. F. (1970). *J. Mol. Spectrosc.*, **36**, 114
108. Brown, J. M. (1969). *Can. J. Phys.*, **47**, 233
109. Christoffersen, J., Hollas, J. M. and Wright, R. A. (1969). *Proc. Roy. Soc. (London), A,* **308**, 537
110. Innes, K. K., Byrne, J. P. and Ross, I. G. (1967). *J. Mol. Spectrosc.*, **22**, 125
111. Jortner, J., Rice, S. A. and Hochstrasser, R. M. (1969). *Adv. in Photochem.*, **7**, 149

6
Molecules in Space

L. E. SNYDER
University of Virginia, Charlottesville, Virginia

6.1   INTRODUCTION

At least 24 molecular species have been found to date in the interstellar clouds of gas and solid particles, or interstellar 'grains', which pervade our galaxy, the Milky Way. The chemical processes which generate interstellar molecules are efficient over large regions of the clouds and appear to be intimately related to the presence of the grains and stars enveloped in the interstellar clouds. Little progress has been made in determining directly the chemical composition of the grains in the past 40 years but many of their properties as submicrometre size particles have been established from observations of extinction, polarisation and reflection of starlight[195]. In very dense clouds, the grains may provide catalytic surfaces which enhance the production of polyatomic molecules whereas in more tenuous regions there is evidence that radiative association could be the predominant mechanism for molecular formation. The usual restrictions of molecular stability placed on the laboratory spectroscopist by wall collisions are negligible in the interstellar clouds. The total gas content of the galaxy is quite large, perhaps 3 per cent of the total mass, but cloud dimensions are of the order of light years; the resulting densities are very low and may only reach 10^6 hydrogen molecules cm^{-3} in very dense clouds[67]. Low densities and average kinetic temperatures between 10 and 100 K are indications that mean free paths between 10^9 and 10^{15} cm are not unusual. The resulting times between collisions (10^4–10^{10} s) are of the order of relaxation times for rotationally-excited molecules; thus electronic and vibrational relaxation times (typically of the order of 10^{-8} and 10^{-2} s, respectively) usually are too short to influence collisional formation processes. In addition, there are indications that the central regions of dense clouds provide sufficient shielding from photodestructive processes to allow fairly complex molecular species, once formed, to survive for years. Hence it is evident that the chemical processes required to explain the formation and relative abundances of many of the interstellar molecules will require models which will be difficult to test in the laboratory. As a result, it is quite possible that astronomers are compiling evidence which offers new insight to the chemistry of molecular formation.

6.1.1   Early detections

As recently as 1968, only 4 molecules were known to exist in the interstellar medium. Three of these, CH, CH^+ and CN had been identified in the period between 1937 and 1941 by means of their absorption spectra superimposed on the optical spectra of bright stars. The detection and identification of these three diatomic molecules helped to establish many of the modern concepts of the basic nature of typical galactic molecular clouds, such as low densities and temperatures, and raised the problem of molecular formation in the interstellar medium. The OH molecule, discovered in 1963, was the first ever detected by radio astronomers. The strong intensities, anomalous line ratios and unusual profiles of OH were to provide evidence of interstellar excitation mechanisms which are not completely understood today.

6.1.2 Recent discoveries

In 1968, the detection of interstellar NH_3 was the first of a remarkable series of discoveries. Following ammonia in rapid progression were detections of interstellar $H_2O(1968)$; $H_2CO(1969)$; CO, H_2, HCN, X-ogen, HC_3N, CH_3OH, HCOOH(1970); NH_2CHO, SiO, OCS, CS, CH_3CN, HNCO, CH_3C_2H, 'HNC', $HCOCH_3$ and H_2CS(1971). In addition to the new molecular detections, the lowest ground-state rotational transition of CN was detected astronomically far in advance of laboratory frequency measurements of the line. Tables 6.1 and 6.2 list the 24 known interstellar molecules. The inorganic interstellar molecules are listed in Table 6.1 and the organic in Table 6.2. The first column in each Table gives the chemical formula and name. The formulae, taken from astronomical detection reports, usually follow the molecular structure and hence may deviate from the I.U.P.A.C. rules of nomenclature. The second column lists the spectroscopic classification of the transition; optical and ultraviolet transitions are classified by electronic states followed by vibrational states and rotational assignments (when resolved). Radio-frequency transitions are classified by ground electronic state (when known) followed by rotational and hyperfine assignments. The third column gives the detection wavelength (in Å) for optical and ultraviolet transitions or the frequency (in MHz) for radio transitions. The fourth column indicates whether the spectral line has been observed in absorption (A) or emission (E) and the last column lists detection references[1-59]. The reference to the initial detection of each molecular isotope is underlined. When more than one reference is listed, usually the first reference is to the detection and the following contain important supplementary information. The electronic-state notation used for polyatomic molecules generally follows modern convention[60]. The earlier optical detections of diatomic molecules, listed elsewhere[61, 63] have been incorporated in Tables 6.1 and 6.2 for completeness. The last category in Table 6.2, labelled 'Other', lists those interstellar molecules which have no laboratory spectra. These molecules are placed in Table 6.2 because they are currently believed to be organic. A positive molecular identification requires the detection of more than one spectral line: hence it is easily seen from the Tables that some interstellar identifications have been established more conclusively than others.

As a direct consequence of the interstellar molecular spectroscopy of the past 2 years a number of phenomena have been discovered which are of direct interest to chemists. These include the non-thermal population statistics

Table 6.1 Inorganic interstellar molecules

Molecule	Classification	Frequency or wavelength	Absorption or emission	Detection reference
	DIATOMIC			
H_2; molecular hydrogen	$B^1\Sigma_u^+ - X^1\Sigma_g^+$, (0,0)	1108 Å	A	(1)
	(1,0)	1092 Å	A	(1)
	(2,0)	1077 Å	A	(1)
	(3,0)	1063 Å	A	(1)
	(4,0)	1049 Å	A	(1)
	(5,0)	1037 Å	A	(1)
	(6,0)	1024 Å	A	(1)
	(7,0)	1013 Å	A	(1)
^{16}OH; hydroxyl	$^2\Pi_{3/2},\ J=\tfrac{3}{2},\ F=1\text{-}2$	1612.231 MHz	A/E	(2)
	1-1	1665.401 MHz	A/E	(3)
	2-2	1667.358 MHz	A/E	(3)
	2-1	1720.533 MHz	A/E	(2)
	$^2\Pi_{3/2},\ J=\tfrac{5}{2},\ F=2\text{-}2$	6030.739 MHz	E	(4)
	3-3	6035.085	E	(4)
	$^2\Pi_{3/2},\ J=\tfrac{7}{2},\ F=4\text{-}4$	13 441.371 MHz	E	(5)
	$^2\Pi_{1/2},\ J=\tfrac{1}{2},\ F=1\text{-}0$	4765.562 MHz	E	(6)
	0-1	4660.242	E	(7)
	$^2\Pi_{1/2},\ J=\tfrac{5}{2},\ F=2\text{-}2$	8135.868 MHz*	E	(8, (9)
^{18}OH	$^2\Pi_{3/2},\ J=\tfrac{3}{2},\ F=1\text{-}1$	1637.53 MHz	A	(11)
	2-2	1639.48	A	(10)
SiO; silicon monoxide	$X^1\Sigma^+,\ J=3\text{-}2$	130 246 MHz	E	(12)
	TRIATOMIC			
H_2O; water	$\tilde{X}^1A_1, J_{K_-K_+} = 6_{16}\text{-}5_{23}$	22 235.08 MHz	E	(13)
	FOUR-ATOMIC			
NH_3; ammonia (inversion doublets)	$\tilde{X}^1A_1, J, K = 3,2\text{-}3,2$	22 834.17 MHz	E	(14)
	2,1-2,1	23 098.79 MHz	E	(14)
	1,1-1,1	23 694.48 MHz	E	(15)
	2,2-2,2	23 722.71 MHz	E	(15)

3,3–3,3	23 870.11 MHz	E	(16)
4,4–4,4	24 129.39 MHz	E	(16)
6,6–6,6	25 056.04 MHz	E	(16)

*The detection of this line was reported in reference 8 but not confirmed by the observations reported in reference 9. It is included here because of the possibility of time-varying intensity. Following publication of reference 9, the rest frequency of this transition was re-measured in the laboratory and determined to have the value given here[22a].

Table 6.2 Organic interstellar molecules

Molecule	Classification	Frequency or wavelength	Absorption or emission	Detection reference
	DIATOMIC			
$^{12}CH^+$	$A^1\Pi$–$X^1\Sigma$, (0,0), R(0)	4232.54 Å	A	(17), (18), (19)
	(1,0), R(0)	3957.70 Å	A	(17), (18), (19)
	(2,0), R(0)	3745.31 Å	A	(20), (18), (19)
	(3,0), R(0)	3579.02 Å	A	(20), (22)
	(4,0), R(0)	3447.08 Å	A	(23), (22)
$^{13}CH^+$	$A^1\Pi$–$X^1\Sigma$, (0,0), R(0)	4232.08 Å	A	(24)
CH	$A^2\Delta$–$X^2\Pi$, (0,0), $R_2(1)$	4300.30 Å	A	(17), (25), (26)
	$B^2\Sigma^-$–$X^2\Pi$, (0,0), $^PQ_{13}(1)$	3890.21 Å	A	(20), (25)
	(0,0), $Q_2(1) + {}^QR_{12}(1)$	3886.41 Å	A	(20), (25)
	(0,0), $R_2(1)$	3878.77 Å	A	(20), (25)
	$C^2\Sigma^+ - X^2\Pi(0,0)$, $^PQ_{13}(1)$	3146.01 Å	A	(23), (27)
	(0,0), $Q_2(1) + {}^QR_{12}(1)$	3143.20 Å	A	(28), (27), (29)
	(0,0), $R_2(1)$	3137.53 Å	A	(23), (27)
CN; cyanogen radical	$X^2\Sigma^+$ $N = 1-0$	113 492 MHz	E	(30)
	$B^2\Sigma^+$–$X^2\Sigma^+$, (0,0), P(3)	3876.84 Å	A	(31)
	(0,0), P(2)	3876.30 Å	A	(31)
	(0,0), P(1)	3875.77 Å	A	(20), (32)
	(0,0), R(0)	3874.61 Å	A	(21), (20), (25)
	(0,0), R(1)	3874.00 Å	A	(20), (32)
$^{12}C^{16}O$; carbon monoxide	$X^1\Sigma^+$, $J = 1-0$	115 271.2 MHz	E	(33)
	$A^1\Pi$–$X^1\Sigma^+$, (1,0)	1509.65 Å	A	(34)

Table 6.2 (continued)

Molecule	Classification	Frequency or wavelength	Absorption or emission	Detection reference
	(3,0)	1447.26 Å	A	(34)
	(4,0)	1418.97 Å	A	(34)
	(6,0)	1367.56 Å	A	(34)
	(8,0)	1322.10 Å	A	(34)
	(9,0)	1301.37 Å	A	(34)
	(10,0)	1281.83 Å	A	(34)
$^{13}C^{16}O$	$X^1\Sigma^+, J = 1-0$	110 201.4 MHz	E	(35)
	$A^1\Pi-X^1\Sigma^+$, (2,0)	1478.8 Å	A	(34)
	(3,0)	1449.3 Å	A	(34)
	(4,0)	1421.4 Å	A	(34)
	(6,0)	1370.8 Å	A	(34)
$^{12}C^{18}O$	$X^1\Sigma^+, J = 1-0$	109 782.2 MHz	E	(35)
CS; carbon monosulphide	$X^1\Sigma^+, J = 3-2$	146 969.16 MHz	E	(36)
TRIATOMIC				
$H^{12}C^{14}N$; hydrogen cyanide	$\tilde{X}^1\Sigma^+, J = 1-0, F = 1-1$	88 630.4157 MHZ*	E	(37)
	2-1	88 631.8473	E	
	0-1	88 633.9360 MHz	E	
$H^{13}C^{14}N$	$\tilde{X}^1\Sigma^+, J = 1-0, F = 1-1$	86 338.75 MHz	E	(37)
	2-1	86 340.05	E	
	0-1	86 342.16 MHz	E	
OCS; carbonyl sulphide	$\tilde{X}^1\Sigma^+, J = 9-8$	109 462.8 MHz	E	(38)
FOUR-ATOMIC				
$H_2^{12}C^{16}O$; formaldehyde	$\tilde{X}^1A_1, J_{K_-K_+} = 1_{10}-1_{11}$	4 829.660 MHz	A/E	(39), (40)
	$2_{11}-2_{12}$	14 488.65	A	(41)
	$3_{12}-3_{13}$	28 974.85	A	(42)
	$2_{12}-1_{11}$	140 839.53	E	(43)
	$2_{02}-1_{01}$	145 602.97	E	(44)
	$2_{11}-1_{10}$	150 498.36	E	(44)
$H_2^{13}C^{16}O$	$\tilde{X}^1A_1, J_{K_-K_+} = 1_{10}-1_{11}$	4593.089 MHz	A	(45)
$H_2^{12}C^{18}O$	$\tilde{X}^1A_1, J_{K_-K_+} = 1_{10}-1_{11}$	4 388.797 MHz	A	(46)

Molecule; name	Transition	Frequency		Ref.
HNCO; isocyanic acid	$J_{K_-K_+} = 4_{04}-3_{03}$	87 925.45 MHz	E	(47)
H$_2$CS; thioformaldehyde	$J_{K_-K_+} = 1_{01}-0_{00}$	21 981.7 MHz	E	(48)
	$J_{K_-K_+} = 2_{11}-2_{12}$	3139.38 MHz	A	(59)
FIVE-ATOMIC				
HCOOH; formic acid	$\tilde{X}^1A', J_{K_-K_+} = 1_{10}-1_{11}$	1638.805 MHz	E	(49)
HC$_3$N; cyanoacetylene	$\tilde{X}^1\Sigma^+, J = 1-0, F = 1-1$	9097.07 MHz	E	(50)
	$2-1$	9098.36 MHz	E	(50)
SIX-ATOMIC				
CH$_3$OH; methyl alcohol	$J_{K_-K_+}(A) = 1_{10}-1_{11}$	834.267 MHz†	E	(51)
	$J_K(E_1) = 4_2-4_1$	24 933.47 MHz	E	(52)
	5_2-5_1	24 959.08 MHz	E	(52)
	6_2-6_1	25 018.14 MHz	E	(52)
	7_2-7_1	25 124.88 MHz	E	(52)
	8_2-8_1	25 294.41 MHz	E	(52)
	$J_K(E_1) = 5_1-4_0$	85 521.5 MHz	E	(53)
CH$_3$CN; methyl cyanide	$\tilde{X}^1A_1, J_K = 6_5-5_5$	110 330.7 MHz	E	(54)
	6_4-5_4	110 349.7 MHz	E	(54)
	6_3-5_3	110 364.5 MHz	E	(54)
	6_1-5_1	110 381.4 MHz	E	(54)
	6_0-5_0	110 383.5 MHz	E	(54)
HCONH$_2$; formamide	$J_{K_-K_+} = 1_{10}-1_{11}$	1539.543 MHz‡	E	(55)
	$2_{11}-2_{12}, F = 2-2$	4617.118 MHz	E	(56)
	$F = 3-3$	4618.970 MHz	E	(56)
	$F = 1-1$	4619.988 MHz	E	(56)
SEVEN-ATOMIC				
CH$_3$C$_2$H; methylacetylene	$\tilde{X}^1A_1, J_K = 5_0-4_0$	85 457.29 MHz	E	(47)
HCOCH$_3$; acetaldehyde	$\tilde{X}^1A', J_{K_-K_+}(A) = 1_{10}-1_{11}$	1065.075 MHz	E	(57)
OTHER				
X-ogen		89 190 MHz	E	(58)
'HNC'; hydrogen isocyanide	$J = 1-0$	90 665 MHz	E	(47)

*The hyperfine components are not resolved in reference 37.
†Laboratory measurement after the detection was reported[26a].
‡The value listed is the calculated unsplit frequency given in reference 211.

of the water vapour maser and the formaldehyde inverse maser, time variations in water vapour spectra, anomalous isotopic abundance ratios of
interstellar carbon, the apparent over-abundance of some complex molecules
relative to simpler molecules and the failure to detect several small molecules
composed of cosmically abundant constituent atoms. A proper interpretation
of many of the observational results is difficult because it requires recognition
of the limitations placed on astronomical measuring techniques. For
example, ambiguous estimates of temperatures, limited information about
population statistics, clumping effects and spectral line saturation tend to
introduce uncertainties in density determinations which could easily lead to
erroneous conclusions about the chemistry of the interstellar medium. To
compensate for some of these difficulties, the detections of unidentified
molecules and the ground-state CN lines emphasise the role of a typical
interstellar cloud as a unique chemical laboratory—a sample cell which
would be difficult if not impossible to duplicate in its entirety in terrestrial
laboratories.

Earlier reviews[62, 63] have discussed our knowledge of the interstellar
medium and the possible reaction mechanisms invoked to explain the
formation of those interstellar molecules found before 1968. In this Review,
we will discuss the new detections and the interpretations of the data which
are currently offered. With the exception of molecular hydrogen, all of the
new molecules were initially detected by radio astronomers. Hence an
introductory description of radio observational techniques is included to aid
those who are not familiar with the radio branch of astronomy.

6.2 OBSERVATIONAL CONSIDERATIONS

6.2.1 Comments on optical v. radio observations

From the late 1930s until 1968, it was generally assumed by most astronomers
and astrophysicists (with a few exceptions discussed elsewhere[64]) that only
diatomic species could exist in the interstellar clouds. This assumption
probably was a direct result of the failure of both optical searches for new
molecules and radio searches for a few diatomic species which either are very
unstable or have poorly determined radio frequencies. In addition, the various
formation mechanisms proposed for diatomic molecules were severely
limited by the lack of systematic observations and appropriate molecular
data[63].

Now that it is becoming apparent that many of the new molecules appear
to be associated with heavily obscured regions of the galaxy, a plausible
explanation for the failure of optical searches for complex molecules is
beginning to emerge. The energy required for excitation of an optical
absorption line may be written as

$$E = E_{el} + E_{vib} + E_{rot} \qquad (6.1)$$

where E_{el} is the energy for electronic excitation; E_{vib} and E_{rot} are the much
smaller energies for vibrational and rotational excitation of the molecule.

E_{el} may be on the order of $13\,000\,\text{cm}^{-1}$ or greater, while E_{vib} is typically 1000 to $5000\,\text{cm}^{-1}$ above the zero-point vibrational level and E_{rot} is generally much less than $500\,\text{cm}^{-1}$ for most of the interstellar molecules detected to date with the marked exception of interstellar H_2O ($447\,\text{cm}^{-1}$). Hence it is well known that to detect optical absorption lines of interstellar molecules the background source should have a temperature of 15 000 K or more. Thus an interstellar cloud through which a bright star with few stellar lines, preferably O- or B-type, can be spectroscopically observed is an ideal candidate for optical searches. As a result, the heavily obscured galactic regions currently associated with the presence of both grains and complex molecules are necessarily excluded from observation. It is interesting to contrast briefly these optical limitations imposed on interstellar spectroscopy with the more favourable situation found in cometary spectroscopy. In a comet, parent molecules in a region of high opacity receive more than the necessary excitation energy in the form of dissociating ultraviolet radiation from the sun. The resulting molecular fragments produce fluorescent spectra which are easily observed[65]. Hence optical spectroscopists have detected many more cometary molecules (e.g. CH, CH^+, CN, C_2, C_3, NH, NH_2, N_2^+, CO^+ and OH) than interstellar molecules.

Radio observations of interstellar molecules do not require particularly special configurations such as a bright star with intervening molecular clouds of low opacity. The excitation conditions needed for detections are not as stringent as those required for observations of optical transitions. For example, an inspection of Tables 6.1 and 6.2 shows that pure rotational transitions are commonly detected. The required excitation energies are often less than $100\,\text{cm}^{-1}$. Energy can be provided by such diverse background sources as supernova remnants[66], radio galaxies, quasars, ionised gaseous nebulae, certain types of infrared stars and, in some cases, by the 2.7 K isotropic background radiation. In very dense clouds, collisions between molecules and neutral particles or electrons must be considered as an important source of excitation for molecular emission lines[67]. Molecular clouds observed by radio techniques are also permitted to have much higher opacities than their optical counterparts because gas and grains attenuate radio waves much less than optical waves. Presumably, the greater number of grains along the line of sight indirectly enhances the detectability of interstellar molecules.

These basic differences between optical and radio spectroscopy have been largely responsible for the remarkable success of interstellar molecular spectroscopy. The remainder of the credit must go to receiver development and other instrumental improvements which we shall discuss next.

6.2.2 Astronomical radio spectroscopy: telescopes and instrumentation

Many of the problems encountered in detecting new interstellar molecules and subsequently in interpreting the data are unique to astronomy. The following description is intended to give the laboratory scientist some insight into current observational equipment and practices used in astronomical radio spectroscopy. The cited references have been chosen to lead the interested reader to more detailed publications and technical reports.

6.2.2.1 Radio telescopes

The simplest radio telescope in terms of antenna gain and resolution is the filled-aperture antenna[68, 69]; it is preferred for molecular searches because of the relative ease of operation and the minimal electronic complexity required for data acquisition. An example of the filled-aperture design most commonly used is the single-element parabolic reflector illustrated by Figure 6.1.

Figure 6.1 The 36 ft radio telescope of the National Radio Astronomy Observatory is located on Kitt Peak near Tuscon, Arizona. The telescope is enclosed by an astrodome which allows spectral line observations to proceed during inclement weather. The radio receiver can be seen near the end of the support legs at the top of the photograph and a portion of the azimuth-elevation mount is visible beneath the reflector.
(Photograph by courtesy of the National Radio Astronomy Observatory)

The half-power beam width (HPBW) of the parabolic reflector may be very critical both for successful molecular detections and for interpreting observations. Theoretically the HPBW is given in radians by

$$HPBW = 1.22 \, \lambda/D \qquad\qquad (6.2)$$

where λ is the wavelength corresponding to the observed frequency and D is the diameter of the reflector. Thus an 11 m parabolic reflector operating at 3.5 mm wavelength has a HPBW of approximately 70 seconds of arc which, in practice, means that weak molecular signals from sources of small angular diameter (<0.01 degrees) may not be detectable because of beam dilution. For example, typical single-element reflectors limit radio spectroscopy of stellar molecules to observations of relatively intense emission lines from a few of the more abundant species.

One way to overcome the inherent difficulties while retaining the advantages of the parabolic reflector is to increase the value of D in equation (6.2); this will be accomplished shortly in the centimetre-wavelength region by the new Bonn 100 m radio telescope[70]. Another way is to use telescopes and receivers which operate in the shorter-wavelength or higher-frequency ranges. This method is being used now as shown by the third column in Tables 6.1 and 6.2. When applicable, interferometric (unfilled-aperture) techniques can be used to obtain much smaller beam widths with correspondingly higher spatial resolution but data acquisition is currently very tedious[71, 72]. With improved technology in the future, interferometric techniques will be used to gather significant new data from both interstellar and stellar molecules which cannot be obtained with single-element reflectors.

6.2.2.2   Receivers and data processors

The design and quality of radio receivers and data processors have been important factors in the new molecular detections[73]. Modern spectral line systems send the collected antenna signal into a front-end section which usually is a Dicke-switched superheterodyne mixer radiometer[74]. Following amplification, filtering, conversion and appropriate calibration, the spectral line is displayed by means of the back-end data processor which is either a multichannel filter bank or an autocorrelation receiver[75]. The minimum signal temperature detectable is called the sensitivity, ΔT_{min}, and is related to the system noise temperature, T_{sys}, by the proportionality

$$\Delta T_{min} = C T_{sys} \left[\frac{N}{Bt} \right]^{\frac{1}{2}} \qquad (6.3)$$

where T_{sys} is the sum of T_a, the antenna noise contribution, and T_r, the receiver noise temperature (including the telescope feed). B is the total frequency band width and N the number of channels in the multichannel receiver. The integration time, t, is usually given in seconds and C, a proportionality constant, is a function of the particular system and observing mode used.

There are three techniques commonly used to improve the sensitivity of a spectral line system. The first method is to merely increase the integration time t. However, system instabilities usually impose a practical limit of a few hours on t. The time limit can be considerably less when filter banks must be used instead of the more stable autocorrelation receiver. The second method is to use the maximum band width consistent with the spectral line width, thereby minimising N/B in equation (6.3). Autocorrelation receivers are particularly versatile for this approach because the band width may be changed easily. For example, the 413 channel Model II Autocorrelation Receiver of the National Radio Astronomy Observatory, Green Bank, West Va., can have channel spacings (B/N) ranging from 0.1017 to 26.0417 kHz or from 0.2034 to 52.0834 kHz with seven intermediate values in either case which depend on the observing configuration selected[76]. At present, band-width limitations restrict autocorrelators to centimetre-wave observations and filter banks must be used to provide the larger band widths necessary for observations of millimetre-wave spectral lines. The third method, currently most promising for the millimetre-wave region, is to reduce T_{sys} by designing receivers with the lowest possible T_r. In the centimetre-region, systems with T_r less than 100 K are common, but at 3 mm, for example, the current value of T_r is close to 4000 K for spectral line observations[73]. The usual technique for reducing the overall system temperature is to insert a preamplifier, either a maser or parametric amplifier, between the antenna feed and the mixer radiometer. Using this technique, improvements in millimetre wave sensitivity between one and two orders of magnitude are expected within the next 5 years.

In the past year, small on-line computers have been utilised successfully to process spectral line data from millimetre wave observations. Although the ability to interact with the spectral line system does not improve the theoretical limit for receiver sensitivity, it has been found to be invaluable for obtaining maximum sensitivity from a given receiver. In practice, the successful detections of relatively weak signals from several new interstellar molecules were greatly expedited through on-line data processing.

6.3 APPLICATIONS OF RADIATIVE TRANSFER

Radiative transfer is a very active and important area of investigation for the spectroscopist who studies interstellar molecules because he can only observe, analyse and interpret the radiation reaching him from a distant sample cell over which he has no control. Our current limited understanding of the dominant interstellar chemical processes, for example, is built upon abundance determinations which are critically dependent upon proper interpretation of temperature and density effects on spectral line radiation. Recently, advances have been made in the application of radiative transfer to astronomical molecular spectroscopy, most notably in the case of maser emission, but often the basic equations are obscured by the specialised terminology. In this Section we will briefly outline those elements of radiative transfer which are central to understanding current abundance determinations in radio astronomy.

6.3.1 The transfer equation

The equation of transfer may be written in the form

$$\frac{dI_\nu}{ds} = -K_\nu I_\nu + J_\nu \tag{6.4}$$

where K_v is the volume absorption coefficient, J_v the volume emission coefficient and $I_v \, dv$ the specific radiation intensity in the frequency range dv[77, 78]. Unit refractive index is assumed. An integrating factor, $e^{\tau_v(s)}$, may be used to find

$$I_v(s_0) = I_v(0)e^{-\tau_v(s_0)} + e^{-\tau_v(s_0)} \int_0^{s_0} J_v e^{\tau_v(s)} \, ds \qquad (6.5)$$

as the solution to equation (6.4); $\tau_v(s)$, the optical depth, is defined by

$$\tau_v(s) = \int_0^s K_v \, ds' \qquad (6.6)$$

The integration path s is taken along the line of sight from an initial point $s = 0$ (which may coincide with the position of a background source) to the position of the observer s_0.

6.3.2   Idealised cloud configuration

Applications of equation (6.5) to both atomic and molecular spectral lines are often discussed for the case of the uniform homogeneous cloud observed with an ideal antenna[77, 79]. In Figure 6.2(a), source radiation of intensity I_0 passes through a uniform cloud with absorption coefficient $K_0(s)$ which corresponds to an atomic or molecular transition of frequency v_0. Any continuum contributions to the absorption and emission coefficients usually can be ignored; thus the intensity of the spectral line with respect to the continuum may be treated directly. Evaluated at the resonance frequency v_0, equation (6.5) becomes

$$I_0(s_0) = I_0(0) \, e^{-\tau_0(s_0)} + (J_0/K_0) \, (1 - e^{-\tau_0(s_0)}) \qquad (6.7)$$

where in equation (6.6) the effective integration path for the optical depth is taken between the cloud boundaries. The first term represents the attenuation of the incident intensity as it passes through the cloud; the second term is the emission contribution of the cloud corrected for absorption. At a second frequency, which is off resonance but very close to v_0, equation (6.5) gives only the continuum contribution due to the background source as $I_0'(s_0) \simeq I_0(0)$. Thus the solution for the excess intensity due to the spectral line is found to be

$$\Delta I_0 \equiv I_0(s_0) - I_0'(s_0) = \left[\frac{J_0}{K_0} - I_0(0) \right] [1 - e^{-\tau_0(s_0)}] \qquad (6.8)$$

The volume absorption coefficient, K_v, is related to the Einstein coefficient for absorption, B_{1u}, and stimulated emission, B_{ul}, by

$$K_v I_v = \frac{hv}{4\pi} f(v) (n_1 B_{1u} - n_u B_{ul}) I_v \qquad (6.9)$$

where h is Planck's constant; the usual subscripts u and l are for the upper and lower levels of a two level system. n_u and n_l are the molecular densities of the upper and lower levels and $f(v)$ is the line-shape function defined by $\int_v f(v) dv = 1$. For simplicity, $f(v)$ is often expressed in terms of the full width at half maximum as $f(v_0) = (\Delta v)^{-1}$. Equation (6.9) represents the intensity contribution due to the net number of radiation induced transitions per cubic centimetre. The well known relations between the Einstein B values

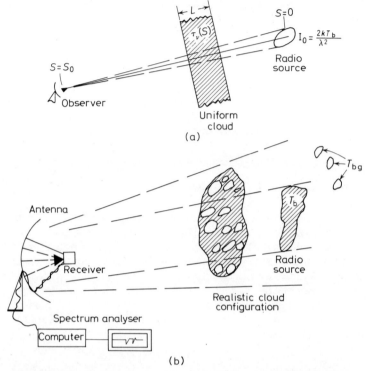

(a)

(b)

Figure 6.2 (a) An observer at s_0 uses an ideal antenna to record the signal from a background radio source (at $s = 0$) through a uniform, homogeneous cloud of depth L and opacity $\tau_v(s)$

(b) The realistic radio observing configuration must take into account beam dilution and efficiency, cloud inhomogeneities and galactic background contributions other than the main radio source.

and the Einstein coefficient for spontaneous emission, A_{ul},

$$A_{ul} = \frac{2hv^3}{c^2} B_{ul} = \frac{2hv^3}{c^2} \cdot \frac{g_l}{g_u} \cdot B_{lu} \tag{6.10}$$

may be used to write the absorption coefficient in the more compact form

$$K_v = \frac{A_{ul}\lambda^2 f(v)}{8\pi}\left[\frac{g_u}{g_l}\cdot n_1 - n_u\right] \tag{6.11}$$

where g_u and g_l are the degeneracy factors for the upper and lower energy levels.

The volume emission coefficient, J_v, is related to A_{ul} by

$$J_v = \frac{A_{ul}\lambda^2}{8\pi} f(v)\left(\frac{hv}{k}\right)\left(\frac{2kv^2}{c^2}\right)n_u \tag{6.12}$$

At this point, the ratio J_0/K_0 in equation (6.8) may be evaluated.

$$J_0/K_0 = \left(\frac{2kv_0^2}{c^2}\right)\left(\frac{hv_0}{k}\right)\left[\frac{g_u}{g_1}\cdot\frac{n_1}{n_u}-1\right]^{-1} \tag{6.13}$$

In radio astronomical research, intensities are almost always expressed in terms of an equivalent blackbody temperature (or 'brightness' temperature) by the Rayleigh–Jeans equation

$$I = \frac{2kv^2}{c^2}\,T_b = \frac{2kT_b}{\lambda^2} \tag{6.14}$$

where k is Boltzmann's constant and c is the speed of light. Equation (6.13) may be conveniently expressed in temperature units by using

$$\frac{n_1}{n_u} = \frac{g_1}{g_u}e^{+hv/kT_e} \approx \frac{g_1}{g_u}\left[1+\frac{hv}{kT_e}\right] \tag{6.15}$$

for $hv \ll kT_e$. In astronomical parlance, T_e is called an excitation temperature. Strictly speaking, T_e is not a state temperature describing the molecular population distribution over the entire ensemble of energy levels, because the interstellar distribution statistics are rarely described by a Boltzmann distribution law. The interstellar deviation from thermal equilibrium may be caused by either radiative or collisional interactions which are very difficult to measure directly. Thus T_e is only a temperature describing the relative populations of the two levels for which it is defined, but in certain cases (see equation (6.35) in Section 6.4.3.1) T_e may be an acceptable first approximation to the kinetic temperature for the molecular ensemble.

Substitution of equations (6.13), (6.14) and (6.15) into equation (6.8) gives the solution of the radiative-transfer equation which radio astronomers use as a starting point for the analysis of spectral line data.

$$\Delta T = (T_e - T_b)(1-e^{-\tau_0(s_0)}) \tag{6.16}$$

ΔT is the temperature difference, measured between the resonant frequency v_0 and the spectral region adjacent to it, as recorded by an ideal antenna for a uniform, homogeneous cloud (Figure 6.2(a)). ΔT is positive for an emission line ($T_e > T_b$) and negative for an absorption line ($T_e < T_b$).

6.3.3   Realistic cloud configuration

A more realistic cloud configuration consists of a number of small condensations or cloudlets in the telescope beam as illustrated in Figure 6.2(b). This configuration, often called 'clumping', was initially treated in detail for interstellar hydrogen absorption and later modified to take into account the effect of galactic background radiation on the excitation temperature of OH[80, 81]. The latter treatment is unchanged when adapted to spectral lines of other molecules as long as the remaining physical assumptions in Section 6.3.2 are still valid. Thus the appropriate modification of equation

(6.16) for a distribution of $(M+N)$ cloudlets observed with a real antenna is

$$\Delta T \frac{\Omega_B}{\eta} = \Omega_C(T_e - T_b - T_{bb} - F(D)T_{bg}) \sum_{j=1}^{M} \frac{\Omega_j}{\Omega_C} G(\theta_j)(1 - e^{-\tau_j})$$

$$+ (T_e - T_{bb} - F(D)T_{bg}) \sum_{i=1}^{N} \Omega_i G(\theta_i)(1 - e^{-\tau_i}) \qquad (6.17)$$

where the additional terms are defined as follows:
Ω_B = solid angle subtended by the main beam of the antenna; Ω_j = solid angle of jth molecular cloud contained in the solid angle Ω_C subtended by the discrete background source; Ω_i = solid angle of ith molecular cloud contained in Ω_B but not in Ω_C; T_{bg} = brightness temperature of any galactic background contribution; $F(D)$ = fraction of T_{bg} absorbed at distance D beyond the molecular cloud group; T_{bb} = brightness temperature of the isotropic background radiation (c. 2.7 K); $G(\theta)$ = antenna gain function[74] normalised to unity on the beam axis (where $\theta = 0$); η = telescope beam efficiency.

It is assumed in equation (6.17) that the antenna is pointed toward the discrete source, and Ω_B is much greater than Ω_i, Ω_j or Ω_C. Furthermore, all molecular clouds are assumed to be clustered in the same general spatial region, which allows $F(D)$ to be constant, and the $(M+N)$ clouds are all assigned the same excitation temperature (which is not necessarily true). Also, any segment of the jth cloud falling outside Ω_C is treated as an ith cloud. Brightness temperature contributions from the galactic background, T_{bg}, and from the isotropic background radiation, T_{bb}, are taken to be uniform over a solid angle greater than Ω_B.

The first term in equation (6.17) represents the contribution of M clouds in front of the background source and the second term the contribution of the remaining N clouds. In practice, T_b is often so much larger than T_e, T_{bb} and T_{bg} that the second term in equation (6.17) can be completely ignored for absorption observations.

6.3.4 Maser emission

The high brightness temperatures and narrow line widths which characterise all emission lines from interstellar H_2O and the majority of interstellar OH emission lines can be explained by a maser mechanism. The standard astronomical treatment of maser emission may be illustrated by explicitly considering only a two-level system although current models are somewhat more sophisticated because the pumping processes introduce other levels. The customary starting equation is found by combining equations (6.4), (6.11) and (6.12) to obtain

$$\frac{dI_v}{ds} = + \frac{A_{ul}f(v)\lambda^2}{8\pi}(\Delta n)I_v + \varepsilon_v \qquad (6.18)$$

where the emissivity, ε_v, is just J_v (equation (6.12)) and Δn, the inversion measure or degree of inversion, is given by

$$\Delta n = n_l \left[\frac{n_u}{n_l} - \frac{g_u}{g_l} \right] = -g_u \left[\frac{n_l}{g_l} - \frac{n_u}{g_u} \right] \tag{6.19}$$

where the terms have been defined in Section 6.3.2. Maser emission occurs when the inversion measure is greater than zero.

The standard maser equation differs from the previous solutions to the radiative transfer equation (equations (6.16) and (6.17)) in that the underlying assumption of a constant value for (J_0/K_0) throughout the cloud depth is replaced by a more general condition found from requiring overall statistical equilibrium between the two levels,

$$n_u \left(A_{ul} + \frac{W_m}{g_u} + \frac{W_c}{g_u} + W_p' \right) = n_l \left(\frac{W_m}{g_l} + \frac{W_c}{g_l} + W_p \right) \tag{6.20}$$

where $W_m = g_u B_{ul} I_v$ and $W_c = g_u C_{ul}$. As a first approximation the collisional de-excitation rate, C_{ul}, may be related to the collisional excitation rate, C_{lu}, through $C_{ul} \simeq g_l/g_u \cdot C_{lu}$ [82]. In the radio-frequency range the spontaneous emission rate, A_{ul}, is usually small enough to ignore in comparison to the induced emission rate W_m/g_u. W_p', the rate between the upper and lower levels ('downpumping' rate) and W_p, the reverse rate, represent the effects of the available pump energy which maintains the necessary population inversion in the masering region of the interstellar cloud; both W_p' and W_p implicitly include other levels. The source of the pump energy might be continuum radiation at an appropriate frequency (ultraviolet, infrared or microwave) or collisions; the radiation initiating maser action might be spontaneous emission or emission from the masering line itself. A typical form for the inversion measure is found by substituting equation (6.20) into equation (6.19) to get

$$\Delta n = \frac{\Delta n_p}{1 + (W_m + W_c)/(g_u W_p')} \tag{6.21}$$

where A_{ul} has been ignored. The term in the numerator, Δn_p, is defined as

$$\Delta n_p = n_l \left[\frac{W_p}{W_p'} - \frac{g_u}{g_l} \right] \tag{6.22}$$

and, by analogy with the definition of Δn in equation (6.19), it may be considered to be the inversion measure due to the radiative pumping process alone[83].

Perhaps a more common method of expressing the inversion measure[84, 85] is to use the total number density for both levels, $n = n_u + n_l$ and equation (6.20) to obtain

$$\Delta n = \frac{\Delta n_0}{1 + \frac{(W_m + W_c)}{W} \left[\frac{1}{g_u} + \frac{1}{g_l} \right]} \tag{6.23}$$

where $W = W_p + W'_p$ and Δn_0 is given by

$$\Delta n_0 = \frac{nW'_p}{W}\left[\frac{W_p}{W'_p} - \frac{g_u}{g_l}\right] = \frac{n}{n_1}\left[\frac{W'_p}{W}\right]\Delta n_p \tag{6.24}$$

Combining equations (6.18) and (6.23), the simplified form of the radiative-transfer equation for describing OH and H_2O emission is given as

$$\frac{dI_v}{ds} = \frac{\alpha I_v}{c + bI_v} + \varepsilon_v \tag{6.25}$$

where for $g_u = g_l$, the terms α, b, and c are defined as follows:

$$\alpha = \frac{\lambda^2 A_{ul}f(v)\Delta n_0}{8\pi}$$

$$b = \frac{2B_{ul}}{W}\left[\frac{\Omega_m}{4\pi}\right]$$

$$c = 1 + \frac{2C_{ul}}{W} \tag{6.26}$$

The gain factor for small stimulated emission and collisional rates, α, is reduced by $(c + bI_v)^{-1}$ when stimulated emission and collisions are not negligible. The factor $\Omega_m/4\pi$ is included in the definition of b to account for non-isotropic emission.

The solution to equation (6.25) for constant α and ε_v has been found to be

$$I_v - I_0 = \left[\frac{c\varepsilon_v}{(\alpha + \varepsilon_v b)} + I_0\right]\left[\exp\left(\frac{(\alpha + \varepsilon_v b)L - b(I_v - I_0)}{c\alpha(\alpha + \varepsilon_v b)^{-1}}\right) - 1\right] \tag{6.27}$$

where L is the maser path length[86, 87]. Equation (6.27) cannot be solved explicitly for I_v; hence two limiting cases, the unsaturated and the saturated maser, are used to interpret the observational data. Unsaturated means that the pumping rate, W, or the collisional rate, W_c, or both greatly exceed W_m, the microwave transition rate. For this limiting case, $c \gg bI_v$ and equation (6.27) becomes

$$I_v = I_0 e^{\alpha L/c} + (\varepsilon_v c/\alpha)(e^{\alpha L/c} - 1) \tag{6.28}$$

where the first term on the right-hand side shows the amplification of the incident radiation I_0 and the second the amplification of the spontaneous emission along the maser path L. A small change in the gain produces an exponential intensity change as well as line narrowing in proportion to $(\alpha L)^{-\frac{1}{2}}$. Saturated, on the other hand, means that W_m is much greater than either W_c or W or both. Thus $bI_v \gg c$ and equation (6.27) becomes

$$I_v = I_0 + \left(\varepsilon_v + \frac{\alpha}{b}\right)L \tag{6.29}$$

where now the first term is the incident radiation and the second is the sum of both spontaneous and stimulated emission along the maser path L. In this case, neither the collisional nor the radiative rates can energetically approach the capacity of the maser and any gain change is reflected as a linear change in the intensity.

Detailed comparison of observation with theory is generally very difficult. Interstellar OH emission investigations require extensive computer programs for equations of the form of equation (6.27); up to seven parameters (I_0, n_u or n_l, Δn_0, $f(v)$, W, W_c and Ω_m) may be required to fit the data obtained from observations of one transition in order to predict the properties of another observed transition (see Table 6.1). Several parameters of the fit may have to be determined by rather indirect means which may limit the usefulness of the results. The present situation for interstellar H_2O is even less satisfactory because only the $6_{16}-5_{23}$ transition has been detected to date. This difficulty is compounded in some sources by time-varying H_2O intensities (see Section 6.4.3).

6.3.5 Number densities

The discussions in the preceding Sections serve to introduce the difficulties encountered in obtaining accurate abundance measurements of interstellar molecules. The solution of the radiative-transfer equation for the simplest case, the idealised cloud configuration in Figure 6.2(a), the assumptions inherent in determining molecular abundances or number densities for the interstellar clouds; any necessary modifications of the following discussion required by more complex cloud geometries may be made using equation (6.17). The following discussion clearly does not apply to abundance determinations in the case of maser action. further illustrates

Two extreme cases are easily considered. First of all, if the cloud is optically thick and T_b is known, the excitation temperature T_e can be determined directly from equation (6.16), thus ruling out a conventional number density determination. On the other hand, if the cloud is optically thin, ΔT in equation (6.16) can be integrated over the spectral line to give

$$\int_v \Delta T dv = (T_e - T_b) \int_v \langle \tau_0(s_0) \rangle dv = (T_e - T_b) \int_v \langle \int_0^{s_0} K_v ds' \rangle dv \quad (6.30)$$

The left-hand side of equation (6.30) is the measurable area contained by the spectral line in the frequency antenna temperature domain. The mean optical depth $\langle \tau_0(s_0) \rangle$ is easily integrated over the line width by using equation (6.11) and the definition $A_{ul} = 64\pi^4 |\mu_{ul}|^2/3h\lambda^3$ to find

$$\int_v \langle \tau_0(s_0) \rangle dv = \frac{8\pi^3}{3ck} \cdot \frac{v_0^2 g_u |\mu_{ul}|^2}{T_e} \cdot \frac{|\mu_{ul}|^2}{g_1} \cdot (n_1 L) \quad (6.31)$$

where $|\mu_{ul}|^2$ is the square of the dipole moment matrix element for the transition from the upper to the lower molecular energy level. The projected density (number density integrated along the line of sight) or column density for the lower energy level is found from equations (6.30) and (6.31) to be

$$n_1 L = \frac{\int_v \Delta T dv}{\left(1 - \frac{T_b}{T_e}\right)\left(\frac{8\pi^3}{3ck}\right)|\mu_{1u}|^2 v_0^2} \quad (6.32)$$

where the substitution $(g_u/g_1)|\mu_{ul}|^2 = |\mu_{1u}|^2$ was used. In cgs units, n_1L is the number of molecules in level 1 contained in a column of length L with a cross-sectional area of 1 cm^2 (number cm^{-2}). To find the volume density, it is necessary to estimate the cloud depth L. It is customary, for example, to map a cloud region and assume a convenient geometry such as spherical symmetry but L can never be directly measured.

An alternate method for arriving at equation (6.32), which perhaps offers more physical insight, is to initially define K_v in terms of the well known Van Vleck–Weisskopf absorption coefficient

$$K_v = \frac{8\pi^2(Nf)|\mu_{1u}|^2v_0^2}{3ckT_e}\left[\frac{\Delta v}{(v-v_0)^2+(\Delta v)^2}+\frac{\Delta v}{(v+v_0)^2+(\Delta v)^2}\right] \quad (6.33)$$

where N is the molecular volume density and f, the fractional number in the lower energy level, is given by n_1/N [88]. The half-width at half-maximum intensity, Δv, is related to the mean time t between collisions by $\Delta v = (2\pi t)^{-\frac{1}{2}}$. When v is near the resonance frequency v_0, and for $\Delta v \ll v_0$, the right-hand side of the bracketed term in equation (9.33) can be neglected in favour of the Lorentzian contribution on the left-hand side. Then

$$\int_v \langle\tau_0(s_0)\rangle dv = \frac{8\pi^2(Nf)|\mu_{1u}|^2v_0^2(\pi L)}{3ckT_e} \quad (6.34)$$

and the result is identical to equation (6.31).

Even when the optically-thin case applies to the simplified geometry of Figure 6.2(a), it may be impossible to find the projected density for the total number of molecules in an interstellar cloud because the distribution function over all molecular energy levels may not be known. Hence n_1L, the lower limit to the total projected molecular density, is the quantity most reliably measured. The number density problem is complicated for maser emission because either n_uL or n_1L must be determined by the indirect means discussed previously.

6.4 ASTRONOMICAL OBSERVATIONS

Observations of all the interstellar molecules found to date, with the exception of OH, are restricted to our own galaxy, the Milky Way; weak OH absorption signals from two external galaxies, NGC 253 and M82, were recently detected with a radio interferometer and thus have the distinction of being the first extragalactic molecular spectra ever observed[89]. Identifications and observations of the newer interstellar molecules in different regions of our galaxy have been greatly expedited through the use of known velocity offsets or Doppler shifts which were determined by earlier work on the 21 cm line of atomic hydrogen (HI), hydrogen recombination lines, or OH. However, the velocity agreement between spectral lines in the same interstellar cloud is not always expected to be exact due to the possibility of different excitation mechanisms which may have different locations.

6.4.1 H$_2$

Indirect evidence (discussed elsewhere[90-93]) has led astronomers to the conclusion that molecular hydrogen, possibly formed by atomic hydrogen recombination on grain surfaces, must be a major constituent of dense interstellar clouds but until recently there was no direct spectroscopic evidence for the existence of H$_2$ in the interstellar medium. Now the first interstellar H$_2$ has been detected[1] in absorption against the far-ultraviolet spectrum of the O-type star ξ-Persei. A rocket-mounted spectrograph recorded eight vibrational transitions of the Lyman resonance series (B$^1\Sigma_u$–X$^1\Sigma_g$) between 1000 and 1120 Å (see Table 6.1); hence H$_2$ is the only interstellar molecule found in the last three decades which was not detected first with a radio telescope. The H$_2$ column density is $c.$ 1.3×10^{20} cm^{-3} for the ξ-Persei cloud and comparison with the atomic hydrogen column density determined from the same spectrum indicates that almost half of the total hydrogen in this region may be in molecular form. It is interesting to note that the H$_2$ volume density between ξ-Persei and a nearby star, ε-Persei, is estimated to be 0.6 cm^{-3} whereas there are indications that the H$_2$ density may be as great as 10^6 cm^{-3} in very dense clouds[67].

Near-infrared searches for the S(1), Q(2) and O(2) transitions of the ground-state (3–0) band in spectra of the Orion nebula, IC 5146, NGC 2264, IC 1499 and near ξ-Persei have not led to a positive detection of interstellar H$_2$ [94]. In addition, radio astronomers have searched for interstellar H$_2^+$ with negative results[95, 96].

6.4.2 NH$_3$

Ammonia was the first interstellar polyatomic molecule ever found[15, 16] and until recently it had been detected and mapped only in the galactic centre clouds[97, 98]. The Sagittarius observations of emission from five NH$_3$ inversion transitions ($J, K = (1, 1), (2, 2), (3, 3), (4, 4)$ and $(6, 6)$ in Table 6.1) have been useful for an extensive study of collisional excitation mechanisms. For example, an intensity versus position map[97] of the (1, 1), (2, 2) and (3, 3) lines in the direction of Sgr B2 revealed distinct regions within the cloud where the (3, 3) emission is strongest, other regions where the (1, 1) emission is strongest, regions where the (2, 2) is very weak and finally, regions where the intensities of all three lines are described by a common temperature of $c.$ 50 K (corresponding to thermal equilibrium). The (3, 3) line was the most intense in the majority of the positions mapped; thus excitation temperatures determined from (3, 3) and (1, 1) line ratios (for example, 100 K) were found to be much higher than excitation temperatures determined from (2, 2), and (1, 1) line ratios (typically $c.$ 25 K). The mapping results clearly show that most of the ammonia in the Sgr B2 cloud distribution is not in thermal equilibrium and hence general population distributions calculated using Boltzmann statistics, for example, cannot precisely describe molecular densities in such regions.

It has been suggested that the NH$_3$ intensity ratios are caused by collisions with molecules (requiring a density of 10^3 cm^{-3} for H$_2$) or with electrons (at 10^{-1} cm^{-3})[97]. Thus collisionally-induced transitions via the (1, 1), (2, 1) and (2, 2) levels could explain the intensities of the (1, 1) and (2, 2) inversion transitions. This explanation appears to be in agreement with recent laboratory studies of NH_3 transitions induced by collisions with H_2 but the NH_3–H_2 results do not explain the (3, 3) line intensities of interstellar NH_3[99, 100]. In addition, an experiment simulating NH_3 evaporation from interstellar grain surfaces indicated no preference for *ortho*-ammonia ($K = 3n$) and hence could not explain the excess population of the (3, 3) levels[99].

The possibility of radiative transitions other than electric–octopole has been considered and it is found that $\Delta K = \pm 3$ electric dipole transitions are allowed between the (2, 2) and (1, 1) levels with radiative probabilities which are not much smaller than collisional probabilities[101]. This leads to the interpretation that radiative transitions could be more important along the rim of the Sgr B2 cloud where the (1, 1) emission tends to dominate and collisions are more important toward the core of the cloud where the (3, 3) intensity dominates. Recently, the (2, 1) and (3, 2) inversion transitions were detected[14] in this cloud (see Table 6.1) and preliminary mapping shows the (2, 1) emission to be coming from a dense core, smaller than and well within the spatial extent of the (1, 1) emission[102]. Interstellar ammonia is an ideal molecule for the study of excitation mechanisms because a number of inversion frequencies (see reference 88 for an extensive list) can be observed from a given radio telescope with little change in half-power beam-width (equation (6.2)) as more sensitive receivers are developed. Thus it is expected that future maps of the (2, 1), (3, 2) and other NH_3 transitions will clarify the question of whether collisional or radiative mechanisms are dominant in different regions of interstellar clouds.

Estimated ammonia column densities for the various Sagittarius clouds range from 5 to 10×10^{15}[55] cm^{-3}. It has been suggested that the abundance of NH_3 molecules in sources outside the galactic centre region may be an order of magnitude less, and recent detections of weak NH_3 emission from Orion, W3(OH), W43, W51 and DR 21(OH) seem to confirm this[98, 102].

6.4.3   OH and H$_2$O

Since the initial detection in 1963[3], radio astronomers have searched for OH in many different regions of the galaxy in order to gain distribution statistics which are not biased by observational selection effects. The four transitions of the ground-state lambda-doublet (1612, 1665, 1667 and 1720 MHz) have been used extensively to locate *c*. 180 OH absorption and 70 emission sources; about one-half of the emission sources have absorption associated with them[103]. Galactic surveys of OH transitions associated with higher lambda-doublets are not as complete because detections of these lines are much more recent (see Table 6.1); for example, the 4765 MHz transition has been reported in four sources to date[118].

Much less is known about the true galactic distribution of interstellar water vapour (extragalactic H_2O has yet to be detected[130]) and the general spatial correlation between OH and H_2O sources. Before H_2O was detected, it was suggested that OH clouds in the vicinity of regions of ionised hydrogen (HII regions) would be important places to search because it was expected that OH and H_2O would be spatially correlated; it was also recognised that minimal detection of the 6_{16}–5_{23} transition of H_2O would require either an abnormally high state temperature and projected densities in the range 10^{16}–10^{17} molecules cm^{-3} in the case of thermal population statistics or

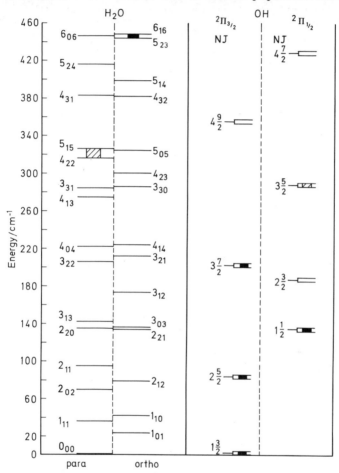

Figure 6.3 Comparative energy level diagrams for the lowest energy levels of H_2O and OH. The solid bars represent astronomically-observed transitions. The dashed bar for H_2O is the maser transition predicted by selective pre-dissociation (see Section 9.4.3.4 and reference 153) while the dashed bar for OH is an unconfirmed detection (see Table 6.1)

non-thermal population anomalies[105]. The differences between the excitation energy requirements of the detected interstellar OH and H_2O lines are evident from an examination of Figure 6.3. Interstellar H_2O was subsequently found in maser emission near three OH clouds which are associated with HII regions[13]. To date, c. 20 H_2O emission sources have been detected in northern sky surveys; all are in maser emission and approximately half are connected with HII regions[83]. The remaining sources can not be classified as interstellar as they are associated with the circumstellar envelopes of infrared stars[129]; typical circumstellar sources are much weaker in intensity[106]. While the general velocity pattern of interstellar OH and H_2O emission often covers the same range, there is no detailed agreement of individual velocity features which would confirm that both molecular species exist in the same clouds[83]. The most obvious interpretation of this apparent chemical anomaly is not that OH and H_2O exist in mutually exclusive clouds but rather that some regions of excitation enhance OH intensities while others enhance the intensity of 6_{16}–5_{23} transition of H_2O (see Figure 6.3); these regions of preferred excitation may often be spatially independent since they are thought to depend on local densities and radiation fields.

It is well known that if OH were observed in an optically thin cloud under conditions of thermal equilibrium, the relative intensities of the 1612, 1665, 1667 and 1720 MHz lines would obey the ratios 1:5:9:1. This ideal ratio is seldom observed for interstellar OH although the deviation is usually less severe for OH absorption; several observations of 5 : 9 ratios have been reported for the 1665 and 1667 MHz lines in emission from dark nebulae and neutral hydrogen (HI) clouds[104, 116]. Because the observational data are strongly correlated with particular modes of excitation (see Section 6.3), it has been customary to treat OH absorption and emission separately, although from the chemist's viewpoint this distinction may be quite artificial. Currently, OH sources may be classified according to the following four categories[86]: class I sources have all four ground-state lambda-doublet lines in emission but the main lines at 1665 and 1667 MHz (particularly 1665 MHz) are anomalously strong with respect to the satellite lines at 1612 and 1720 MHz; class II(a) sources have the 1720 MHz satellite line in emission and the remaining lines at 1612, 1665 and 1667 MHz in absorption; class II(b) sources have the 1612 MHz satellite line in anomalous emission, the two main lines in emission, and the 1720 MHz line in absorption; class III may be used to describe sources which have all four lines in absorption[117].

6.4.3.1 OH *absorption*

Virtually every galactic plane continuum source which has been searched shows 18 cm OH absorption due to dense clouds of gas and dust which lie in and near the plane[81, 107]. Velocities found from OH absorption measurements often agree with the 21 cm absorption measurements of atomic hydrogen (HI), which suggests that HI and OH commonly occupy the same spatial regions. The problem of interpreting the excitation temperature of OH in neutral hydrogen clouds (HI regions) has been treated extensively by several authors for the ground-state lambda-doublet[108, 109] and extended to describe the excitation of higher-lying states of OH[110]. The latter treatment is general enough to be applied to proton collisional excitation of the $N = 1$–0 transition of CN, carbon ion excitation of the CH ground-state lambda-doublet and proton excitation of the $2^2S_{\frac{1}{2}}$–$^2P_{\frac{1}{2}}$ transition of atomic hydrogen. The condition of overall statistical equilibrium in the absence of pumping radiation ($W_p = W'_p = 0$ in equation (6.20)) and the definition of the excitation temperature (equation (6.15)) lead to the following first-order approximation for the excitation temperature, T_e, of the ground-state lambda-doublet of OH in an HI cloud when the molecule is subject to both 18 cm radiation and ion collisions:

$$T_e = T_K \left[\frac{T_R + T_0}{T_K + T_0} \right] \tag{6.35}$$

In equation (6.35), T_K is the kinetic temperature corresponding to the average velocity of a colliding particle relative to an OH molecule; T_R is the radiation temperature corresponding to 18 cm radiation in the Rayleigh–Jeans approximation (equation (6.14)); and the ratio of the collisional de-excitation to spontaneous emission rate is expressed by $T_0 = (hv/k)(C_{ul}/A_{ul})$. Thus, when T_0 is very small, radiation dominates over collisions and the excitation temperature may be interpreted as a radiation temperature T_R. When collisions dominate, T_0 becomes large and T_e may be interpreted as a kinetic temperature in analogy with the interpretation given the observed excitation temperature of the 21 cm line of atomic hydrogen[111]. However, it may be seen from equation (6.35) that the coupling between T_e and T_K generally will be weaker for the OH spectral lines than for the 21 cm line because the transition moments of the former are electric dipole while the moment of the latter is magnetic; thus the matrix element contribution to the spontaneous emission rate will be about four orders of magnitude smaller for HI than for OH.

Due to line saturation, projected densities determined from OH ground-state absorption measurements often represent only lower bounds to the true values. Typical values[81] of $n_1 L$ for an assumed excitation temperature of 3 K are: Orion A, 1.6×10^{13} cm^{-3} ($\pm 20\%$) for the 5.5 km s^{-1} line; Taurus A (Crab nebula), 5×10^{12} cm^{-3} ($\pm 18\%$) for the 2.3 km s^{-1} line and 8×10^{12} cm^{-3} ($\pm 18\%$) for the 13.2 km s^{-1} line. These projected densities give OH to HI ratios of 5.4×10^{-9} for Orion A and 10^{-8} for Taurus A. The OH projected density in the direction of the galactic centre is impossible to determine accurately due to line saturation but it may be estimated as 10^{16} cm^{-2} or greater for the absorption feature at 60 km s^{-1} in Sagittarius B2[112]. The high abundance of ^{16}OH in the galactic centre clouds greatly aided the initial detection and subsequent observations of the less common molecular isotope ^{18}OH [11, 113]. The most recent estimate for the ^{16}O:^{18}O ratio is 390:1, based on the 40 km s^{-1} velocity component in Sagittarius A[11]. Given the assumptions involved, this ratio is not appreciably different from the terrestrial ratio of 490:1. The geometry of the 40 km s^{-1} cloud has been studied extensively by lunar occultation techniques and the cloud has been found to be 6×4 arc minutes angular size which corresponds to linear dimensions of approximately 59×39 light years if the distance to the galactic centre is taken to be 33 000 light years[114, 115]. Hence if the 40 km s^{-1} cloud is at least as deep as it is wide, with OH uniformly distributed throughout, the ^{16}OH volume density is 2.5×10^{-4} cm^{-3} (or less) and the ^{18}OH density is 6×10^{-7} cm^{-3}. These represent lower bounds in the sense that the molecular distribution over all energy levels is not taken into account but even with the most optimistic population schemes the average density per cubic centimetre probably will not exceed one OH molecule cm^{-3} for the galactic centre clouds.

6.4.3.2 OH *emission*

The first OH emission spectra were detected at 1665 MHz by three independent groups of astronomers who were observing the galactic centre clouds[119–121]. The unusually narrow intense lines observed in the initial detections were so unexpected that two of the three groups could not believe they were observing OH [122]. The first group to detect OH emission thought the spectrum was due to an instrumental flaw[119] while another group preferred to refer to the 1665 MHz emission as 'mysterium' until its identity could be established. The subsequent detection of other spectral lines belonging to the ground-state lambda-doublet proved that the spectrum was due to OH in maser emission. OH emission, always in varying degrees of anomalous excitation, has been found to be associated with HII regions, supernova remnants[81, 123], infrared stars[124, 125, 128], dark nebulae[104], planetary nebulae[126], galactic nebulae, and HI clouds[116]. Strong right- and left-hand circular polarisation as well as linear polarisation is commonly found in anomalous OH emission spectra[107, 122, 127] but all absorption spectra examined to date have been found to be unpolarised.

A number of OH emission sources, usually class I, have been observed to undergo temporal variations in intensity where the time scale for noticeable change ranges from a few days to several months[118, 127]. An analysis of emission from two sources, NGC 6334 and Orion A, indicates that intensity fluctuations may be as great as an order of magnitude; spectral elements separated by 0.4 km s^{-1} may vary independently or may show either positive or negative correlation; no discernable periodicity is found in the variation of a given spectral line; and the polarisation properties appear to remain constant throughout the intensity variations[107]. In addition, corresponding velocity features in 1665 and 1667 MHz spectra from the same source may vary independently. A recent survey of 60 infrared stars at 1612 MHz led to the detection of temporal changes in intensity which may be easier to explain[125]. The survey yielded seven OH emission sources associated with either red-giant or supergiant stars; the observed emission generally followed the Class II(b) pattern in that 1612 MHz emission was strongest and no 1720 MHz emission was found. Time variations found for two of the Mira-type variable stars, IRC + 10011 and NML Tau, seemed to be correlated with an accompanying change in the infrared flux. Thus, in contrast to Class I variations, these changes probably can be explained in terms of an expanding stellar atmosphere.

Very accurate measurements of the positions and sizes of several strong OH sources associated with HII regions have been made using radio interferometry. Early measurements of four sources (W3, W24, W49 and NGC 6334) revealed that no OH emission position measured coincided with an optical object and the emission points were found to be adjacent to continuum maxima but not coincident with them[131]. Each OH emission feature, corresponding to a different radial velocity, originated from a source of small angular size (less than 25 seconds of arc in most cases). A very small angular source size means that the associated radio brightness temperature will be very high (see equations (6.14) and (6.17)). For example, early size measurements of the OH emission points associated with the W3 HII region placed upper limits of less than 2 seconds of arc which corresponds to a linear dimension of 0.052 light years or less[132]. The corresponding brightness temperatures were placed at no less than 10^9 K; subsequent observations reduced the size limits to 0.1 seconds of arc or smaller with associated brightness temperatures of 10^{11} K or greater[133]. A more detailed examination using very long baseline interferometry (VLBI) revealed each velocity component to be associated with an emission source of 0.02–0.005 seconds of arc in diameter and some sources were found to be composed of two or more elements[134]. Thus the brightness temperatures had to be revised upward again to 10^{12} K or higher. Such high brightness temperatures give a striking example of non-thermal emission and illustrate why maser models must be invoked to explain OH emission.

6.4.3.3 H₂O emission

The initial observations of interstellar water vapour established that all known sources exhibit maser emission and it was recognised that the emission sources were smaller than the beam-widths of the largest single-element telescopes[135–139]. The H_2O emission regions were determined to be less than 0.5 minutes of arc in angular extent and generally agreed with OH emission positions to within 0.5 minutes of arc. The spectral features of some sources were observed to have linear polarisation but no circular polarisation has ever been found. Furthermore, it was found that individual lines often undergo pronounced short-term non-periodic variations on a time scale much shorter than that observed for OH. H_2O spectral features are not correlated with those of anomalous OH emission from the same source except that they may have the same overall velocity range.

The H_2O emission spectrum from W49 is perhaps one of the most interesting ever observed by radio astronomers. In 1969 the spectrum was found to exhibit Doppler-shifted features covering an extraordinarily large range from -180 to $+210$ km s^{-1} with respect to the local standard of rest (see Figure 6.4); this corresponds to extreme deviations from the rest frequency of 13.4 and -15.6 MHz, respectively. In contrast, OH emission from the planetary nebula NGC 2438 covers a velocity range from -20.6 to 90.0 km s^{-1} which is the widest known for any OH emission source[126]. The measured antenna temperature of the central H_2O emission feature in W49 was found to be 2000 K (see Figure 6.4) and several of the spectral features were observed to change intensity in periods of less than 2 weeks.

Two different types of observational programmes, periodic monitoring and very long baseline interferometry (VLBI), were started in 1969 in order to learn more about the nature of the anomalous H_2O emission sources. The H_2O sources W3, Orion A, VY Canis Majoris, NGC 6334, Sgr B2, W49, W51, ON1 and W75 were periodically monitored from January 1969, until April 1970[83, 140]. It was found that all of the main emission features

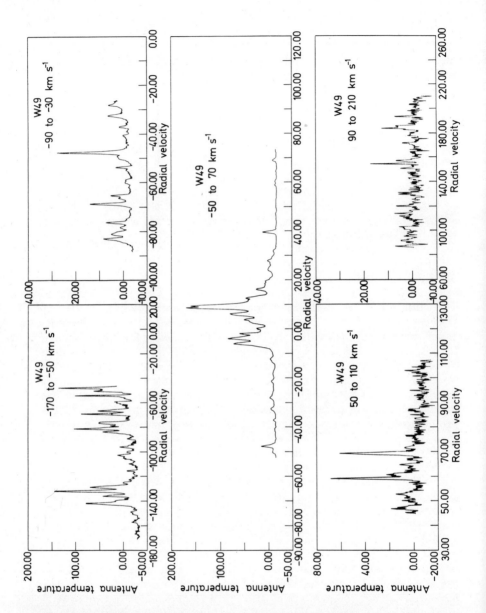

Figure 6.4 Interstellar water vapour emission from W49. The ordinate is antenna temperature (K) and the abscissa is radial velocity in km s^{-1} with respect to the local standard of rest. The spectra were taken with the 140 ft radio telescope of the National Radio Astronomy Observatory on the following dates in 1969: (a) April 8, -170 to -50 km s^{-1}; (b) May 3, -90 to -30 km s^{-1}; (c) April 26, -50 to 70 km s^{-1}; (d) April 17, 50 to 110 km s^{-1}; (e) April 25, 90 to 210 km s^{-1}. Note that the four features centred at -80 km s^{-1} changed relative intensities between April 8 and May 3.
(From Buhl et al.[138], by courtesy of University of Chicago Press)

changed during the monitoring period but the changes followed no periodic pattern and occurred on time scales ranging from several days to months. Most features underwent small radical velocity ($c.\ 1.0\,\mathrm{km\,s^{-1}}$ or less) and width changes along with the intensity changes. Spectral features separated by as little as $1.0\ \mathrm{km\ s^{-1}}$ usually varied independently which indicates association with different emission regions. Linear polarisation, when present, was found to vary in the same sense as the intensity of the feature. Emission spectra from several sources were found to be extremely variable. For example, the intensity of the $+6.5\ \mathrm{km\ s^{-1}}$ H_2O feature in W49 flared from an antenna temperature of 1400 to 4700 K in the short period from September to November 1969, and returned to 1000 K by January 1970. Thus the current W49 emission pattern is very different from the April–May 1969 spectrum shown in Figure 6.4.

The first VLBI measurements were started in December 1969, and January 1970, in order to accurately measure the positions and sizes of the H_2O emission regions associated with the strong sources W3(OH), Orion A, W49 and VY Canis Majoris[141]. At baseline lengths of 1.7×10^7 and 5×10^7 wavelengths no source was spatially resolved, which meant that all except W3(OH) were smaller than 3×10^{-3} seconds of arc; a limit of 10^{-2} seconds of arc was placed on W3(OH). These upper limits on source size were used to place lower limits on brightness temperature of 10^{13} K for W49 and 10^{11}–10^{12} K for the remaining sources. In June 1970, subsequent measurements were made on the W49 emission source using baselines of 6.3×10^7, 2.2×10^8 and 2.7×10^8 wavelengths[142]. The source was found to be constructed of many small emission centres, all less than 5×10^{-4} seconds of arc in angular size, and contained in a circle of 1.5 seconds of arc angular diameter. In W49, the measured H_2O emission centres were found to coincide with known OH emission centres to within the measurement uncertainty but they were at least 100 times smaller in angular size. The most recent VLBI measurements of W49 utilised a baseline of 5.47×10^8 wavelengths (from Westford, Mass., U.S.A. to Semeiz, U.S.S.R.) to determine an apparent brightness temperature of 1.4×10^{15} K for the very intense emission feature at $-2\ \mathrm{km\ s^{-1}}$ [143].

In view of the strong emission from $H_2^{16}O$ it is reasonable to expect $H_2^{18}O$ to be detectable[144], but at least two searches for the 6_{16}–5_{23} transition at 5625 MHz have been unsuccessful[145, 146]. If the $H_2^{16}O$ were undergoing saturated maser emission (see Section 6.3.4) and if the excitation temperatures of $H_2^{16}O$ and $H_2^{18}O$ were the same, then $H_2^{18}O$ should have been detected in at least six sources. Since it was not found, two possible conclusions are that either the H_2O pumping mechanism is dependent on isotopic species or the H_2O maser is not fully saturated at the two frequencies[146]. Thus the identification of $H_2^{16}O$ has yet to be confirmed by the detection of a second transition (see Table 6.1).

6.4.3.4 Agreement with theory

During the past few years there have been several proposals and some indirect observational evidence that the intense interstellar OH and H_2O

emission sources, especially those associated with HII regions, are proto-stars or regions where star formation is occurring[147-150]. Thus maser emission could provide an important mechanism for cooling hot masses of gas under-going gravitational contraction and, as a result, the formulation of a quanti-tative theory of maser emission could be central to understanding an im-portant phase of stellar evolution. Attainment of such a theoretical formula-tion appears to be growing more difficult, however, because as the obser-vational data becomes more·sophisticated the conditions placed on the source parameters become more stringent. For example, it is possible to use time variations to place an upper limit on the length L of an amplifying column by assuming that the propagation velocity of the disturbance is less than or equal to the velocity of light. If a given spectral feature changes intensity in 4 days, then the upper limit on L is approximately 10^{16} cm whereas a 40 day change corresponds to 10^{17} cm. Time variations give an L of 10^{16} cm for H_2O emission[138] from W49 but the source sizes measured by VLBI tech-niques are of the order of 10^{13} cm for H_2O and 10^{15} cm for OH. The dis-agreement between source sizes determined by the two techniques pre-sumably could be resolved by incorporating propagation velocities several orders of magnitude less than that of light into maser emission models. A quantitative model also must explain the fairly rapid H_2O time variations versus the characteristically slower OH changes and, when polarisation is observed, the predominantly circular polarisation of OH spectral features versus the linear polarisation of H_2O features.

Most of the current theories for OH and H_2O maser emission invoke either collisional or radiative pumping and relaxation processes (see Section 6.3.4) to explain the observed properties of the spectrum and it is usually straightforward to find qualitative agreement between theory and obser-vation. For instance, small optical depth changes such as density fluctuations could produce large intensity changes in a short time period if the maser is unsaturated (see equation (6.28)). Thus the relatively short time scale of H_2O intensity variations suggests that the emission is unsaturated while the somewhat longer OH time scale suggests saturated emission[151]. An un-saturated H_2O maser model also agrees with one of the reasons which was given for not finding interstellar $H_2^{18}O$. As another example, one possible model for H_2O emission associated with the HII regions W3, Orion A and W49 can explain the observations for a maser gain of 10^{10}, kinetic temperature of 500 K, amplifier dimensions of $1 \times 1 \times 100$ astronomical units (where 1 a.u. $= 1.496 \times 10^{13}$ cm) and H_2O density of 10^3 molecules cm^{-3} [140]. Even though the present constraints placed on theory by the data are by no means unique, a model which invokes far infrared pumping radiation from a few M5 stars appears to be able to explain unpolarised OH emission from dark nebulae in that it is the only known model reported to be capable of reason-ably correct quantitative predictions[86]. Given the intensities of the 1665 and 1667 MHz lines, it is possible to assume reasonable values for kinetic tem-perature, source brightness temperature, emission solid angle, and product of density by amplifier length which are consistent with a dark nebula and hence predict the observed intensity of the 1720 MHz emission line. A model pro-posed for W3 and W49 maser emission treats the OH–H_2O source sizes and polarisation properties by suggesting that the signals emerge from small mass condensations which are hot spots of unsaturated amplification distributed along the edges of two large, massive, co-rotating parallel discs; polarisation would be the consequence of the non-linear properties of an associated weak magnetoplasma[152].

Recently a model has been suggested which uses selective predissociation as the mechanism to trigger H_2O maser emission in regions of intense ultraviolet radiation[153]. Vacuum ultraviolet radiation is known to selectively predissociate H_2O molecules having non-zero angular momentum components, K_-, corresponding to rotation about the a axis. Thus H_2O molecules which absorb vacuum ultraviolet photons are excited to upper electronic states, but only those which have $K_- = 0$ in the excited states survive and return to the ground-state by spontaneous emission. The result is a depopulation of the ground state levels which is in direct proportion to the value of K_- or a relative overpopulation of levels for which $J_{K_-K_+}$ equals either J_{0J} or J_{1J}. The first orthostate transition which can transfer this relative overpopulation from the J_{0J} and J_{1J} levels is the 6_{16}–5_{23} transition at 22.235 GHz. Once maser emission starts, induced emission would keep the process going. This model also predicts that the 5_{15}–4_{22} transition at 325 GHz will be seen in maser emission since it is the first transfer transition encountered by $para$-H_2O as it cascades down through the ground state following ultraviolet excitation (see Figure 6.3). The only special requirements of the selective predissociation model are an efficient production process to replace the dissociated H_2O and exciting ultraviolet radiation of energy 2×10^{-10} erg cm^{-2}.

6.4.4 H_2CO and CO

6.4.4.1 H_2CO observations

Formaldehyde, the first polyatomic organic molecule ever detected in the interstellar medium, was discovered in absorption through the 1_{10}–1_{11} transition at 4830 MHz[39]. At present, five other transitions of $H_2^{12}C^{16}O$ have been found as well as the 1_{10}–1_{11} transitions of $H_2^{13}C^{16}O$ and $H_2^{12}C^{18}O$ (see Table 6.2) but the 4830 MHz line undoubtedly has been the most important to radio astronomy. The 4830 MHz transition has been detected in so many different regions of the galaxy that H_2CO is believed to be a common constituent of the spiral arms of the Milky Way. Galactic surveys have located more than 60 different H_2CO sources and each source is typically composed of multiple clouds or condensations along the line of sight[154–156] H_2CO surveys are expected to be used in conjunction with hydrogen recombination line and continuum surveys to help resolve the twofold distance ambiguity which arises when velocity–distance relations are applied to galactic radio sources which are closer than the sun to the galactic centre. The overall velocity agreement between OH and H_2CO spectral features is good even though the OH surveys are usually made with larger beam-widths due to the lower frequencies of the ground-state transitions (see equation (6.2) and Table 6.1): e ch OH velocity component often has a recognisable H_2CO counterpart. However, possibly as a result of different excitation for the two molecules, the small scale velocity agreement may be poor. An extended survey has determined that the H_2CO velocity distribution in the galactic centre clouds shows greater similarity to OH than to HI (atomic hydrogen). Furthermore, the H_2CO distribution is composed of massive (10^6 solar masses), discrete clouds which suggests that molecular clouds represent extreme condensations within smoother HI density distributions[157]. Lunar occultation measurements show that the position of the $+40$ km s^{-1} H_2CO cloud in Sgr A coincides with the previous occultation positions for the $+40$ km s^{-1} feature in the 1665, 1667 and 1720 MHz OH lines and thus gives further evidence for the presence of OH and H_2CO in the same clouds[158]. If it can be assumed that the excitation temperature for OH is about three times greater than that for H_2CO, then typical H_2CO column densities are found to be 30 times less than OH column densities in the same cloud[154].

Shortly after the initial detection of interstellar H_2CO, one of the most interesting effects in astronomical spectroscopy was found. A search for H_2CO emission in several dark nebulae where OH emission is observed

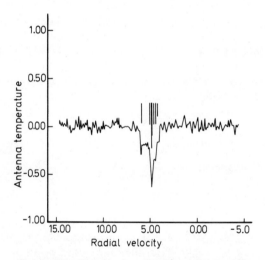

Figure 6.5 The $1_{10}-1_{11}$ transition of interstellar formaldehyde observed in absorption against the 2.7 K microwave background. This spectrum is from Cloud 2, a dark nebula with very little turbulence. The ordinate is antenna temperature in K and the abscissa is radial velocity in km s^{-1} with respect to the local standard of rest. The vertical bars represent the positions and relative intensities of the hyperfine components (determined by laboratory measurement[208]) which have an extreme frequency separation of only 29.61 kHz. (From Palmer *et al.*[225])

yielded weak spectra of H_2CO in absorption[159]. This was unexpected at the time because the particular clouds observed have no known associated continuum sources at 4830 MHz and detailed observations failed to reveal the necessary background sources of energy which must be present for absorption to occur. Thus it was concluded that in these dense nebulae H_2CO is absorbing radiation from the universal 2.7 K microwave background. In order for this to happen, the excitation temperature must be less than 1.8 K. Figure 6.5 shows a spectrum of H_2CO in absorption against the microwave background.

Another interesting measurement resulted from the detection of the 1_{10}–1_{11} transition of $H_2^{13}C^{16}O$ at 4593 MHz in the galactic centre clouds[160]. It was found that the lower limit on the ^{12}C:^{13}C ratio for the Sgr A velocity feature at $+41.4\,km\,s^{-1}$ was 10.5 and the limit for the Sgr B2 feature at $+61.6\,km\,s^{-1}$ was 8.6. These ratios were surprisingly low when compared with the terrestrial abundance ratio of 89:1 and could only be regarded as lower bounds because of possible saturation of the $H_2^{12}C^{16}O$ spectral features and cloud inhomogeneities which could not be spatially resolved by a single-element telescope. Recent interferometric measurements provide much higher spatial resolution and indicate that these limits should be moved to 25 or greater for both clouds[161], but observations of $H_2^{12}C^{18}O$ indicate that the ^{13}C:^{18}O ratio is approximately twice the terrestrial value[162]. Thus it appears that the galactic centre contains clouds which by terrestrial standards have an abnormally high concentration of $H_2^{13}C^{16}O$. Of course, these abundance determinations are dependent on the reasonable assumption that in a given region the same excitation temperature applies for the three molecular isotopes of formaldehyde.

6.4.4.2 H_2CO *pumping models*

The detection of H_2CO absorption in dark nebulae implied an excitation temperature which is lower than both the isotropic background radiation temperature and the expected kinetic temperature. Just as maser action requires a non-thermal excitation mechanism, the overpopulation of the lower 1_{11} level of H_2CO which is implied by the low excitation temperature (see equation (6.15)) requires a non-thermal cooling mechanism.

The first mechanism to be proposed was H_2CO cooling caused by collisions with molecular hydrogen[163]. An examination of the effect of classical hard-sphere collisions with H_2CO shows that, on the average, rotation about the c-axis (perpendicular to the plane of the molecule) results more often than rotation about the b-axis (perpendicular to the a or figure axis and in the plane of the molecule). Preferential rotation about the c-axis corresponds to preferential population of the lower components of the H_2CO *ortho*-state K doublets (i.e. the 1_{11}, 2_{12}, 3_{13}, 4_{14} and higher levels). Following collisions between H_2 and H_2CO, the resulting overpopulation would be rapidly transferred by $\Delta J = 1$ transitions to the 1_{11} level and thus cause the observed low excitation temperature of H_2CO in dark nebulae. The collisional pump requires H_2 densities of $10^3\,cm^{-3}$ or greater in regions where H_2CO is absorbing radiation.

A second pumping model predicts that if the 2.7 K isotropic background radiation departs slightly from a black-body spectrum at frequencies corresponding to H_2CO transitions connecting the lower K-doublets (c. 140

and 150 GHz), then anomalous cooling can result in an overpopulation of the 1_{11} level[164]. A consequence of this radiative pumping model is the prediction that absorption spectroscopy of the 4830 MHz transition of H_2CO should be possible in any direction of the galaxy provided that beam dilution and velocity smearing are not appreciable and the optical depth is sufficient. A second consequence is that the 2_{11}–2_{12} transition at 14 489 MHz is expected to be observed in emission.

A third set of calculations demonstrates that newly formed H_2CO molecules will undergo 'adiabatic cooling' as they follow dipole selection rules to make transitions to lower levels by spontaneous emission[165]. The resulting low temperature will last for c. 10^7 s before H_2CO equilibrates with either the black-body radiation or the kinetic environment by collisions. Thus this model suggests the low excitation temperature observed in dark nebulae may be a natural consequence of recent H_2CO formation.

Finally, a fourth pumping model invokes shock-heated layers in gravitationally unstable dust clouds to form hot H_2CO molecules which in turn emit infrared radiation to be absorbed by cooler molecules far behind the shock front[166]. The radiation, corresponding to one of the perpendicular bands (v_4, v_5, or v_6), will act as a pump between the 1_{10} level in the ground state and the 1_{01} level in the vibrationally excited state, and hence effectively anti-invert the populations of the 1_{10} and 1_{11} ground-state levels of the cooler molecules. It is interesting to note that OH emission from the same dark nebulae also appears to be explained by an analogous infrared pumping model (see Section 6.4.3.4).

Extensive computations and observations have been undertaken to try to learn which of the proposed pumping mechanisms are applicable. It is not currently known whether H_2CO absorption is anomalous in clouds associated with continuum sources because no examples have yet been found outside of dark nebulae where the temperature of the 4830 MHz absorption line exceeds the measured continuum temperature[157]. On the other hand the predominance of absorption and the absence of 4830 MHz emission in all clouds observed (except Orion where a spectrum reversal is interpreted as weak emission[167]), argues strongly for the presence of an effective anti-inverting mechanism associated with this transition of H_2CO in all molecular clouds. A recent quantum mechanical treatment of the collisional pumping mechanism for the optically thin case predicts that the required anti-inversion of the 1_{11} and 1_{10} populations occurs only when the kinetic temperature exceeds 45 K; population inversion (maser emission) is predicted to occur at the lower kinetic temperatures which generally characterise dark nebulae[168]. Another computation shows that the effects of optical thickness do not alter the preceding restrictions placed on collisional pumping; furthermore, radiation trapping of millimetre wavelength H_2CO lines would serve to modify the 2.7 K black-body spectrum and hence lead to anomalous 4830 MHz absorption as predicted by the radiative pumping model[169]. Observations of the 2_{11}–2_{12} transition in absorption at 14 489 MHz indicate that H_2CO in the Sgr B2 region does indeed trap a significant amount of its own 2 mm radiation but the implied intensity of the 2_{12}–1_{11} transition is not the most favourable for radiative pumping[41]. On the other hand, observations of H_2CO emission at 140.8, 145.6 and 150.5 GHz (and possibly 4830 MHz) from the dense infrared nebula in Orion have been interpreted as tentative evidence for the radiative and against the collisional pumping process[170, 171]. However, this same result may not hold for dark nebulae due to different physical conditions.

Finally it should be noted that two of the suggested models, infrared and collisional pumping, in principle can explain both OH emission and anomalous H_2CO absorption in dark nebulae whereas the radiative pump utilising deviations from the black-body spectrum (possibly in the form of trapped H_2CO millimetre wavelength radiation) can explain anomalous 4830 MHz absorption but not the OH emission spectrum.

6.4.4.3 CO observations

Three molecular isotopes of carbon monoxide, $^{12}C^{16}O$, $^{12}C^{18}O$, and $^{13}C^{16}O$, have been detected through their $J = 1-0$ emission spectra[35, 36] (see Table 6.2); following the radio detections, a rocket-mounted spectrograph was used to detect ultraviolet band elements of the $(A\Pi - X^1\Sigma^+)$ fourth positive system of $^{12}C^{16}O$ and $^{13}C^{16}O$ in the interstellar absorption spectrum of ζ-Ophiuchi[34]. At present, radio detections and observations of carbon monoxide emission have been reported from nine galactic regions, an infrared star[172, 173], and several dark nebulae[174]. Currently there is no observational evidence that CO is undergoing maser emission[35]. The most extensive observations reported to date have been made in the direction of the HII region associated with W51. There the total projected number density is found to be approximately 3×10^{19} cm^{-3} which corresponds to a CO volume density of 1 cm^{-3} for a spherical-cloud model. From preliminary reports of other radio observations, this large CO density appears to be typical for dense interstellar clouds; thus it appears that CO is at least two orders of magnitude more abundant than any other interstellar molecule observed to date except molecular hydrogen. In comparison, the column densities found from ultraviolet observation of the interstellar cloud associated with the O-type star ζ-Ophiuchi are 1.8×10^{15} cm^{-3} and 2.3×10^{13} cm^{-3} for $^{12}C^{16}O$ and $^{13}C^{16}O$, respectively (after recomputation of the observed data using recently measured values for the oscillator strengths[175]).

The lifetime of CO against photodissociation by an ambient interstellar radiation field of 4×10^{-17} erg cm^{-3} Å^{-1} is known to be of the order of 1000 years in the absence of shielding; this lifetime is considerably longer than those obtained for other observed molecules such as OCS (10 years) H_2CO and CH_3C_2H (30 years) NH_3 (40 years) and H_2O (65 years)[176]. Grain shielding of the ambient radiation field equivalent to three magnitudes or more of visual extinction can raise the lifetime against photodissociation to over one million years[177]. Because CO is very stable in comparison with other interstellar molecules, it is expected that much of the observed interstellar CO corresponds to the residue remaining after the photodestruction of more complex molecules. For example, CO has been found to be a primary product of the photolysis of formaldehyde at 1470 and 1236 Å [178]. The stability of CO should allow it to exist in regions of lower obscuration than other interstellar molecules and this appears to be borne out by CO emission extending beyond the visible boundaries of those cloud regions which can be observed optically such as Orion[33].

Lower limits for the $^{12}C:^{13}C$ abundance ratio have been determined from observations of $^{12}C^{16}O$, $^{13}C^{16}O$ and $^{12}C^{18}O$ in selected sources. The limits are ten for W51 [35]; 79, which is nearly the terrestrial value, for the cloud in front of ζ-Ophiuchi[34]; and four for the envelope of the infrared star IRC + 10216 [173]. The latter value would be expected for material processed through the C—N—O cycle.

6.4.5   CN and HCN

The reported radio observations of the $N = 1$–0 transition of CN and the $J = 1$–0 of HCN are currently very limited. Optical transitions of CN, detected in optically-thin interstellar clouds in front of hot stars, have been studied for many years (see Table 6.2 and Section 6.2.1) and successfully used as a thermometer to measure the temperature of the microwave background radiation[179, 180]. The recent radio observations of CN are associated with dense regions of gas and dust and hence sample a different environment[30] but eventually CN will present astronomers with the rare opportunity to correlate radio and optical data from the same molecule. The radio identification is based on agreement between the astronomically-measured rest frequency and a predicted frequency which was calculated from optically-determined molecular constants; it is supported by the astronomical measurement of two hyperfine components associated with the upper spin doublet of the $N = 1$–0 transition. Despite several attempts this transition has yet to be measured directly in the laboratory; thus accurate ground-state molecular parameters of CN probably will be determined initially from astronomical observations if the transitions associated with the lower-spin doublet can be detected. At present, the $N = 1$–0 transition has been reported in emission from Orion, W51, the Sagittarius region (galactic centre) and the infrared star IRC + 10216 [30, 173].

HCN was the second polyatomic organic molecule to be detected in the interstellar clouds[37]. At present, emission from the $J = 1$–0 transition of $H^{12}C^{14}N$ has been observed in W3, Orion, Sgr A, Sgr B2, W49, DR 21 and IRC + 10216; $H^{13}C^{14}N$ has been detected in Orion, Sgr A, and IRC + 10216 [37, 181, 182]. There is no observational evidence that maser action is significant for either CN or HCN and, although it is probable that neither molecule is in thermal equilibrium, rough abundance comparisons for Orion, W51 and IRC + 10216 indicate that HCN may be as much as three times more abundant than CN in these clouds[35, 182]. Exact comparisons must be regarded as tentative until questions involving line saturation, beam dilution, excitation temperature and antenna calibration are resolved, since small errors in any of these quantities could change the apparent abundance values. However, the basic inferences about the relative abundances are likely to remain unchanged. Current lower limits to the $^{12}C:^{13}C$ ratio have been found from HCN measurements to be 8.9 for Orion and 4.7 for Sgr A [37]; at present these limits are affected more severely by data quality than by uncertainties in excitation temperature.

6.4.6   CS and OCS

The $J = 3–2$ transition of CS has been detected in emission from Orion, W51, IRC + 10216, and W51 [36, 173] but emission from the $J = 9–8$ transition of OCS has been found only in Sgr B2 [38] (see Table 6.2). Although it is not yet possible to compare observations of CS and OCS from the same region, preliminary abundance comparisons with other molecules have been made. OCS is estimated to have a column density greater than 3×10^{15} cm^{-2} in Sgr B2; this density is comparable with H_2CO, HC_3N and NH_3 but about one order of magnitude less than OH and four orders less than CO in this region [38]. The peak column density of CS in Orion has been determined to lie between 53 and 2.3×10^{13} cm^{-3} and between 210 and 4.3×10^{13} cm^{-3} for W51. Thus CS appears to be more abundant than CN but less abundant than HCN in these two regions [36]. In contrast, observations of HCN emission from IRC + 10216 indicate the HCN:CS ratio is $c.$ 24:1 but the CN:CS ratio is $c.$ 8:1 [182]. These estimates depend upon the value used for the excitation temperature and assumptions about cloud opacity (see equation (6.32)); hence some modifications are expected as other transitions are found.

6.4.7   CH₃OH

CH_3OH was the first molecule with hindered internal rotation to be found in the interstellar medium. The A level $1_{10}–1_{11}$ transition (i.e. connecting the K-type doublet components of the $J_K = 1_1$ level) at 834 MHz was the first line found and it has been observed in emission only in the galactic centre region [51]. No abundance estimate has been made for this transition due to the lack of information about the background continuum brightness temperature and excitation temperature. However, five CH_3OH rotational transitions corresponding to the $J = 4, 5, 6, 7$ and 8 levels of E_1 symmetry have been detected in emission from the direction of the infrared nebula in Orion [52]. All five frequencies are near 25 GHz (see Table 6.2) which minimises any uncertainties in receiver calibration and essentially eliminates errors in relative intensities caused by unequal beam dilution (see equation (6.2)). Maser action does not appear to be significant because the measured line intensities are within a factor of two of those predicted from thermal equilibrium considerations but the angular size of the emission region is 1 arc minute or less. If the region is exactly coincident with the infrared nebula in Orion, the projected density is 5×10^{16} cm^{-3} which in turn implies a CH_3OH volume density of 0.25 cm^{-3}. The size of the CH_3OH emission region is much smaller than that found for simpler molecules such as H_2CO, CO, CS and HCN. It has been suggested that this small size may be an indication that the more complex molecules in Orion are formed in the dense central region of the infrared nebula [52].

6.4.8   Spectra without laboratory identification: X-ogen and 'HNC'

Radio astronomers have reported detections of two different spectral lines which have no corresponding laboratory identification. The first line was found accidentally when a superheterodyne radio receiver which was being used to observe $H^{13}CN$ with the lower sideband recorded an unidentified emission line in the upper sideband at the astronomically-determined rest frequency of 89.190 GHz. The new line was named X-ogen because it is generated by an unidentified molecule of unknown extraterrestrial origin[58]. It has been suggested that X-ogen may be the molecular ion HCO^+ because 89.190 GHz is well within the estimation error of the calculated $J = 1-0$ transition[183]. An astronomical detection of the theoretically-predicted $J = 1-0$ transition of $H^{13}CO^+$ at 86.708 GHz would support this proposed identification. Two reported $H^{13}CO^+$ searches have yielded negative results but neither one reached the limits needed for detection if the $H^{12}CO^+$: $H^{13}CO^+$ ratio happens to have the terrestrial value[184, 185]; thus HCO^+ has not been ruled out as a possible identification.

The X-ogen emission line has been detected in eight sources (W3, Orion, L134, Sgr A, W51, NGC 2024, K3-50 and NGC 6334) and mapped in Orion over an angular extent of 8 arc minutes[58, 185]. The detection of X-ogen in a number of regions which are not particularly outstanding molecular sources and its fairly widespread emission pattern over Orion tend to argue against a complex molecule (see Section 6.4.7) and in favour of a simple molecule with cosmically abundant constituent atoms.

The X-ogen detection led radio astronomers to search for another spectral line which has yet to be seen in the laboratory. Hydrogen isocyanide was originally suggested as a possibility for the identification of the 89.190 GHz emission line[186] but the terrestrial existence of HNC has been confirmed only in the solid state[187]; thus the frequency of the $J = 1-0$ transition of gaseous HNC had to be deduced from models based on the most probable structure. A number of calculations (outlined elsewhere[188]) indicated that this transition should fall slightly above the X-ogen frequency but below 91 GHz. An ensuing search of W51 and DR 21 yielded an emission line with an astronomically determined rest frequency of 90.665 ± 0.001 GHz, which is well within the estimation error of the calculated $J = 1-0$ transition of HNC. Thus the 'HNC' identification listed in Table 6.2 is based on a theoretical structure and hence there may be better alternatives which have yet to be found. For example, it is not yet possible to rule out currently unassigned or unmeasured transitions of molecules with complex energy-level patterns. To check some of the possibilities, NH_2CHO, NO_2, H_2O, CH_3OH, H_2CO and NH_3 have been examined in the laboratory for previously unmeasured transitions near 90.665 GHz and it has been established that none of these molecules can be responsible for the new interstellar line[189]. Gaseous HNC is very difficult to generate and measure in the laboratory; thus the most promising identification test may be an interstellar search for the predicted $J = 1-0$ transition of $HN^{13}C$ at 87.112 GHz.

6.4.9 Molecules detected only in the galactic centre clouds

The remaining molecules reported by radio astronomers have been observed only in the galactic centre regions Sgr A and Sgr B2. Due to the weak signal strengths found in many cases, the determination of galactic distribution and abundances relative to the molecules discussed previously have yet to be established conclusively and may have to await the development of more sensitive radio receivers.

HC$_3$N (cyanoacetylene) has been detected in emission from Sgr B2 and identified by observation of two of the quadrupole-split lines of the $J = 1$–0 transition[50] (see Table 6.2). The projected density is estimated to be 2.1×10^{16} cm^{-3} which is approximately an order of magnitude greater than the estimated H$_2$CO density for the same source. HC$_3$N has not been detected in a search of 16 other galactic regions.

A weak emission line of HCOOH at 1639 MHz has been detected in the direction of Sgr B2 and possibly Sgr A but it has not been found elsewhere in the galaxy[49, 190]. Unfortunately, the HCOOH line is somewhat masked by emission lines from ^{18}OH; hence the detection of other HCOOH transitions which are not obscured would be helpful for confirmation of this identification.

The $J = 3$–2 transition of SiO was detected in emission from Sgr B2 and identified on the basis of a calculated frequency[12]. The estimated projected density is 4×10^{13} cm^{-3}, which is more than an order of magnitude less than the corresponding H$_2$CO column density. The $J = 3$–2 transition was not detected in six other sources and an earlier search for the $J = 2$–1 transition at 86 846.9 MHz was not successful[191].

The $J = 6$–5 transition of CH$_3$CN has been observed in both the Sgr A and Sgr B2 regions[54]. Definite population anomalies have been observed; for example, the $K = 2$ line is missing (see Table 6.2) and a kinetic temperature of 150 K is required to explain the intensities of the $K = 4$ and 5 lines in Sgr B2. Column densities of the order of 2×10^{14} cm^{-3} are comparable with those of CN and hence may be an indication that complex stable molecules are at least as abundant as chemically unstable diatomic species in dense interstellar clouds.

Both the 1_{10}–1_{11} and 2_{11}–2_{12} rotational transitions of formamide (HCONH$_2$) have been observed in emission[55, 56]. A conservative lower limit to the column density of 4×10^3 cm^{-2} was derived from observations of the 2_{11}–2_{12} transition in Sgr B2[56]. This transition possibly was found in Sgr A also but not in the eight other galactic sources surveyed.

The remaining molecules have recently been identified and only preliminary reports are available. Isocyanic acid (HNCO) has been detected in emission and identified in Sgr B2 by observations of both the 4_{04}–3_{03} and 1_{01}–0_{00} transitions[47, 48]. An emission feature at 85 457 MHz has been identified as the 5_0–4_0 transition of methylacetylene (CH$_3$C$_2$H)[47]. A weak emission feature from the 1_{10}–1_{11} transition of acetaldehyde (HCOCH$_3$) was reported from both Sgr A and Sgr B2[57]. Finally, a very recent detection of the 2_{11}–2_{12} transition of H$_2$CS in absorption against Sgr B2 has ended a search which began shortly after the discovery of interstellar formaldehyde[59]. Interstellar H$_2$CO was detected in so many different galactic regions (see Section 6.4.4.1) that it was soon recognised that the sulphur-containing analogue of H$_2$CO might be important to astronomy. Since no spectral information was available in the literature, gaseous H$_2$CS was successfully generated in the laboratory and characterised from its gas-phase rotational spectrum[192]. In a subsequent search of six interstellar regions the 1_{10}–1_{11} transition was not detected[193] but the search for the $2_{11}-2_{12}$ transition was successful[59]. The typical reactive half-life of H_2CS is 6 min in a laboratory cell at pressures of 10–50 μm Hg [194]; however, this short laboratory lifetime does not appear to be an important consideration in the interstellar clouds (see Section 6.1).

6.4.10   Recent searches for other molecules

Searches for a number of other interstellar molecules have been conducted but at present only limited observational results have been reported. The ground-state transitions of NO have been thoroughly analysed[197] and the expected optical depths have been estimated for transitions in the decametre, centrimetre and millimetre wavelength ranges[198]. Several searches for the ground-state lambda-doublet transitions of NO at 206, 226, 411 and 431 MHz have given negative results which are consistent with expected interstellar abundances[199, 200]. Two searches have been conducted for ground-state transitions of SH at 111 and 112 MHz with no detection reported[201, 202]. In addition, searches for low-frequency transitions of OD, HDO, ClO and H_2C_2O have not been successful[199]. It should be recognised that negative results obtained at low frequencies may not be conclusive because the beam-dilution factor is usually large (see equation (6.2) and Section 6.3.3), and man-made interference is often present. In addition, it can be shown from equations (6.15) and (6.32) that for reasonable excitation temperatures the detection of normal emission is relatively more difficult at low frequencies (e.g. 400 MHz) than at appreciably higher frequencies (e.g. 40 GHz). Hence several of the low-frequency negative results will become more definitive as the search range is extended to higher frequencies; at present they tend to rule out maser emission or unexpectedly large abundances but do not prove that a given molecular species is significantly underabundant in the interstellar clouds.

At higher frequencies, an initial search for the $1_{01}-0_{00}$ transition of HCO at 86 796 MHz has not yielded a detection[184]. Recently, a negative result was reported for the $J = 1-0$ transition of N_2O at 25 123 MHz [52]. A possible detection of interstellar methane (CH_4) has been observed in the high-resolution infrared spectrum of α-Orionis[196].

Information about the relative abundances of molecular isomers is expected to lead to a better understanding of the predominant molecular formation processes; hence several searches have been initiated for isomers of previously-detected interstellar molecules. At present two isomers of formamide, CH_3NO and CH_2NOH, have not been detected in an initial search[203].

Sensitive searches for radio-frequency transitions of interstellar molecules are conducted at a fixed frequency with a typical maximum band-width of 10 MHz in the centrimetre wavelength spectral region and 200 MHz in the millimetre region. Limited spectrum scans can be performed by using short integration periods and laboriously stepping the fixed frequency through adjoining intervals over the band-pass of the receiver. The short integration periods, on the order of 40 s, required to cover a typical tuning range limit the possible detections to very intense spectral lines (see equation (6.3)). To date, no new interstellar molecules have been found by frequency scanning techniques[204, 205].

6.5 RELATED STUDIES OF ASTRONOMICAL INTEREST

6.5.1 Frequency measurements

Radio astronomers often rely heavily on accurate laboratory frequency measurements for conducting meaningful interstellar searches for new molecules, computing radial velocities and assigning correct identifications. Hence recent compilations of measured molecular transitions have proved extremely useful for the study of molecules in space[206]. Precise laboratory measurements of H_2CO and HC_3N have been used to remove radial velocity uncertainties and to study hyperfine spectra in several interstellar clouds[207-210]. A critical review series, recently started, will contain a complete list of frequencies (measured or calculated) and energy levels up to several hundred wavenumbers for many of the known interstellar molecules[211]. The completion of this series is expected to facilitate the future identification of interstellar lines which may arise from transitions between higher rotational levels.

Radio-frequency transitions of many molecules of potential astronomical importance either have not been or cannot be directly measured in the laboratory. For example, the ground-state lambda-doublet transition of CH was recently determined to be 3374 ± 15 MHz using optical methods[212]. However, measurements of ground-state transitions are lacking for known stellar molecules such as TiO, VO, HF and SiC_2 as well as for simple metallic species which should have reasonable interstellar abundances such as MgO and FeO. Finally, on a slightly more complex scale, laboratory efforts to measure and characterise gaseous amino acids and their derivatives should be undertaken. Two promising candidates, glycine and alanine, are both composed of cosmically abundant H, C, N and O and their sub-groups are closely related to known interstellar molecules.

6.5.2 Studies of collisional processes

Proper evaluation of the excitation temperature, T_e, is central to the determination of interstellar molecular abundances (see equations (6.15), (6.16) and (6.34)). The approximate expression given in equation (6.35) serves to illustrate the dependence of T_e on both radiative and collisional interactions. Collisional interactions are understood in principle but the relative importance of collisions between interstellar molecules and electrons or ions (e.g. dipole and higher order interactions) v. neutral atoms or molecules (e.g. hard-sphere interactions) is difficult to determine. A combination of collisional and radiative transition rates applied to HCN emission observations leads to approximate density estimates of either 10^6 cm^{-3} for molecular hydrogen or 1 cm^{-3} for electrons in dense interstellar clouds but it is not necessarily clear which is preferred[67]. At present, the available information on collision cross-sections is perhaps more complete for NH_3 than for any other interstellar molecule[88] and collisional selection rules have been studied for NH_3–rare gas and NH_3–H_2 interactions[99, 100]. But accurate numbers for the cross-sections of many other interstellar molecules are not yet available, particularly in the case of hard-sphere collisions. For example, an interesting investigation of several models for collisional excitation of interstellar CO uses a more exact form of equation (6.35) and the rate equations for the ten lowest rotational levels to show that hard collisions (which do not obey any selection rules) are capable of introducing CO maser emission[213]. However, due to the lack of appropriate theoretical and experimental information on the interaction potential for hard collisions between CO and neutral atoms or molecules, it is possible that the cross-sections used may lack the accuracy required for quantitative predictions.

6.5.3   Molecular formation and destruction processes

A number of theories have been presented to explain the interstellar formation processes, particularly for polyatomic molecules, but many are updated versions of theories advanced originally to explain the formation of CH, CH^+, CN and OH. Most of the current ideas fit one of the following three categories: formation on grain surfaces; formation by two-body collision; and formation in stellar atmospheres followed by ejection of molecules, atoms and grains to form interstellar clouds. The merits of many of these molecular formation proposals have been discussed elsewhere[62, 63, 67]; hence the following remarks are confined to those recent models which appear to contribute to a quantitative understanding of the interstellar chemistry. It should be realised that several processes probably are responsible for the observed molecular abundances; thus one of the current goals of observational astronomy is to correlate the observed abundances with the specific physical conditions found in different interstellar regions.

It has been shown that radiative association and two-body chemical reactions can account for the abundances of CH, CH^+, CN and c. 0.13 of the CO observed in the low density cloud associated with the star ζ-Opiuchi[214]. Formation of OH by inverse predissociation can account for approximately half of the OH density in dark nebulae but not in less dense HI clouds[215]. The problem of polyatomic formation is more difficult but photodissociation and photoionisation, which are thought to be the dominant interstellar destruction processes, place some stringent conditions on the possible formation mechanisms for polyatomic molecules in dense clouds. In unshielded interstellar regions, the lifetimes of typical molecules such as OCS, H_2CO, CH_3C_2H, NH_3 and H_2O are known to be less than 100 years[176] (with the exception of CO as discussed in Section 6.4.4.3). Reasonable lifetimes on the galactic time scale can be expected only in the dense clouds where they are protected from the ambient interstellar radiation field[177]. Thus the polyatomic molecules are thought to have formed in the clouds where they are observed and not elsewhere. This restriction, along with the observational correlation between polyatomic molecules and heavily obscured regions, currently gives preference to mechanisms which invoke polyatomic formation on grain surfaces with subsequent release into the relatively shielded regions of an interstellar cloud. However, the agreement between theory and observation is still not well confirmed. For example, it has been amply demonstrated that simple precursor atoms or molecules deposited on a variety of surfaces, which are subsequently irradiated with ultraviolet light, lead to the synthesis of a wide variety of polyatomic organic molecules. The types of surfaces used for laboratory synthesis or discussed for interstellar synthesis include nickel[216], powdered Vycor glass[217], a quartz cold finger[218], brass[219], iron grains[220] and even sterilised soil[217]. Thus, it does not seem to matter that the chemical composition of the interstellar grains is somewhat uncertain[195]; as long as an adsorption surface is present to collect the precursor gas and the appropriate energy is supplied by the radiation field, presumably polyatomic formation will proceed in dense clouds. Unfortunately, not much information about molecular ejection rates is available and thus there is no quantitative evaluation of the contribution to the interstellar abundance by molecules formed on grain surfaces.

Finally, it is hoped that the answer to the polyatomic formation problem will lead to the solution of the oldest spectroscopic problem associated with the interstellar clouds — the origin of the interstellar diffuse bands. The initial detection of four diffuse absorption features at 5780, 5797, 6284 and 6614 Å, reported in 1934, was responsible for the first speculation about the possible existence of interstellar molecules[221]. Currently, at least 24 diffuse bands between 3137 and 4300 Å have been observed[222] and, although their origin still has not been established, they are thought to be molecular features[223]. The most recent proposal is that the diffuse features are due to a particular porphyrin molecule, bispyridylmagnesium-tetrabenzoporphine ($MgC_{46}H_{30}N_6$), which has a laboratory spectrum very similar to that observed for the interstellar diffuse bands[224]. However, this proposal is not supported strongly by a recent observational analysis which suggests at least three molecular species are responsible for the 24 observed diffuse features[221].

Note added in proof

The $J = 2 \to 1$ rotational transition of silicon monoxide, SiO, at 86 847 MHz has been detected in Orion A and Sgr B2 [227]. An earlier search for this line was not successful[191].

References

1. Carruthers, G. R. (1970). *Astrophys. J. (Lett.)*, **161**, L81
2. Gardner, F. F., Robinson, B. J., Bolton, J. G. and van Damme, K. J. (1964). *Phys. Rev. Lett.*, **13**, 3
3. Weinreb, S., Barrett, A. H., Meeks, M. L. and Henry, J. C. (1963). *Nature (London)*, **200**, 829
4. Yen, J. L., Zuckerman, B., Palmer, P. and Penfield, H. (1969). *Astrophys. J. (Lett.)*, **156**, 127
5. Turner, B. E., Palmer, P. and Zuckerman, B. (1970). *Astrophys. J. (Lett.)*, **160**, L125

6. Zuckerman, B., Palmer, P., Penfield, H. and Lilley, A. E. (1968). *Astrophys. J. (Lett.)*, **161**, L69

7. Thacker, D. L., Wilson, W. J. and Barrett, A. H. (1970). *Astrophys. J. (Lett.)*, **161**, L191

8. Schwartz, P. R. and Barrett, A. H. (1969). *Astrophys. J. (Lett.)*, **157**, L109

9. Ball, J. A., Gottlieb, C. A., Meeks, M. L. and Radford, H. E. (1971). *Astrophys. J. (Lett.)*, **163**, L33

10. Rogers, A. E. E. and Barrett, A. H. (1966). *Astron. J.*, **71**, 868

11. Wilson, W. J. and Barrett, A. H. (1968). *Research Laboratory of Electronics, M.I.T., Quarterly Progress Report No. 90*, 9; also (1970). *Astrophys. Lett.*, **6**, 231

12. Wilson, R. W., Penzias, A. A., Jefferts, K. B., Kutner, M. and Thaddeus, P. (1971). *Astrophys. J. (Lett.)*, **167**, L97

13. Cheung, A. C., Rank, D. M., Townes, C. H., Thornton, D. D. and Welch, W. J. (1969). *Nature (London)*, **221**, 626

14. Morris, M., Zuckerman, B., Turner, B. E. and Palmer, P. (1971). *B.A.A.S.*, **3**, 499

15. Cheung, A. C., Rank, D. M., Townes, C. H., Thornton, D. D. and Welch, W. J. (1968). *Phys. Rev. Lett.*, **21**, 1701

16. Cheung, A. C., Rank, D. M., Townes, C. H. and Welch, W. J. (1969). *Nature (London)* **221**, 917

17. Dunham, T., Jr. (1937). *Publ. Astr. Soc. Pacific*, **49**, 26

18. Douglas, A. E. and Herzberg, G. (1941). *Astrophys. J.*, **94**, 381

19. Douglas, A. E. and Herzberg, G. (1942). *Can. J. Res.*, **20**, 71

20. Adams, W. S. (1941). *Astrophys. J.*, **93**, 11

21. Adams, W. S. (1938–1939). *Ann. Rep. Dir. Mt. Wilson Obs.*

22. Douglas, A. E. and Morton, J. R. (1960). *Astrophys. J.*, **131**, 1

23. Herbig, G. H. (1960). *Astron. J.*, **65**, 491

24. Bortolot, V. J., Jr. and Thaddeus, P. (1969). *Astrophys. J. (Lett.)*, **155**, L17

25. McKellar, A. (1940). *Publ. Astr. Soc. Pacific*, **52**, 187

26. Swings, P. and Rosenfeld, L. (1937). *Astrophys. J.*, **86**, 483

27. Feast, M. W. (1955). *Observatory*, **75**, 182

28. Spitzer, L., Jr. and Field, G. B. (1955). *Astrophys. J.*, **121**, 300

29. McKellar, A. and Richardson, E. H. (1955). *Astrophys. J.*, **122**, 196

30. Jefferts, K. B., Penzias, A. A. and Wilson, R. W. (1970). *Astrophys. J. (Lett.)*, **161**, L87

31. Münch, G. (1964). *Astrophys. J.*, **140**, 107

32. McKellar, A. (1941). *Publ. Dom. Astrophys. Obs., Victoria, B.C.*, **7**, 251

33. Wilson, R. W., Jefferts, K. B. and Penzias, A. A. (1970). *Astrophys. J. (Lett.)*, **161**, L43

34. Smith, A. M. and Stecher, T. P. (1971). *Astrophys., J. (Lett.)*, **164**, L43

35. Penzias, A. A., Jefferts, K. B. and Wilson, R. W. (1971). *Astrophys. J. (Lett.)*, **165**, L229

36. Penzias, A. A., Solomon, P. M., Wilson, R. W. and Jefferts, K. B. (1971). *Astrophys. J. (Lett.)*, **168**, L53

37. Snyder, L. E. and Buhl, D. (1971). *Astrophys. J. (Lett.)*, **163**, L47

38. Jefferts, K. B., Penzias, A. A., Wilson, R. W. and Solomon, P. M. (1971). *Astrophys. J. (Lett.)*, **168**, L111

39. Snyder, L. E., Buhl, D., Zuckerman, B. and Palmer, P. (1969). *Phys. Rev. Lett.*, **22**, 679

40. Kutner, M. and Thaddeus, P. (1971). *Astrophys. J. (Lett.)*, **168**, L67

41. Evans, N. J., II, Cheung, A. C. and Sloanaker, R. M. (1970). *Astrophys. J. (Lett.)*, **159**, L9

42. Welch, W. J. (1970). *B.A.A.S.*, **2**, 355

43. Kutner, M., Thaddeus, P., Jefferts, K. B., Penzias, A. A. and Wilson, R. W. (1971). *Astrophys. J. (Lett.)*, **164**, L49

44. Thaddeus, P., Wilson, R. W., Kutner, M., Penzias, A. A. and Jefferts, K. B. (1971). *Astrophys. J. (Lett.)*, **168**, L59

45. Zuckerman, B., Palmer, P., Snyder, L. E. and Buhl, D. (1969). *Astrophys. J. (Lett.)*, **157**, L167

46. Gardner, F. F., Ribes, J. C. and Cooper, B. F. C. (1971). *Astrophys. Lett.*, **9**, 181

47. Snyder, L. E. and Buhl, D. (1971). *B.A.A.S.*, **3**, 388

48. Buhl, D., Snyder, L. E. and Edrich, J., to be published

49. Zuckerman, B., Ball, J. A. and Gottlieb, C. A. (1971). *Astrophys. J. (Lett.)*, **163**, L41

50. Turner, B. E. (1971). *Astrophys. J. (Lett.)*, **163**, L35

51. Ball, J. A., Gottlieb, C. A., Lilley, A. E. and Radford, H. E. (1970). *Astrophys. J. (Lett.)*, **162**, L203

52. Barrett, A. H., Schwartz, P. R. and Waters, J. E. (1971). *Astrophys. J. (Lett.)*, **168**, L101

53. Zuckerman, B., Turner, B. E., Johnson, D. R., Palmer, P. and Morris, M., to be published
54. Solomon, P. M., Jefferts, K. B., Penzias, A. A. and Wilson, R. W. (1971). *Astrophys. J. (Lett.)*, **168**, L107
55. Palmer, P., Gottlieb, C. A., Rickard, L. J. and Zuckerman, B. (1971). *B.A.A.S.*, to be published
56. Rubin, R. H., Swenson, G. W., Jr., Benson, R. C., Tigelaar, H. L. and Flygare, W. H. (1971). *Astrophys. J. (Lett.)*, **169**, L39
57. Ball, J. A., Gottlieb, C. A., Lilley, A. E. and Radford, H. E. (1971). *Inter. Astrom. Union Circular No. 2350*
58. Buhl, D. and Snyder, L. E. (1970). *Nature (London)*, **228**, 267
59. Sinclair, M. W., Ribes, J. C., Fourikis, N., Brown, R. D. and Godfrey, P. D. (1971). *Inter. Astrom. Union Circular No. 2362*
60. Herzberg, G. (1967). *Molecular Spectra and Molecular Structure Vol. III*, 580. (Toronto: D. Van Nostrand Co., Inc.)
61. Herbig, G. H. (1963). *J. Quant. Spectry. Rad. Transfer*, **3**, 529
62. Dieter, N. H. and Goss, W. M. (1966). *Rev. Mod. Phys.*, **38**, 256
63. McNally, D. (1968). *Advances in Astronomy and Astrophysics*, ed. Kopal, Z., 173. (New York: Academic Press)
64. Snyder, L. E. and Buhl, D. (1970). *Sky and Telescope*, **40**, 267
65. Herzberg, G. (1971). *The Spectra and Structures of Simple Free Radicals*, 5. (Ithaca: Cornell University Press)
66. Scheuer, P. A. G. and Williams, P. J. S. (1968). *Annual Review of Astronomy and Astrophysics*, 321. (Palo Alto: Annual Reviews Inc.)
67. Rank, D. M., Townes, C. H. and Welch, W. J. (1971). *Science*, **174**, 1083
68. Findlay, J. W. (1971). *Annual Review of Astronomy and Astrophysics*, 271. (Palo Alto: Annual Reviews Inc.)
69. Cogdell, J. R., McCue, J. J., Kalachev, P. D., Salomonovich, A. E., Moiseev, I. G., Stacey, J. M., Epstein, E. E., Altshuler, E. E., Feix, G., Day, J. W. B., Hvatum, H., Welch, W. J. and Barath, F. T. (1970). *IEEE Trans. Antenna Propag.*, **AP-18**, 515
70. Hachenberg, O. (1970). *Sky and Telescope*, **40**, 338
71. Heeschen, D. S. (1967). *Science*, **158**, 75
72. Swenson, G. W., Jr. and Mathur, N. C. (1968). *Proc. IEEE*, **56**, 2114
73. Buhl, D. and Snyder, L. E. (1971). *Nature, Physical Science*, **232**, 161
74. Kraus, J. D. (1966). *Radio Astronomy*. (New York: McGraw-Hill Book Co.)
75. Weinreb, S. (1963). *Tech. Rept. AD 418-413, Clearinghouse for Federal Scientific and Technical Information*. (Springfield, Va.)
76. Shalloway, A. M., Mauzy, R., Greenhalgh, J. and Weinreb, S. (1968). *Electronics Division Internal Report No. 75, Nat. Rad. Astr. Obs.* (Green Bank, West Va.)
77. Milne, E. A. (1966). *Selected Papers on the Transfer of Radiation*, ed. Menzel, D. H., 77. (New York: Dover)
78. Chandrasekhar, S. (1950). *Radiative Transfer*. (Fair Lawn, N.J.: Oxford University Press)
79. Barrett, A. H. (1958). *Proc. I.R.E.*, **46**, 250
80. Hagen, J. P., Lilley, A. E. and McClain, E. F. (1955). *Astrophys. J.*, **122**, 361
81. Goss, W. M. (1967). *Astrophys. J. Suppl. 133*, **15**, 131
82. Hummer, D. G. and Rybicki, G. (1971). *Annual Review of Astronomy and Astrophysics*, 237. (Palo Alto: Annual Reviews Inc.)
83. Sullivan, W. T., III (1971). *Ph.D. Dissertation*, University of Maryland
84. Litvak, M. M., McWhorter, A. L., Meeks, M. L. and Zeiger, H. J. (1966). *Phys. Rev. Lett.*, **17**, 821
85. Litvak, M. M. (1969). *Astrophys. J.*, **156**, 471
86. Turner, B. E. (1970). *P.A.S.P.*, **82**, 996
87. Turner, B. E. (1970). *R.A.S.C. J.*, **64**, 221
88. Townes, C. H. and Schawlow, A. L. (1955). *Microwave Spectroscopy*. (New York: McGraw-Hill)
89. Weliachew, L. (1971). *Astrophys. J. (Lett.)*, **167**, L47
90. Field, G. B., Somerville, W. B. and Dressler, K. (1966). *Annual Review of Astronomy and Astrophysics.*, 207. (Palo Alto: Annual Reviews Inc.)
91. Carruthers, G. (1970). *Space Sci. Rev.*, **10**, 459
92. Heiles, C. (1969). *Astrophys. J.*, **156**, 493
93. Cartwright, D. C. and Drapatz, S. (1970). *Astron. and Astrophys.*, **4**, 443

94. Gull, T. R. and Harwit, M. O. (1971). *Astrophys. J.*, **168**, 15
95. Penzias, A. A., Jefferts, K. B., Dickinson, D. F., Lilley, A. E. and Penfield, H. (1968). *Astrophys. J.*, **154**, 389
96. Jefferts, K. B., Penzias, A. A., Ball, J. A., Dickinson, D. F. and Lilley, A. E. (1970). *Astrophys. J. (Lett.)*, **159**, L15
97. Cheung, A. C., Rank, D. M., Townes, C. H., Knowles, S. H. and Sullivan, W. T., III (1969). *Astrophys. J. (Lett.)*, **157**, L13
98. Knowles, S. H. and Cheung, A. C. (1971). *Astrophys. J. (Lett.)*, **164**, L19
99. Oka, T. (1970). *Seventeenth International Astrophysical Conference.* (Liege, Belgium)
100. Daly, P. W. and Oka, T. (1970). *J. Chem. Phys.*, **53**, 3272
101. Oka, T., Shimizu, F. O., Shimizu, T. and Watson, J. K. G. (1971). *Astrophys. J. (Lett.)*, **165**, L15
102. Zuckerman, B., Morris, M., Turner, B. E. and Palmer, P. (1971). *Astrophys. J. (Lett.)*, **169**, L105
103. Turner, B. (1971)., to be published
104. Heiles, C. (1968). *Astrophys. J.*, **151**, 919
105. Snyder, L. E. and Buhl, D. (1969). *Astrophys. J.*, **155**, L65
106. Schwartz, P. R. (1971). *Ph.D. Dissertation*, M.I.T.
107. Weaver, H. (1968). *Interstellar Ionized Hydrogen Gas*, ed. Terzian, Y., 645. (New York: W. A. Benjamin, Inc.)
108. Barrett, A. H. and Lilley, A. E. (1957). *Astron. Jour.*, **62**, 4
109. Rogers, A. E. E. and Barrett, A. H. (1968). *Astrophys. J.*, **151**, 163
110. Goss, W. M. and Field, G. B. (1968). *Astrophys. J.*, **151**, 177
111. Purcell, E. M. and Field, G. B. (1956). *Astrophys. J.*, **124**, 542
112. Palmer, P. and Zuckerman, B. (1967). *Astrophys. J.*, **148**, 727
113. Gardner, F. F., McGee, R. X. and Sinclair, M. W. (1970). *Astrophys. Lett.*, **5**, 67
114. Kerr, F. J. and Sandqvist, A. (1968). *Astrophys. Lett.*, **2**, 195
115. Sandqvist, A. (1971). *Ph.D. Dissertation*, University of Maryland
116. Verschuur, G. L. (1971). *Astrophys. Lett.*, **7**, 217
117. Turner, B. E. (1969). *Astrophys. J.*, **157**, 103
118. Zuckerman, B., Ball, J. A., Dickinson, D. F. and Penfield, H. (1969). *Astrophys. Lett.*, **3**, 97
119. McGee, R. X., Robinson, B. J., Gardner, F. F. and Bolton, J. G. (1965). *Nature (London)*, **208**, 1193
120. Gunderman, E. J. (1965). *Ph.D. Dissertation*, Harvard (recording the observations of Gunderman, E. J., Goldstein, S. J., Jr. and Lilley, A. E.)
121. Weaver, H., Williams, D. R. W., Dieter, N. H. and Lum, W. T. (1965). *Nature (London)*, **208**, 29
122. Barrett, A. H. (1967). *Science*, **157**, 881
123. Weaver, H. F., Dieter, N. H. and Williams, D. R. W. (1968). *Astrophys. J. Suppl.*, **16**, 219
124. Raimond, E. and Eliasson, B. (1967). *Astrophys. J. (Lett.)*, **150**, L171
125. Wilson, W. J. and Barrett, A. H. (1968). *Science*, **161**, 778
126. Turner, B. E. (1971). *Astrophys. Lett.*, **8**, 73
127. Manchester, R. N., Robinson, B. J. and Goss, W. M. (1970). *Aust. J. Phys.*, **23**, 751
128. Wilson, W. J., Barrett, A. H. and Moran, J. M. (1970). *Astrophys. J.*, **160**, 545
129. Schwartz, P. R. and Barrett, A. H. (1970). *Astrophys. J. (Lett.)*, **159**, L123
130. Dickinson, D. F. and Chaisson, E. J. (1971). *Astrophys. J.*, **169**, 207
131. Rogers, A. E. E., Moran, J. M., Crowther, P. P., Burke, B. F., Meeks, M. L., Ball, J. A. and Hyde, G. M. (1967). *Astrophys. J. (Lett.)*, **147**, L369
132. Moran, J. M., Barrett, A. H., Rogers, A. E. E., Burke, B. F., Zuckerman, B., Penfield, H. and Meeks, M. L. (1967). *Astrophys. J. (Lett.)*, **148**, L69
133. Davies, R. D., Rowson, B., Booth, R. S., Cooper, A. J., Gent, H., Agdie, R. L. and Crowther, J. H. (1967). *Nature (London)*, **213**, 1109
134. Moran, J. M., Burke, B. F., Barrett, A. H., Rogers, A. E. E., Ball, J. A., Carter, J. C. and Cudaback, D. D. (1968). *Astrophys. J. (Lett.)*, **152**, L97
135. Knowles, S. H., Mayer, C. H., Cheung, A. C., Rank, D. M. and Townes, C. H. (1969). *Science*, **163**, 1055
136. Meeks, M. L., Carter, J. C., Barrett, A. H., Schwartz, P. R., Waters, W. J. and Brown, W. E. (1969). *Science*, **165**, 180
137. Knowles, S. H., Mayer, C. H., Sullivan, W. T. and Cheung, A. C. (1969). *Science*, **166**, 221
138. Buhl, D., Snyder, L. E., Schwartz, P. R. and Barrett, A. H. (1969). *Astrophys. J. (Lett.)*, **158**, L97

139. Turner, B. E., Buhl, D., Churchwell, E. B., Mezger, P. G. and Snyder, L. E. (1970). *Astron. and Astrophys.*, **4**, 165
140. Sullivan, W. T., III (1971). *Astrophys. J.*, **166**, 321
141. Burke, B. F., Papa, D. C., Papadopoulos, G. D., Schwartz, P. R., Knowles, S. H., Sullivan, W. T., Meeks, M. L. and Moran, J. M. (1970). *Astrophys. J. (Lett.)*, **160**, L63
142 Johnston, K. J., Knowles, S. H., Sullivan, W. T., III, Moran, J. M., Burke, B. F., Lo, K. Y., Papa, D. C., Papadopoulos, G. D., Schwartz, P. R., Knight, C. A., Shapiro, I. I. and Welch, W. J. (1971). *Astrophys. J. (Lett)*, **166**, L21
143 Burke, B. F., Matveyenko, L. I., Moran, J. M., Moiseyev, I. G., Knowles, S. H., Clark, B. G., Efanov, V. A., Johnston, K. J., Kogan, L. R., Kostenko, V. I., Lo, K. Y., Papa, D. C., Papadopoulos, G. D., Rogers, A. E. E. and Schwartz, P. R. (1971). *B.A.A.S.*, **3**, 468
144. Powell, F. X. and Johnson, D. R. (1970). *Phys. Rev. Lett.*, **24**, 637
145. Snyder, L. E., Buhl, D. and Hemenway, P. D. (1970), private communication
146. Ball, J. A., Gottlieb, C. A. and Radford, H. E. (1971). *Astrophys. J.*, **163**, 429
147. Shklovsky, I. S. (1966). *Astr. Tsirk.*, No. 372; (1967). ibid., No. 424
148. Menon, R. K. (1967). *Astrophys. J. (Lett.)*, **150**, L167
149. Mezger, P. G. and Robinson, B. J. (1968). *Nature (London)*, **220**, 1107
150. Litvak, M. M. (1969). *Science*, **165**, 855
151. Litvak, M. M. (1970). *Phys. Rev. A.*, **2**, 2107
152. Litvak, M. M. (1971). *Astrophys. J.*, **170**, 71
153. Oka, T. (1971). *Proc. Conf. on Int. Molecules.* (Charlottesville, Va., U.S.A.)
154. Zuckerman, B., Buhl, D., Palmer, P. and Snyder, L. E. (1970). *Astrophys. J.*, **160**, 485
155. Whiteoak, J. B. and Gardner, F. F. (1970). *Astrophys. Lett.*, **5**, 5
156. Wilson, T. L. (1971). *B.A.A.S.*, **3**, 18
157. Scoville, N. Z., Solomon, P. M. and Thaddeus, P. (1972). *Astrophys. J.*, **172**, 335
158. Kerr, F. J. and Sandqvist, A. (1970). *Astrophys. Lett.*, **5**, 59
159. Palmer, P., Zuckerman, B., Buhl, D. and Snyder, L. E. (1969). *Astrophys. J. (Lett.)*, **156**, L147
160. Zuckerman, B., Palmer, P., Snyder, L. E. and Buhl, D. (1969). *Astrophys. J. (Lett.)*, **157**, L167
161. Formalont, E. B. and Weliachew, L. N., to be published
162. Gardner, F. F., Ribes, J. C. and Cooper, B. F. C. (1971). *Inter. Astrom. Union Circular No. 2354*
163. Townes, C. H. and Cheung, A. C. (1969). *Astrophys. J. (Lett.)*, **157**, L103
164. Solomon, P. M. and Thaddeus, P. (1971)., to be published
165. Oka, T. (1970). *Astrophys. J. (Lett.)*, **160**, L69
166. Litvak, M. M. (1970). *Astrophys. J. (Lett.)*, **160**, L133
167. Kutner, M. and Thaddeus, P. (1971). *Astrophys. J. (Lett.)*, **168**, L67
168. Thaddeus, P. (1972)., to be published
169. White, R. E. (1971). *Ph.D. Dissertation*, Columbia University
170. Kutner, M., Thaddeus, P., Jefferts, K. B., Penzias, A. A. and Wilson, R. W. (1971). *Astrophys. J. (Lett.)*, **164**, L49
171. Thaddeus, P., Wilson, R. W., Kutner, M., Penzias, A. A. and Jefferts, K. B. (1971). *Astrophys. J. (Lett.)*, **168**, L59
172. Solomon, P., Jefferts, K. B., Penzias, A. A. and Wilson, R. W. (1971). *Astrophys. J. (Lett.)*, **163**, L53
173. Wilson, R. W., Solomon, P. M., Penzias, A. A. and Jefferts, K. B. (1971). *Astrophys. J. (Lett.)*, **169**, L35
174. Solomon, P. (1971)., private communication
175. Smith, A. M. and Stecher, T. P. (1971)., private communication
176. Stief, L. J. (1971). *Proc. Conf. on Int. Molecules.* (Charlottesville, Va., U.S.A.)
177. Stief, L. J. Donn, B., Glicker, S., Gentieu, E. P. and Mentall, J. E. (1972). *Astrophys. J.*, **171**, 21
178. Glicker, S. and Stief, L. J. (1971). *J. Chem. Phys.*, **54**, 2852
179. Field, G. B. and Hitchcock, J. L. (1966). *Phys. Rev. Lett.*, **16**, 817
180. Thaddeus, P. and Clauser, J. F. (1966). *Phys. Rev. Lett.*, **16**, 819
181. Snyder, L. E. and Buhl, D. (1971). *Inter. Astrom. Union Circular No. 2330*
182. Morris, M., Zuckerman, B., Palmer, P. and Turner, B. E. (1971). *Astrophys. J. (Lett.)*, **170**, L109
183. Klemperer, W. (1970). *Nature (London)*, **227**, 1230

184. Jefferts, K. B., Penzias, A. A., Wilson, R. W., Kutner, M. and Thaddeus, P. *Astrophys. Lett.,* **8,** 43
185. Snyder, L. E. and Buhl, D. (1971). *B.A.A.S.,* **3,** 251
186. Herzberg, G. (1970)., private communication
187. Milligan, D. E. and Jacox, M. E. (1963). *J. Chem. Phys.,* **39,** 712
188. Snyder, L. E. and Buhl, D. (1971). *Proc. N.Y. Acad. Sciences,* to be published
189. Johnson, D. R. (1971)., private communication
190. Cato, M., Cato, T., Landgren, P. and Sume, A. (1970). *Astrophys. J. (Lett.),* **160,** L131
191. Dickinson, D. F. and Gottlieb, C. A. (1971). *Astrophys. Lett.,* **7,** 205
192. Johnson, D. R. and Powell, F. X. (1970). *Science,* **169,** 679
193. Evans, N. J., II, Townes, C. H., Weaver, H. F. and Williams, D. R. W. (1970). *Science,* **169,** 680
194. Johnson, D. R., Powell, F. X. and Kirchhoff, W. H. (1971). *J. Mol. Spectrosc.,* **39,** 136
195. Greenberg, J. M. (1968). *Stars and Stellar Systems, Vol. VII: Nebulae and Interstellar Matter,* 221 (Middlehurst, B. M. and Aller, L. H., editors). (Chicago: University of Chicago)
196. Herzberg, G. (1970). *Symp. Int. Molecules, 13th Meeting Amer. Assoc. Advan. Science,* Chicago
197. Neumann, R. M. (1970). *Astrophys. J.,* **161,** 779
198. Kotsev, I. N., Georgiev, G. I. and Kontorovich, V. M. (1969). *Sov. Phys.-Astron.,* **12,** 1030
199. Paschoff, J. M., Gottlieb, C. A., Snyder, L. E., Buhl, D., Palmer, P., Zuckerman, B. and Dickinson, D. F. (1970). *B.A.A.S.,* **2,** 213
200. Turner, B. E., Heiles, C. E. and Scharlemann, E. (1970). *Astrophys. Lett.,* **5,** 197
201. Meeks, M. L., Gordon, M. A. and Litvak, M. M. (1969). *Science,* **163,** 173
202. Heiles, C. E. and Turner, B. E. (1971). *Astrophys. Lett.,* **8,** 89
203. Flygare, W. H. (1971)., private communication
204. Howard, W. E., III, and Hvatum, H. (1969). *Astrophys. J. (Lett.),* **157,** L161
205. Hvatum, H. and Howard, W. E., III (1970). *Astrophys. J. (Lett.),* **162,** L167
206. (1968). *Microwave Spectral Tables,* V. (Washington, D.C.: National Bureau of Standards, U.S. Government Printing Office)
207. Kukolich, S. G. and Ruben, D. J. (1971). *J. Molec. Spectrosc.,* **38,** 130
208. Tucker, K. D., Tomasevich, G. R. and Thaddeus, P. (1971). *Astrophys. J.,* **169,** 429
209. Johnson, D. R. and Lovas, F. (1971). *Astrophys. J.,* **169,** 617
210. de Zafra, R. L. (1971). *Astrophys. J.,* **170,** 165
211. Johnson, D. R., Kirchhoff, W. H. and Lovas, F. J. (1971)., to be published
212. Baird, K. M. and Bredohl, H. (1971). *Astrophys. J. (Lett.),* **169,** L83
213. Goldsmith, P. F. (1971)., to be published
214. Solomon, P. M. and Klemperer, W. (1971)., to be published
215. Julienne, P. S., Krauss, M. and Donn, B. (1971). *Astrophys. J.,* **170,** 65
216. Breuer, H. D. (1971). *Proc. Conf. on Int. Molecules.* (Charlottesville, Va., U.S.A.)
217. Hubbard, J. S., Hardy, J. P. and Horowitz, N. H. (1971). *Proc. Nat. Acad. Sci.,* **68,** 574
218. Khare, B. N. and Sagan, C. (1971). *Proc. Conf. on Int. Molecules.* (Charlottesville, Va., U.S.A.)
219. Greenberg, J. M. (1970). *Symp. Int. Molecules, 13th Meeting,* Amer. Assoc. Advan. Sci., Chicago
220. Brecher, A. and Arrhenius, G. (1971). *Nature, Physical Science,* **230,** 107
221. McNally, D. (1970). *Highlights in Astronomy.* (14th Inter Astrom. Union Meeting, Brighton, England)
222. Herbig, G. (1970)., private communication cited in ref. 221)
223. Herzberg, G. (1968). *Proc. XIII Coll. Spec. Int.,* 3. (London: Adam Hilger)
224. Johnson, F. M. (1971). *Proc. N.Y. Acad. Sci.,* to be published
225. Palmer, P., Snyder, L. E., Zuckerman, B., Snider, D. and Buhl. D. (1972)., to be published
226. Radford, H. E. (1972)., to be published
227. Dickinson, D. F. (1972). Paper 12.05.07, 137th meeting, Amer. Astron. Soc., Seattle, Washington

7
Millimetre Wave Spectroscopy

G. WINNEWISSER
University of British Columbia

and

M. WINNEWISSER and B. P. WINNEWISSER
Mississippi State University

7.1   INTRODUCTION

The development of microwave (MW) gas phase absorption spectroscopy in the millimetre wave (MMW) region has reached a state of maturity such that it seems at this time justified to review this field separately from either microwave spectroscopy in the centimetre wave region or the rapidly developing field of far infrared spectroscopy.

The term microwave refers to that portion of the electromagnetic spectrum which extends from *c.* 30 cm to 0.3 mm, thus covering a frequency range from *c.* 1000 MHz to 1000 GHz. It encompasses the gap between the radio-frequency region at longer wavelengths and the infrared region at shorter wavelengths. Microwaves may be classified according to their wavelength as centimetre waves (30–1 cm), millimetre waves (10–1 mm) or sub-millimetre waves (shorter than 1 mm).

Spectroscopy in the centimetre wave region is a well-established research field with a well-documented history of continuous growth beginning in earnest in 1946, involving many laboratories[1-3]. The extension of MW spectral measurements into the MMW region, on the other hand, was mainly accomplished by Gordy and co-workers at the Duke microwave laboratories. The early period of this development of MMW spectroscopy has been reviewed by Gordy[4,5], and a more thorough treatment may be found by Gordy and Cook in the book on *Chemical Applications of Spectroscopy*[6].

In 1954 Burrus and Gordy[7] achieved the link between microwave and infrared spectroscopy by making MW spectral measurements in the sub-MMW region at 0.78 mm overlapping with infrared grating measurements made by Genzel and Eckhardt[8]. In 1964 Jones and Gordy[9] succeeded in extending the operational frequency range to 691 472.70 MHz by measuring the $J = 6 \leftarrow 5$ transition of $^{12}C^{16}O$, and in 1970 Helminger, De Lucia and Gordy[10] extended these measurements considerably further employing a photoconducting detector operated at 1.6 K. So far the highest frequency measured with microwave techniques is the $J = 67 \leftarrow 66$ transition of $^{16}O^{12}C^{32}S$ at 813 353.706 MHz[10]. The MMW region remains, however, the least explored region of the electromagnetic spectrum.

The information gained from MW and MMW measurements include molecular structure parameters, electric and magnetic dipole moments, quadrupole coupling tensors, spin–rotation coupling tensors, barriers to internal rotation, and centrifugal stretching parameters which are directly related to the molecular force field. Measurements of the rotational spectrum of molecules in the MMW region up to high J quantum numbers reveal the full impact of centrifugal distortion and rotation–vibration interaction. Although most polar molecules show pure rotational transitions in the MW region, light molecules such as HCl, HCN, H_2O, H_2S, H_2O_2, H_2S_2, O_2, O_3 and OH have few if any transitions below 60 GHz. In addition, molecules which can be studied in the centimetre region have strong absorption lines in the MMW region, since the total absorption of electromagnetic radiation is proportional to the square of the absorption frequency. This frequency factor makes the MMW region suitable to measure the weak absorption of molecules in excited vibrational states or of isotopic species. Furthermore, certain types of rotation–vibration transitions in heavy molecules fall in this spectral region[11].

Both rotational spectra and low-lying vibration–rotation transitions can also be measured with far-infrared techniques[12] as low as 10 cm^{-1} or 300 GHz. The distinction between far-infrared and MMW spectroscopy does not lie any longer in the frequency range covered, but in the experimental techniques used. The difference has to be seen in the use of monochromatic sources, and thus in the high resolution and accurate frequency determination of MMW spectroscopy, versus the broad banded sources and dispersive elements employed currently in far-infrared spectroscopy.

Common to both fields is the possibility of focusing and directing the radiation by the use of optical or semi-optical techniques[4,13]. For example, conditions for observing the Lamb-dip in the Doppler profile of a MMW absorption line[14,15] may be achieved utilising these techniques with a simple free-space cell[15]. The now widely used free-space cells lend a remarkable experimental flexibility to a MMW spectrometer. In particular, unstable or corrosive species are amenable to spectral measurements with such cells.

Microwave spectroscopy has been reviewed recently by Morino and Saito[16] and Rudolph[17]. In this Review, we shall concentrate on experimental techniques for measurement in the region from c. 100 GHz (3 mm) to 1000 GHz (0.3 mm) and the results which have been obtained by absorption spectroscopy in this frequency region. However, much of the significant work has been incorporated by Gordy and Cook into a book[6], so that emphasis here will be given to the developments of the last 3 years. Because of the limitations put upon such a review, the policy has been adopted to list only the major publications relating to each topic, and in some cases this may mean the only publication. In order to supplement this Review we would like to refer to several general reviews of MW spectroscopy[18–23] and some books[24–26], in addition to those already mentioned. Attention should also be directed to the forthcoming supplement volume in the Landoldt–Börnstein series[27, 28], which will contain an extensive tabulation of molecular constants derived from MW spectroscopy, gas phase electron spin resonance, molecular beam techniques and related fields, covering the literature until the middle of 1971. Further useful listings are the five volumes of measured line frequencies tabulated by the National Bureau of Standards[29] and the MW gas spectroscopy bibliographies compiled by Guarnieri and Favero[30] and Starck[31]. A far-infrared bibliography compiled by Palik[32] is also available.

Following a section covering experimental techniques and developments for measurements in the MMW region, the results of MMW spectroscopy will be discussed for successive categories of molecules each having common spectral features. Special attention has been directed to the spectroscopy of unstable closed-shell and open-shell molecules, since such ·molecular species are of importance in astrophysics and astrochemistry. Theoretical developments relevant to the interpretation of MMW spectral data will be included as the appropriate molecules illustrating the theory are discussed. Finally, a few topics not limited to any one type of molecule will be touched upon.

7.2 EXPERIMENTAL TECHNIQUES USED IN MILLIMETRE WAVE SPECTROSCOPY

The measurements listed in Table 7.1 are representative of the range of molecules and frequencies which have been investigated in the last few years by MMW gas phase spectroscopy. Most of the frequency measurements listed, and indeed most of those carried out in the MMW region have been made with video-type spectrometers of the type originally developed by Gordy and co-workers[4, 33]. The characteristics of such spectrometers were reviewed and analysed in detail by Baker[34]. In the last few years two major improvements have been made for this type of spectrometer:

(a) Extension of the frequency range and improvement in spectrometer sensitivity by use of a photoconducting detector[10].

(b) Application of a signal-averaging computer to increase the sensitivity

Table 7.1 Some representative ground-state rotational transitions measured by millimetre wave techniques

Molecule	Transition $J' \leftarrow J''$	Frequency*/MHz	Reference	Molecule	Transition $J' \leftarrow J''$	Frequency*/MHz	Reference
Diatomic molecules				*Diatomic molecules*			
CO	$7 \leftarrow 6$	806 651.719	10	DF	$1 \leftarrow 0$	651 099.393	10
$H^{35}Cl$	$1 \leftarrow 0$ $(F' = \frac{3}{2} \leftarrow F'' = \frac{1}{2})$	625 932.007	108	$T^{35}Cl$	$1 \leftarrow 0$ $(F' = \frac{1}{2} \leftarrow F'' = \frac{3}{2})$	222 160.50	109
$H^{127}I$	$1 \leftarrow 0$ $(F' = \frac{3}{2} \leftarrow F'' = \frac{1}{2})$	770 540.543	108	$D^{79}Br$	$3 \leftarrow 2$ $(F' = \frac{9}{2} \leftarrow F'' = \frac{7}{2})$	763 853.201	108
$^6Li^{19}F$	$6 \leftarrow 5$	538 072.65	118	AlF	$14 \leftarrow 13$	461 329.74	124
$Na^{35}Cl$	$23 \leftarrow 24$	312 109.88	119	$Cs^{127}I$	$139 \leftarrow 138$	195 432.52	74
Linear molecules				*Linear molecules*			
$^{16}O^{12}C^{32}S$	$67 \leftarrow 66$	813 353.706	10	$^{16}O^{12}C^{80}Se$	$39 \leftarrow 38$	313 217.57	347
HCN	$3 \leftarrow 2$	265 886.18	347	HCP	$9 \leftarrow 8$	359 506.14	327
HCNO	$18 \leftarrow 17$	412 786.066	338	HCCD	$4 \leftarrow 3$	237 793.36	348
Symmetric-top molecules				*Symmetric-top molecules*			
PH_3	$2 \leftarrow 1, K = 0$	533 794.32	356	$^{75}AsH_3$	$2 \leftarrow 1, K = 0$	449 789.56	367
$^{121}SbH_3$	$2 \leftarrow 1, K = 0$	352 095.67	367	NF_3	$13 \leftarrow 12, K = 0$	277 580.38	370
CH_3F	$5 \leftarrow 4, K = 0$	255 332.20	391	CHF_3	$15 \leftarrow 14, K = 0$	310 313.76	391
CH_3Br	$10 \leftarrow 9, K = 0$	191 324.10	397	$(CH_2O)_3$	$19 \leftarrow 18, K = 0$	200 347.11	392
Asymmetric-top molecules				*Asymmetric-top molecules*			
H_2O	$2_{02} \leftarrow 2_{11}$	752 033.23	478	D_2S	$5_{24} \leftarrow 5_{15}$	672 337.45	463
HOF	$5_{05} \leftarrow 4_{04}$	261 264.22	447	$HO^{35}Cl$	$6_{06} \leftarrow 5_{05}$	179 025.31	448
NOF	$10_{0,10} \leftarrow 9_{0,9}$	220 912.37	433	HCOOH	$13_{0,13} \leftarrow 12_{0,12}$	282 939.8	436
SO_2	$2_{20} \leftarrow 1_{11}$	192 651.02	424	S_2O	$15_{0,15} \leftarrow 14_{0,14}$	140 456.20	435
HN_3	$9_{09} \leftarrow 8_{08}$	214 316.89	450	H_2C_2O	$9_{09} \leftarrow 8_{08}$	181 826.35	444
H_2S_2	$41_{2,39} \leftarrow 41_{1,41}$	420 321.48	455	H_2O_2	$30_{0,30} \leftarrow 30_{1,30}$	278 691.28	476

*Hypothetical centre frequency given where quadrupole splitting is present and not stated otherwise.

and accuracy of measurement without affecting the basic functions and
characteristics of the spectrometer[15]. Both of these techniques are suitable
for operation in video mode or with source modulation and phase-sensitive
detection.

Some special techniques have been developed, among which are the
application of quasi-optical techniques[35, 36] and the observation of the
Lamb-dip in the MW and MMW regions. Another is the application of
molecular-beam methods to MMW spectroscopy, which has led to the
construction and successful operation of molecular-beam absorption spectrometers, electric resonance spectrometers and molecular-beam maser
spectrometers. Sub-MMW lasers will be briefly mentioned. Further specialised
applications of sub-MM waves are covered in the proceedings of a meeting
at the Brooklyn Polytechnic Institute[37].

7.2.1 Video-type millimetre wave spectrometer

The basic characteristics of the video type spectrometer, a current version
of which is shown in the block diagram in Figure 7.1, are a wide frequency
range and rapid scanning of the chosen spectral region, which provide a
fast search capability. The former property is due to the broad-banded
characteristics of the frequency multiplier, absorption cell and detection
system, while the rapid scanning is determined by the electronic tuning
capability of the source oscillator. Reflex klystrons in the fundamental
region from 20 to 75 GHz and with output power between 100 mW and
500 mW are usually chosen. As can be seen from Figure 7.1, the fundamental
microwave power is fed into a frequency multiplier housing a point contact
rectifying diode for harmonic generation[38]. Despite the fact that the MMW
region may now be reached by tunable electronic oscillators[39] and by
far-infrared lasers[40, 41], used for laser electron paramagnetic resonance
spectroscopy[42], harmonic generation of the fundamental frequency of a
klystron or backward wave oscillator is still the most commonly used
technique today for producing coherent radiation at MM and sub-MM
wavelengths. Currently available MMW electronic oscillators are narrow
banded and quite expensive. The classic design of the harmonic generator
used in MMW spectroscopy was introduced by King and Gordy[38] in 1954
and has remained essentially unchanged since that time, though variations
have appeared[43-46]. The theory of harmonic generation and its efficiency as
a function of the harmonic number have been reviewed[34] and semiconductor
materials for harmonic generation have also been reported[45-49].

As is shown in Figure 7.1 and in most applications of MM or sub-MMW
for spectroscopic purposes, the MMW radiation is collimated with the use
of horns and dielectric lenses through an oversized absorption cell and
focused onto the detector. Such free-space absorption cells are particularly
suited for the study of reactive molecular species, and for substances with
high temperatures of vaporisation[4, 23, 50]. Parallel Stark plates can be easily
inserted for precision dipole moment measurements[51, 52]. The transmission
losses of the cells are of the order of 2dB over a frequency range up to 400 GHz
for a 2 m cell.

The usual detector in the MMW region, a point contact diode, bears close resemblance in its basic design to the multiplier. The sensitivity of semi-conductor diode detectors at frequencies up to the sub-MMW region has been investigated by Bauer *et al.*[49]. In the sub-MMW region the photo-conductive indium antimonide detector, operated at 1.6 K, was found to be

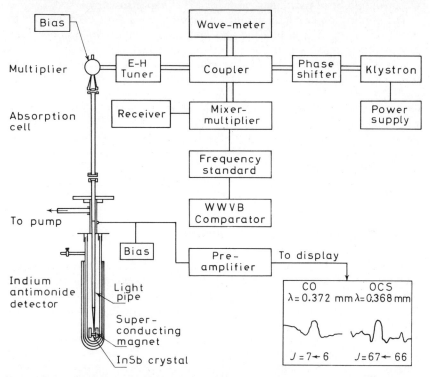

Figure 7.1 Block diagram of a MMW spectrometer.
(From Helminger, De Lucia and Gordy[10, 461], reproduced by courtesy of Academic Press and The American Institute of Physics)

c. ten times more sensitive[10] than the conventional semiconductor diode. The InSb detectors are commercially available[53, 54]. The photoconductive effect in the sub-MMW region and the performance of such devices are discussed by Putley[55]. Other information on sub-MMW and far-infrared detectors can be found in a number of places[37, 56-60].

7.2.2 Comments on spectrometer sensitivity

The sensitivity of the instrument is determined by the overall noise level. An increase in spectrometer sensitivity depends finally on reduction of the product of receiver bandwidth times noise figure, $B \cdot N$, which is achieved by employing the Stark-effect modulation method or the source-modulation method[33]. This latter method is of great importance to MMW spectroscopy where the Stark-effect modulation method becomes less advantageous due to experimental difficulties. The limiting sensitivity of a Stark-effect spectrometer is $c.$ 10^{-10} cm^{-1} for frequencies up to 150 GHz. The sensitivity is further improved by employing a phase-stabilised microwave source which allows a very narrow band-width detection amplifier. Recently, Gilbert[61] and Törring[62] introduced the promising method of saturation-effect modulation, which should be of particular value for MMW spectroscopy.

The application of a data acquisition system using a digital computer for signal averaging, digital filtering and preliminary data reduction can bring considerable improvement in sensitivity. This has been shown for the centimetre wave region by Gwinn and co-workers[63] and for the MMW region by Winnewisser[15]. When signal averaging is applied simultaneously to absorption line and frequency marker, the accuracy of the frequency measurements can be enhanced by a factor of 5–10 over the accuracy obtained with conventional video-techniques which is $c.$ ± 150 kHz at 500 GHz. The signal-averaging technique is particularly useful in its application to video-type spectroscopy. It allows an improvement in the signal-to-noise ratio by a factor of 100 with the accumulation of 10 000 sweeps, which means for a video-type spectrometer without modulation a limiting sensitivity of $c.$ 10^{-8} cm^{-1}.

7.2.3 Millimetre wave interferometers

As we have seen from the discussion of the video-type spectrometer, MMW spectroscopy is performed mainly in oversized waveguide absorption cells or free-space cells. For special applications, such as MMW masers and Lamb-dip investigations, the resonant cavity with high quality factor Q is most suitable and frequently used.

In the centimetre region the low-loss cylindrical cavity mode TE_{01} is usually employed for maximum quality factor Q. At MMW or sub-MMW lengths, however, the wall losses of this mode become appreciable, and it becomes difficult to couple MMW power into the circular waveguide TE_{01} mode from the rectangular waveguide TE_{10} mode over a wide range of frequencies. These problems of loss, coupling efficiency, and tuning range can be greatly alleviated by use of a Fabry–Perot interferometer cavity system[36, 64]. In the centimetre region the early investigations of the basic designs and properties of such cavity interferometers began with the work of Culshaw[65, 66] in 1953. Lichtenstein, Gallagher and Cupp in 1963[67] used the semi-confocal Fabry–Perot interferometer in order to achieve high signal-to-noise ratios in the MMW region where low power is available. Further design considerations are given by Valkenburg and Derr[68].

The advantage of semi-confocal cavities over parallel-plate cavities is that they are considerably less critical to align and that the diffraction losses are substantially reduced. A further advantage is their small physical size in the millimetre and sub-millimetre wave region which permits them, for example, to be placed between the poles of an electromagnet to study Zeeman effects[69]. The open structure of the cavity lends itself to free-radical spectroscopy. The main application, however, is found in MMW and sub-MMW maser and laser systems.

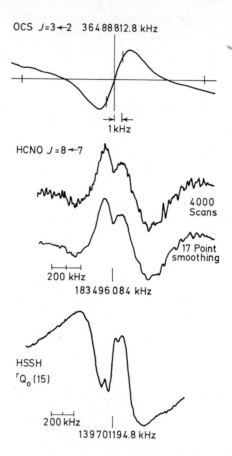

OCS $J=3\leftarrow2$ 364 88 812.8 kHz

1 kHz

HCNO $J=8\leftarrow7$

4000
Scans

17 Point
smoothing

200 kHz

183 496 084 kHz

HSSH
$^rQ_o(15)$

200 kHz

139 701194.8 kHz

Figure 7.2 Examples of the high resolution achieved with the saturation-dip technique
(From references 14, 15 and 73, reproduced by courtesy of Springer, Verlag and National Research Council, Canada)

7.2.4   High-resolution spectroscopy

The resolving power of the video-type spectrometer is essentially limited
by the Doppler width of the particular absorption line to be investigated and
is *c.* 600 000, provided collision broadening is negligible. A substantial
increase in resolution can be achieved by one of two techniques, saturation-
dip spectroscopy or molecular beam spectroscopy.

7.2.4.1   *Saturation-dip spectroscopy*

The saturation dip, predicted by Lamb[70] and observed first in emission in an
He–Ne optical laser[71], was subsequently observed in laser absorption
experiments[72]. Since the frequency source in MW spectroscopy is monochromatic it is logical to apply this technique in the MW and MMW region.
Saturation-dip conditions, as conveniently summarised by Costain[14], are
especially easy to fulfil in the MMW region. In Figure 7.2 the saturation
dip is shown for three different molecules at three widely different frequencies
measured in three different laboratories. The saturation dip with only
1.5 kHz half-width at 36 GHz was recorded by Costain[14] with an elaborate
spectrometer and a stabilised klystron radiation source. The saturation dip
in the Doppler profile of the $J = 8\leftarrow7$ transition in $H^{12}C^{14}N^{16}O$ was obtained with a MMW video-type spectrometer employing a free-running
klystron by reflecting the radiation beam back parallel to the incident beam
through a free-space absorption cell (pressure $\approx 10^{-4}$ torr). The dip width is
c. 35 kHz. The recording was obtained with the aid of a signal-averaging
system[15]. Winton and Winnewisser[73] using a Fabry–Perot cavity spectrometer under saturation conditions were able to observe the torsional
splitting of HSSH in its ground vibrational state. This recording was obtained
employing source modulation in order to enhance the sensitivity.

From these examples we can assume that saturation-dip spectroscopy
will have considerable future application in the MMW range. However,
it appears that the method cannot compete fully with the resolution and
sensitivity achievable with molecular-beam absorption spectrometers,
molecular-beam electric resonance spectrometers, or molecular-beam maser
spectrometers. The advantage of the saturation-dip technique in the MMW
region lies in the simple experimental apparatus and thus in the speed with
which these measurements can be made.

7.2.4.2   *Millimetre wave molecular-beam absorption spectrometers*

The extension of molecular beam techniques to the MMW region was not
made until 1962 due to the difficulties of obtaining sufficient stable MMW
power and of providing adequate molecular beams[74]. A considerable improvement in resolution was obtained by Huiszoon and Dymanus[75] in 1966
through the introduction of a high-resolution millimetre wave molecular-
beam spectrometer, which employs an absorption cell containing two flat
parallel Stark plates. The high resolution of the system is obtained by

measuring the absorption due to a well-collimated molecular beam passing between the plates perpendicular to the direction of propagation of the microwave power. The beam is obtained from a multichannel Zacharias-type oven. A superheterodyne detection scheme has been used because of the very low signal power levels required in order to avoid saturation. In addition, square-wave Stark modulation and phase-sensitive detection have been used for maximum sensitivity. For the recorded $1_{01}-1_{10}$ rotational transition for H_2 ^{32}S reproduced in the upper part of Figure 7.3, the line-width was reported as 9 kHz. The broadening due to the finite transit time of the molecules through the MMW field is $c.$ 2.1 kHz, the Doppler broadening 2.7 kHz and the broadening caused by signal frequency instability 2.8 kHz. This spectrometer reduced the line-width by a factor of 50 over that obtained with a conventional MMW spectrometer. The system was found suitable for precision measurements of the nuclear hyperfine splitting and the electric dipole moments of HI, HBr and DBr[76-78].

7.2.4.3 Millimetre wave electric resonance spectroscopy

In the longer wavelength, radio-frequency and centimetre wave regions, molecular-beam electric and magnetic resonance techniques have been employed very productively for many years[79, 80].

The first electric resonance experiment in the MMW region was performed on $^6Li^{19}F$ at $c.$ 90 GHz by Wharton et al.[81] in 1963. However, this experiment left an important question open: Is the spectral purity of the radiation source (klystron and frequency multiplier output at MMW) sufficient to observe Ramsey patterns[82] in the MMW region, and would they provide further resolution of hyperfine structures over that obtained with single resonator spectrometers? Special effort was directed by Gallagher and co-workers[83] towards answering these questions, and in 1966 the design of a molecular-beam electric resonance spectrometer operating in the MMW region was presented[84]. The first success in recording the Ramsey pattern in the MMW region was reported on the $1_{01}-1_{10}$ rotational transition of H_2S, whose magnetic hyperfine structure was completely resolved with a resolution expressed in spectral line Q as 5×10^8, a value which exceeds that achieved in caesium frequency standards. The hyperfine components have been measured with an accuracy of 2 in 10^{10}, a figure not reached by any other MMW spectroscopy method. This may be seen in Figure 7.3, where the hyperfine structure with six Ramsey interference patterns is compared with the recording of the same transition as observed with a molecular-beam absorption spectrometer.

7.2.4.4 Millimetre wave molecular-beam masers

So far the number of operational millimetre and sub-millimetre molecular-beam masers is limited. Marcuse[85-87] in 1961 achieved beam-maser operation on the 88.6 GHz transition of HCN and Krupnov and Skvortsov[88-90] in 1964 on the 72.8 GHz transition of CH_2O. Since then, De Lucia and Gordy[91,92]

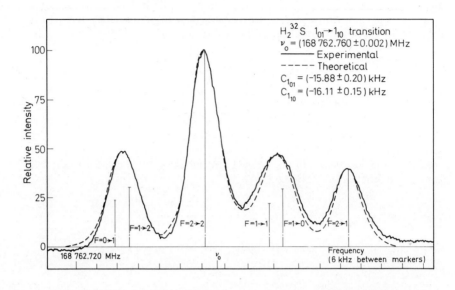

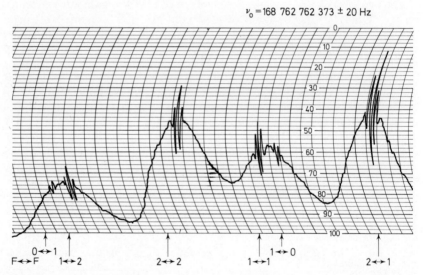

Figure 7.3 High-resolution recording of the $1_{10}-1_{01}$ transition of H_2S by molecular beam absorption spectroscopy (upper recording from Huiszoon and Dymanus[75], reproduced by courtesy of North-Holland) and by electric resonance spectroscopy (lower recording from Cupp *et al.*[84]). (Reproduced by courtesy of The American Institute of Physics)

and De Lucia, Cederberg and Gordy[93] obtained maser action with various transitions of DCN, HCN, D_2O and ND_2H between 57 and 317 GHz.

Constructing a spectrometer similar in principal to the first successful design of a maser by Gordon, Zeiger and Townes[94, 95], De Lucia and Gordy[91, 92] pass a collimated molecular beam through a state selector where the population inversion necessary for maser action is obtained by the selective deflection of molecules according to their particular energy states. This is accomplished by the action of an inhomogeneous electric field acting on the dipole moments

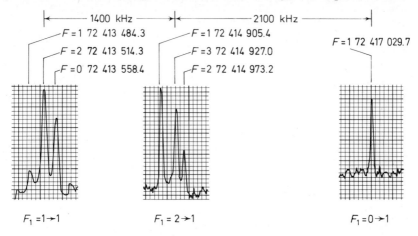

Figure 7.4 Maser emission spectrum of the $J = 1 \rightarrow 0$ transition of DCN showing the deuterium splitting of the nitrogen quadrupole lines in kHz
(From De Lucia and Gordy[91], reproduced by courtesy of The American Institute of Physics)

of the molecules. Figure 7.4 shows the maser output for the $J = 1 \rightarrow 0$ transition of $D^{12}C^{14}N$. The effect of the deuterium coupling is to split into triplets the levels resulting from the $F_1 = 1$ and $F_1 = 2$ states of $J = 1$ arising from the ^{14}N quadrupole nucleus. The line resulting from the $F_1 = 0$ level is not split. For these measurements the cavity Q was c. 300 000 and the experimental line-width of the order of 10 kHz at a frequency of 72 GHz. The Doppler width at this frequency is c. 200 kHz, which in the usual absorption spectrometer would prevent the evaluation of the quadrupole and magnetic fine-structure interaction.

7.2.5 Millimetre and sub-millimetre wave lasers

Although laser devices generally provide higher power than do beam masers, they do not have the stability of frequency, the resolving power, and the low-noise figure of the beam maser which circumvents both Doppler and collision broadening. More detailed information on lasers may be found in the book by Röss[96], who also provides a bibliography complete until 1967. Button and Lax[97] discuss the molecular gas lasers in sub-MMW spectroscopy and give a list of the strongest laser emission lines available from the most common molecular gas lasers with ICN, HCN, DCN, H_2O,

D_2O, SO_2 and CO_2. A complete tabulation of these transitions has been compiled by Button[98].

The HCN laser developed by Gebbie et al.[99–101] in 1964 is still by far the most powerful sub-MMW radiation source. The principal HCN laser emission is centred around 890.760 GHz and is in near coincidence with the tentatively assigned $46_{35, 11}$–$47_{35, 12}$ rotational transition of difluoroethylene which has been measured by Duxbury et al.[102]. At very low pressure this gas was introduced into a cell located inside the laser cavity and the laser cavity was tuned through the gain profile, which showed a dip corresponding to the Doppler profile of the $C_2F_2H_2$ line.

Automatic frequency control and phase locking of the HCN laser line to a multiplied frequency standard to provide a high-power secondary frequency standard in the sub-MMW region has been accomplished by Corcoran, Gallagher and Cupp[103–104]. They achieved a long-term stability of 10^{12}. We might note that a sub-MMW laser locked to the saturation dip of an absorption line would provide an absolute frequency and wavelength standard of unrivalled accuracy. Another exciting experiment should be mentioned: sub-MMW laser action in optically pumped CH_3F. Chang and Bridges[106] observed laser action in six rotational transitions in methyl fluoride gas which is pumped between rotational levels of the ground state and the first excited state of v_3, the C—F stretching vibration at 1048.6 cm^{-1}. Rotational transitions both in the ground state and in the v_3 state could be observed between 550 and 665 GHz in emission.

All laser transitions in the sub-MMW region can in principle be used for laser paramagnetic resonance experiments[42], which recently made possible the first MMW observation of the unstable CH radical[107]. The possible combinations of MMW, sub-MMW and quasi-optical techniques with infrared and laser techniques together with the exploitation of the Zeeman and Stark effects are far from exhausted. The development of even more efficient detectors and radiation sources in the MMW and sub-MMW region will steadily increase the number of useful spectroscopic techniques and lead to further opportunities to investigate molecular systems.

7.3 RESULTS OF MILLIMETRE WAVE SPECTROSCOPY

7.3.1 Closed-shell diatomic molecules

The extent of the data now available for closed-shell diatomic molecules can be conveyed by noting that with the exception of HF, all of the hydrogen halides, including DF[10, 108], have been measured. All of the alkali halides, from LiF to CsI, have also been studied. All stable isotopes of these molecules have been included, and even the tritium halides have been measured. The heavier alkali halides have some absorptions in the MW region, but even these molecules have been measured at frequencies up to at least 180 GHz. The results obtained for these and other diatomic molecules are summarised in Tables 7.2, 7.3 and 7.4.

The analysis of the data where measurements of at least two or more excited vibrational states are available is usually carried out by applying

Table 7.2 Spectral constants and equilibrium bond lengths of the hydrogen halides (from reference 108)

	B_0/MHz	D_0/MHz	eQq_{hal}/MHz	C_{hal}/MHz	r_e/(Å)[†††]
D^{19}F	325 584.98 ±0.300	17.64*	—	—	0.916 914
H^{35}Cl	312 989.297 ±0.020	15.836†	-67.800 ±0.095	0.068 ±0.010	1.274 599 1
H^{37}Cl	312 519.121 ±0.020	15.788†	-53.436 ±0.095	0.056 ±0.010	1.274 599 0
D^{35}Cl	161 656.238 ±0.014	4.196 ±0.003	-67.417 ±0.098	0.026 ±0.011	1.274 599 8
D^{37}Cl	161 183.122 ±0.016	4.162 ±0.003	-53.073 ±0.113	0.024 ±0.013	1.274 599 7
T^{35}Cl	111 075.84‡	1.977††	-67.0 ±0.6§	0.018 ±0.008¶	1.274 598 5
T^{37}Cl	110 601.62‡	1.960††	-53.0 ±0.6§	0.016 ±0.008¶	1.274 598 8
H^{79}Br	250 358.51 ±0.15‖	10.44 ±0.06***	535.4 ±1.4††	0.29 ±0.20††	1.414 469 1
H^{81}Br	250 280.58 ±0.15‖	10.44 ±0.06***	447.9 ±1.4††	0.31 ±0.20††	1.414 470 5
D^{79}Br	127 357.639 ±0.012	2.6529 ±0.0014	530.648 ±0.074	0.148 ±0.009	1.414 469 8
D^{81}Br	127 279.757 ±0.017	2.6479 ±0.0020	443.363 ±0.105	0.158 ±0.012	1.414 469 8
T^{79}Br	86 251.993‡	1.234**	530 ±2§	0.100 ±0.005‡‡	1.414 470 5
T^{81}Br	86 174.078‡	1.232**	443 ±2§	0.106 ±0.006‡‡	1.414 469 1
H^{127}I	192 657.577 ±0.019	6.203 ±0.003	-1828.418 ±0.200	0.349 ±0.010	1.609 018
D^{127}I	97 537.092 ±0.009	1.578 ±0.001	-1823.374 ±0.105	0.165 ±0.006	1.609 018
T^{127}I	65 752.305§§	0.7150¶¶	-1823 +3‖‖	0.115 ±0.005***	1.609 018

* Calculated from data in reference 138.
† Calculated from data in reference 139.
‡ Calculated from the line-frequency measurements in reference 109 and the D_0 listed here.
§ From reference 109. A sign error has been corrected in the TCl values.
¶ Calculated from the microwave values of C_{hal} for DCl.
‖ Calculated from the line-frequency measurements in reference 140 and the D_0 listed here.
** From reference 141.
†† From reference 140.
‡‡ Calculated from the microwave values of C_{hal} for DBr.
§§ Calculated from the line-frequency measurements in reference 110 and the D_0 listed here.
¶¶ Calculated from H^{127}I and D^{127}I microwave data.
‖‖ From reference 110.
*** Calculated from the microwave values of C_{hal} for HI and DI.
††† Calculated by use of $B(\text{MHz}) = (5.053\,76 \times 10^5)/I_b$ (amu Å^2). See reference 6.

Dunham's theory[6, 111] of diatomic rovibrators. This formulation results in a representation of the energy with the Dunham series, an expansion of the rovibrational energy in the vibrational and rotational quantum numbers v and J:

$$E_{v,J} = h \sum_{l,m} Y_{lm}(v+\tfrac{1}{2})^l J^m (J+1)^m \tag{7.1}$$

where l and m are integers. The Y values are known as Dunham coefficients and are obtained from the experimental data. MMW measurements, when sufficient values of v and J are represented, yield up to five Dunham coefficients Y_{01}, Y_{11}, Y_{21}, Y_{02} and Y_{12} which are directly related to the rotational and potential constants including the equilibrium rotational constant B_e. In some cases where insufficient MW data is available infrared data are used, which, although not as accurate, suffice to make possible the evaluation of $B_e = h(8\pi^2 \mu R_e^2)$. If the MMW and infrared data is sufficient to determine

Table 7.3 Spectral constants and equilibrium bond lengths of diatomic molecules

	B_0/MHz	D_0/kHz	eqQ/MHz	r_e/Å	Reference
Group III halides					
^{10}BF	48 022.63(8)	–	−9.5(8)		121
^{11}BF	45 185.77(6)	–	−4.5(4)	1.2625	121
Group IV–Group VI compounds					
$^{12}C^{16}O$	57 635.971(6)	183.59(14)	–	1.12823(5)	10, 116
$^{12}C^{32}S$	24 495.592(6)	42.85(17)	–	1.53492(7)	13

the value of B_e, then the precision with which internuclear distances are given is limited by the accuracy of Planck's constant to six significant figures rather than the experimental accuracy of the measurement. The high experimental accuracy is exploited by applying correction terms in the determination of B_e for some of the light molecules such as CO [112] and LiD [113].

These corrections were introduced by Van Vleck[114] and were needed when it became clear that the experimental accuracy of MW data did not justify the complete neglect of terms dropped from the Hamiltonian in making the Born–Oppenheimer approximation. The detectable contribution of these terms, which describe the interaction of electronic motion with the overall rotation, is known as the breakdown of the Born–Oppenheimer approximation[112, 115]. Bunker[116, 117] in 1970 derived the necessary corrections which must be included in a determination of R_e. As shown by De Lucia et al.[108] in the case of HCl, the functional dependence of the corrections is such that the earlier, less rigorous, approach to the problem leads to the same value of R_e.

7.3.1.1 Hydrogen halides

The present state of the knowledge of precise rotational parameters and bond lengths of the hydrogen halides is summarised in the 1971 paper by De Lucia, Helminger and Gordy[108], which contains the latest measurements, including

Table 7.4 Molecular constants and equilibrium bond lengths of diatomic molecules from a Dunham analysis*

Molecule	$Y_{01} (\approx B_e)$/MHz	α_e/MHz	D/kHz	ω_e/cm^{-1}	$\omega_e x_e$/cm^{-1}	r_e/Å	Reference
Alkali halides							
^{7}LiF	40 329.808(14)	608.182(29)	352.36(17)	910.25(22)	8.104(42)	1.563 856 7	118
Li^{35}Cl	21 181.0004(4)	240.122(5)	102.19(2)	643.31(7)	4.501(19)	2.020 663 5	118
Li^{79}Br	16 650.318(6)	169.09(8)	64.71(25)	563.5(2.2)	3.88(2)	2.170 42(4)	74,142
LiI	13 286.262(7)	122.62(10)	43.41(25)	490(14)	3.1(1)	2.391 91(4)	74,142
^{23}NaF	13 097.951(7)	136.618(14)	34.797(45)	536.10(35)	3.83(14)	1.926 032	50
NaCl	6 537.367(6)	48.711(9)	9.354 0(50)	364.60(10)	1.755(30)	2.360 898(46)	119
NaBr	4 534.460 3(32)	28.209(4)	4.657(5)	298.49(17)	1.16(3)	2.502 01(4)	74,142
NaI	3 531.723 2(43)	19.420(5)	2.918(4)	259.20(16)	0.96(2)	2.711 43(4)	74,142
^{39}KF	8 392.313(5)	69.999(8)	14.493(16)	426.04(24)	2.43(9)	2.171 554	50
KCl	3 856.373(9)	23.681(13)	3.260(3)	279.80(10)	1.167(5)	2.666 772(54)	119
KBr	2 434.947(1)	12.136(1)	1.337 63(34)	219.170(29)	(0.758(5))	2.820 75(5)	74,143
KI	1 824.978 6(13)	8.027(2)	0.777 5(3)	186.53(4)	0.574	3.047 81(5)	74,142
^{85}RbF	6 315.548(3)	45.652(4)	8.046(4)	373.27(8)	1.80(3)	2.270 435	50
RbCl	2 627.393(4)	13.599(4)	1.483(1)	233.34(8)	0.856(8)	2.786 865(55)	119
RbBr	1 424.852 3(12)	5.576(1)	0.448 33(16)	169.46(3)	0.463(7)	2.944 71(5)	74,142
RbI	984.306 6(21)	3.281 56(17)	0.221 239(11)	138.511(35)	0.335(6)	3.176 84(5)	74,142
^{133}CsF	5 527.265(1)	35.247(2)	6.046(1)	352.56(4)	1.62(1)	2.345 462(45)	50
CsCl	2 161.246(2)	10.119(2)	0.979 1(4)	214.22(5)	0.740(4)	2.906 411(57)	119
CsBr	1 081.333 14(20)	3.720 52(40)	0.251 775(22)	149.503(7)	0.360 28(8)	3.072 21(5)	74,142
CsI	708.329 20(89)	2.046 38(57)	0.111 330(25)	119.195(13)	0.254 2(21)	3.315 15(6)	74,142
Group III halides							
^{27}AlF	16 562.930(6)	149.420(9)	31.37(2)	802.85(25)	4.86(6)	1.654 36(2)	124,125
Alkali hydrides							
^{7}LiD	126 905.36(4)	2 744 61(4)				1.594 90(2)	113

*For other isotopic species see references indicated.

more precise values for previously measured transitions. In spite of the thoroughness of these sub-MMW measurements, the data for these molecules is insufficient for a complete Dunham analysis. The incompleteness of the experimental data is due in part to the large B_0 values, which allow the measurement of the lowest rotational transitions only ($J = 1 \leftarrow 0$ for HCl; up to $J = 3 \leftarrow 2$ for DI and DBr) and in part to the large value of ω_e which causes an unfavourable Boltzmann distribution. In combination with the MMW measurements of B_0 the values of ω_e and $\omega_e x_e$ from infrared measurements yield a precise evaluation of B_e and R_e.

The $J = 1 \leftarrow 0$ transitions of HBr, DBr and HI have recently been reinvestigated with molecular-beam techniques by Van Dijk and Dymanus[76, 77]. The resolution inherent to the molecular-beam method made possible the determination of precise hyperfine coupling constants. They also reported accurate dipole moment measurements for these transitions[78].

7.3.1.2 Alkali halides

The recent publication of the measurement of the spectrum of LiF up to $J = 6 \leftarrow 5$ and of LiCl up to $J = 12 \leftarrow 11$ by Pearson and Gordy[118] marks the conclusion of a long series of measurements in the Duke microwave laboratory[50, 74, 119]. The large family of the various isotopes of the alkali halides spans the entire range of microwave frequencies and a formidable range of experimental problems. For the investigation of these molecules and others, high-temperature absorption cells have evolved in several laboratories, so that now substances with temperatures of vaporisation of nearly 1000 °C have been measured in the MMW region[113] and 1700 °C in the MW region[120] using beam-type apparatus in which the waveguide or free-space cell is maintained at a temperature considerably lower than that of the gas oven. At such temperatures the observation of vibrational satellites is greatly facilitated, so that transitions up to $v = 4$ states are common if the fundamental frequency is not too high. For a more detailed discussion of high-temperature microwave spectroscopy the reader is referred to the recent review by Lovas and Lide[23].

For all of the alkali halides, the use of the Dunham formulation was possible without drawing upon infrared data. In fact, the evaluation of ω_e and $\omega_e x_e$ from the Dunham coefficients obtained from the rotational spectrum provide more accurate values than the infrared data for such molecules as CsF, RbF, KF and LiF [50, 118].

7.3.1.3 Other molecules

The lightest of the Group III monohalides, BF, has recently been identified by Lovas and Johnson[121] who produced this very unstable species in the gas phase by an electric discharge in flowing BF_3. The nuclear quadrupole hyperfine structure of the $J = 1 \leftarrow 0$ transition was observed and analysed for both isotopic species yielding an $eqQ = -9.5 \pm 0.8$ MHz for the ^{10}B nucleus ($I = 3$) and $eqQ = -4.5 \pm 0.4$ MHz for the ^{11}B nucleus in BF ($I = \frac{3}{2}$). The dipole moment was found to be 0.5 ± 0.2 D, and its magnitude agrees with the dipole moment predicted by *ab initio* calculations[122, 123] which indicate the polarity to be $B^- F^+$.

One other Group III monohalide, AlF, has been measured in the MMW[124] and MW[125] region with high temperature cells. Other molecules of this family whose rotational spectra have been observed were measured only in the MW region[126].

Carbon monoxide, CO, the lightest of the Group IV–Group VI compounds, may be one of the best understood diatomic molecules. The rotational spectrum has been measured[10] up to $J = 7 \leftarrow 6$ and the vibrational spectrum, using laser emission[127, 128], up to $v = 37 \leftarrow 36$. The evaluation of the constants describing the molecular potential is very complete[128]. However, the new infrared and MMW data have not yet been subjected to a simultaneous fit.

The detection of the unstable carbon monosulphide[13], CS, in the MMW region initiated the use of free-space cells for the observation of transient species. The electric dipole moment of CS was determined in later measurements[51] to be $\mu = 1.966 \pm 0.005$ D for the $J = 1 \leftarrow 0$ transition of the ground state (value is corrected to conform to new value for OCS moment used as standard, see Section 7.4.1). This value is *c.* 20 times larger than that of CO ($\mu = 0.112 \pm 0.005$ D)[112, 129], which strongly favours the observation of the less abundant CS in interstellar space. The rotational g_J-value and the magnetic susceptibility of both CS [69, 130] and CO [112, 130] have been determined from the rotational Zeeman effect.

MMW emission from one transition of SiO has been detected in interstellar space[131]. This transition, $J = 3 \to 2$ at 130 246 MHz, was predicted from MW data[132, 133] and has not yet been measured in the laboratory. For this reason SiO is not included in Table 7.3.

Although the investigations by Hoeft and co-workers[120, 126, 134-136] of a large number of diatomic molecules, including many formed by combination of elements of Group IV with elements of Group VI, have been confined to the region 7–50 GHz, enough experimental data has been obtained in most cases to apply the Dunham formalism.

Since molecular-beam electric resonance provides complementary structural and spectroscopic data on molecules of interest to MMW spectroscopy, the study of $^{31}P^{14}N$ by this technique should be mentioned here[137]. PN is isoelectronic to CS with a dipole moment of 2.7465(1)D. The rotational absorption spectrum of PN has not yet been observed.

The first alkali hydride to be observed in the MMW region is LiD. Pearson and Gordy[113] were able to measure transitions in the ground state and one excited vibrational state of both ^{6}LiD and ^{7}LiD in spite of the rapid dissociation of the molecule. Using isotopic relationships, constants for LiH could be derived from those obtained for LiD from the MMW data. The small number of electrons makes these molecules good subjects for testing theoretical *ab initio* calculations of molecular parameters.

7.3.2 Open-shell diatomic molecules

As opposed to the molecules discussed in the last Section, open-shell molecules have by definition an electronic configuration other than a $^1\Sigma$ ground electronic state. For an introduction to the spectra of free radicals, both diatomic and polyatomic the reader is referred to the book by Herzberg[144].

The first open-shell molecule to be investigated by MW spectroscopy was oxygen, O_2, by Burkhalter, et al.[145] in 1950. Oxygen was a logical candidate due to its natural abundance, its chemical stability, and its effect on the atmospheric transmission of MMW radar[146–148] in the 5 mm region.

The subsequent research effort in this field followed two avenues: (a) The route of pursuing chemically stable open-shell molecules led to the thorough investigation of nitric oxide, NO, which has a $^2\Pi$ electronic ground state[149–152]. Subsequent observation of the MW spectrum of a polyatomic molecule with an unpaired spin, NO_2 [153–156], stimulated a long series of experimental and theoretical investigations of this molecule[157–162] and the other known stable polyatomic open-shell molecule, chlorine dioxide, ClO_2 [163–168].

(b) The observation of the first chemically-unstable free radical came in 1953 when Sanders, Schawlow, Dousmanis and Townes[169–172] observed the Λ-type doubling transitions in the hydroxyl radical, OH, and its deuterated species, OD, both of which have a $^2\Pi$ electronic ground state.

Improvements in spectrometer design such as free-space cells[13, 173], improved discharge generators and chambers for the production of free radicals[174], Stark-effect modulated spectrometers with specially coated Stark septums[175], generally improved sensitivity and above all the search for more efficient chemical methods for producing unstable species[176], led to more precise OH spectra, and finally in 1964 to the detection of another radical, sulphur monoxide, SO, in a $^3\Sigma^-$ electronic ground state. This work was achieved simultaneously by two independent groups. Winnewisser et al.[177] observed and analysed the MMW spectrum of SO, while Powell and Lide[175] reported the lower-frequency MW spectrum. With this encouragement other research groups have detected and analysed several other free radicals. Particularly noteworthy is the detection of chlorine monoxide[178, 179], ClO, and the first observation of the rotational absorption spectrum of SO in the first excited electronic state, $^1\Delta$, by Uehara, Saito and Morino[180] and Saito[181] in 1970.

Parallel to the microwave rotational absorption investigations, Radford[183–187], McDonald[188] and Carrington and co-workers[189–192] conducted a thorough investigation of a series of diatomic and polyatomic radicals by gas-phase electron spin resonance. Their results have been summarised in a number of places[193–196]. Microwave rotational spectroscopy and electron spin resonance spectroscopy are complementary methods in the gas-phase study of molecules with unpaired electrons, and both methods combined yield a wealth of spectroscopic information. It should be emphasised that electron spin resonance and microwave spectroscopy measure in principle the same type of transitions, although the details of the experimental techniques differ.

In order to give an overall view of the development of the study of open-shell diatomic molecules, we have attempted to collect all of those so far detected by various experimental methods, all of which in one way or another are related to microwave spectroscopy. In listing the open-shell diatomic molecules investigated we sum up to 30 species. Negative results are, of course, hard to publish but we can estimate that at least 30 more species have been attempted by these methods. The radicals are grouped according to their electronic states and given in Table 7.5. It should be noted that some radicals are listed two or more times if they have been observed in two or more electronic states.

The molecules and methods included in Table 7.5 refer only to laboratory measurements. The recent detection of CN in the ground electronic state by MMW radio-astronomy[199] constitutes a measurement in radio-astronomy of an identified frequency which has not been determined in the laboratory. The transition observed is the upper component of the spin doublet (doublet splitting ~ 340 MHz) for $N = 1 \rightarrow 0$ in the $X^2\Sigma$ state of CN at 113 492 MHz. The instability of the species BO, CN and CP known to have $^2\Sigma$ ground states has possibly prevented laboratory observation of them in the MW region. It appears then, that MW and MMW radio-astronomy may be the source of further information concerning the spectra of extremely short-lived species, and in particular of further transition frequencies of CN.

The molecular constants of the eight radicals observed by MW and MMW techniques and the other radicals listed in Table 7.5 are presented in detail in reference 27.

For a more detailed theoretical treatment of diatomic molecules, the reader is referred to an excellent recent monograph by Hougen[200]. The various coupling cases of angular momenta which can occur in an open-shell diatomic molecule are discussed by Herzberg[197]. Kovac's work on diatomic molecules may also be consulted[201].

7.3.2.1 Molecules with a $^3\Sigma$ electronic state

(a) *Oxygen* (O_2) — The ground state $X^3\Sigma_g^-$ millimetre wave spectrum exhibited by O_2 in the range from 55 to 118 GHz can be described as the spin-reorientation spectrum, which consist of magnetic dipole transitions between the three spin components. Although these transitions ($\Delta N = 0$, $\Delta J = \pm 1$) of O_2 have been known for some time[202–205] only recently have McKnight and Gordy[206] succeeded in observing rotational transitions of the type $\Delta N = +2$, $\Delta J = 0$, ± 1 by measuring the $N = 3 \leftarrow 1$, $\Delta J = 0$ transition at 424 763.80 $\pm$ 0.20 MHz. By use of combination relations the rotational constant of O_2 can be determined from MMW data alone, since the energy expression of the F_2 levels ($N = J$) is:

$$E(N = J) = B_0 N(N+1) + D_0 N^2 (N+1)^2 + H_0 N^3 (N+1)^3 \qquad (7.2)$$

and does not involve the magnetic fine-structure constants. The use of the distortion constants D_0, H_0 determined by Babcock and Herzberg[208] from optical data using transitions involving high rotational states are sufficiently accurate to determine $B_0 = 43\,100.589 \pm 0.022$ MHz for O_2 from these measurements.

Figure 7.5 shows the three lowest allowed energy levels of O_2 drawn to scale. The observed fine-structure transitions are entered in the diagram together with the observed rotational transition $N = 3 \leftarrow 1$, $\Delta J = 0$. In 1968 Evenson *et al.*[207] observed the first laser electron paramagnetic resonance (LEPR) absorption in a gaseous O_2 sample. Employing an HCN laser with a centre frequency of 890 759 $\pm$ 0.1 MHz the paramagnetic absorption

Table 7.5 Spectroscopic observations of gaseous diatomic open-shell molecules using microwave and related techniques

Number	Radical	Electronic state	Frequency range of observation/MHz	Method of observation*	Year of observation†	References‡
		$^2\Sigma$				
1.	H_2^+	$X^2\Sigma_g^+$	5–1412	RF–OP(D)	1969	242–245
2.	OH	$A^2\Sigma^+$		RF–OP	1969	246
3.	CN	$B^2\Sigma^+$	113–11 504	MW–OP	1964	247–249
4.	NO	$A^2\Sigma^+$		RF–OP(F)	1968	250
		$^3\Sigma$				
5.	N_2	$A^3\Sigma_u^+$		MBER	1970	251
6.	O_2	$X^3\Sigma_g^-$	Spin-reorientation spectrum 53 592–118 750	MMW	1950	145,202–206,252–258
			Rotational spectrum 368 499–775 770	MMW, LEPR	1968	206,207
				ER	1955	205,256–258
7.	SO	$X^3\Sigma^-$	12 524–172 182	MW, MMW	1964	175,177,209,210
				ER	1963	188,191,211–214,259–263
8.	SeO	$X^3\Sigma^-$		ER	1967	260,262,264
		$^1\Pi$				
9.	CS	$A^1\Pi$		RF–OP	1970	265
		$^2\Pi$				
10.	CH	$X^2\Pi_i$	(H₂O laser at 2 527 952.8 MHz)	LEPR	1971	107
11.	OH	$X^2\Pi_i$	1612–36 995	MW	1953	169,172,183–185,219–224,266–271
				ER	1961	183,184,266–271
			(H₂O laser) $F = 5 \rightarrow 5$: 23 826.621 MHz	LEPR	1970	225
			$F = 4 \rightarrow 4$: 23 817.615 MHz			
				MBM	1971	198
12.	SH	$X^2\Pi_i$		ER	1963	241,272–274
13.	SeH	$X^2\Pi_i$		ER	1964	273,275,276
14.	TeH	$X^2\Pi_i$		ER	1964	275
15.	CN	$A^2\Pi_i$	1579–1650	RF–OP	1964	247,248,277,278
16.	CF	$X^2\Pi_i$		ER	1965	273,279

Table 7.5 (continued)

Number	Radical	Electronic state	Frequency range of observation/MHz	Method of observation*	Year of observation†	References‡
17.	NO	$X^2\Pi_i$	150 175–257 867	MMW	1953	151,152,231,234,235,281
				ER	1966	280
18.	NS	$X^2\Pi_i$	0.6–652	MBER	1970	235
			69 000–116 215	MMW	1969	239
19.	OCl	$X^2\Pi_i$	55 064–92 990	ER	1968	237,238
				MMW	1968	178,179
20.	OBr	$X^2\Pi_i$	63 681–64 379	ER	1968	190,241,273,282,283
				MMW	1969	240
21.	OI	$X^2\Pi_i$		ER	1966	273
22.	SF	$X^2\Pi_i$		ER	1967	273,284–286
23.	SeF	$X^2\Pi_i$		ER	1969	273,286,287
		$^3\Pi$		ER	1969	273,287
24.	H_2	$c^3\Pi_u$	4 928–5 898	MBMR	1962	288–290
25.	H_2	$d^3\Pi_u$		RF–OP(F)	1969	291
26.	CO	$a^3\Pi_{1;2}$	5–2 058	MBER	1970	292–294
		$^1\Delta$				
27.	NF	$a^1\Delta_2$		ER	1971	295
28.	O_2	$a^1\Delta_2$		ER	1965	296,297
29.	SO	$a^1\Delta_2$	127 770	MMW	1970	180,181,298
				ER	1971	241,273,299
30.	SeO	$a^1\Delta_2$		ER	1969	273,300

* In referring to the various methods of observation the following abbreviations are used:

RF–OP(F) or (D) : Radiofrequency–optical double resonance with optical detection through (F) fluorescence, or (D) photodissociation.

MW–OP : Microwave–optical double resonance with optical detection.

LEPR : Laser electron paramagnetic resonance.

MW : Microwave spectroscopy.

MMW : Millimetre wave spectroscopy.

ER : Electron spin resonance.

MBER : Molecular-beam electric resonance.

MBMR : Molecular-beam magnetic resonance.

MBM : Molecular-beam maser.

† The year of the appearance of the first major publication is referred to as year of observation.

‡ The references contain only major publications or the first mention of the molecule in question. This list of references should not be considered complete.

between the $(N,J,M) = (3,4,-4)$ and $(5,5,-4)$ levels of the ground state of O_2 was observed in a magnetic field of $16\,418 \pm 1$ G. They were able to determine the zero-field energy difference of $775\,770 \pm 10$ MHz between the $(N,J) = (5,5)$ and $(3,3)$ levels as indicated in Figure 7.5.

(b) *Sulphur monoxide* (SO)—The rotational energy levels of the ground state of the SO radical are compared in Figure 7.5 with those of O_2. It can

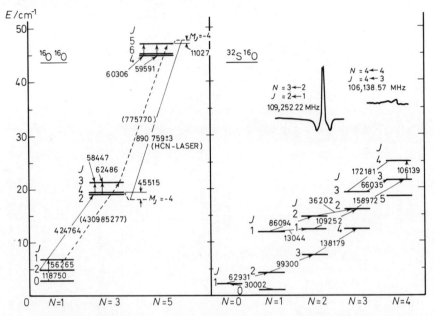

Figure 7.5   Energy level diagram and observed transitions of O_2 and SO

be seen that the triplet splitting of SO in the ground state is almost equal to the separation of adjacent rotational levels. With the exception of one tentatively-assigned weak magnetic dipole transition, all lines are the reported[175, 177] electric dipole transitions (dipole moment = 1.55 D), with the selection rules $\Delta N = +1$, $\Delta J = 0, \pm 1$. Since these first observations were made, considerable spectroscopic information has been accumulated both by MW spectroscopy[209, 210] and electron spin resonance[211-214]. Amano, Hirota and Morino[209] observed rotational transitions of SO in the first vibrationally-excited state, thus in combination with the ground-state data the equilibrium parameters are obtained: $B_e = 21\,609.55 \pm 0.15$ MHz; $\lambda_e = 157\,780.0 \pm 25.0$ MHz and $r_e = 1.481\,08\,(5)$ Å. Furthermore, the spectra of the isotopic species $^{34}S^{16}O$ and $^{33}S^{16}O$ were also studied. The detection of SO in interstellar space could yield interesting results on molecule formation processes.

7.3.2.2   Molecules with a $^2\Pi$ electronic state

The radicals with a $^2\Pi$ ground state whose Λ-type spectra or pure rotational spectrum have been observed by microwave or laser paramagnetic-resonance spectroscopy can be found in Table 7.5. The six molecules which are in this category are CH, OH, NO, NS, ClO and BrO.

(a) *CH radical* — The Λ-type double spectrum of the CH radical in the $^2\Pi_{\frac{1}{2}}$ electronic ground state begins in the 10 cm region with a transition between the two components of the $J = \frac{1}{2}$ level and extends into the shorter wavelength region. Although this transition is of potential interest to astrophysics it has not been possible to observe this line either in interstellar space or by direct measurement in the laboratory. By photographic Fabry–Perot interferometry, however, this transition has recently been determined by Baird and Bredohl[182] to be at 3374 ± 15 Mz. Evenson, Radford and Moran[107] detected an absorption spectrum of the CH radical which can be identified with the pure rotational transition $J,N = (\frac{7}{2},3) \leftarrow (\frac{5}{2},2)$. This transition was brought into coincidence with the frequency of a water-vapour laser at 2527.953 GHz due to the Zeeman shift of the CH rotational energy levels at a field strength of 10 kG. The remarkable feature of this experiment is the sensitivity of the laser magnetic resonance technique. CH concentrations of 10^{12} cm^{-3} in the ground vibrational and electronic state yielded a signal strength 260 times the noise with an absorption cell only 5 cm in length.

(b) *Hydroxyl radical* (OH) — The absorption spectrum of the OH radical has been studied extensively in the MW region[27, 29] since its first detection by Dousmanis, Sanders and Townes[172]. The Λ-doubling transition in the $J = \frac{3}{2}$ rotational state of the $^2\Pi_{\frac{3}{2}}$ ground electronic state was reported in 1959[215] and in 1963 this transition at 18 cm was detected in absorption in interstellar space[216]. Thus OH became the second constituent of the interstellar medium to be identified by radio-astronomical methods after atomic hydrogen was characterised by its 21 cm radio emission spectrum in 1951 by Ewen and Purcell[217]. Intensity anomalies observed in the 18 cm OH interstellar radio lines have stimulated further theoretical and experimental work[198, 220–224].

The paramagnetic resonance absorption of OH was studied by Radford[184]. The low-field Zeeman spectrum of the Λ-type doubling transitions of the $^2\Pi_{\frac{3}{2}}$ state were investigated by Johnson and Lin[218] and the dipole moment was 1.660 D[219]. However, the actual MW measurements have not yet been extended above 40 GHz. In 1970 Evenson, Wells and Radford[225] measured the electric dipole transition of OH between the levels with $(^2\Pi_{\frac{3}{2}}, J = \frac{3}{2}) \leftarrow (^2\Pi_{\frac{1}{2}}, J = \frac{1}{2})$ by the laser paramagnetic resonance absorption method, using a water-vapour laser at 79 μm. A near overlap of a water line on one magnetic dipole transition suggests a possible pumping mechanism for the 18 cm maser emission of interstellar OH[220].

(c) *Nitric oxide* (NO) — The MMW spectrum of the pure rotational transition $J = \frac{3}{2} \leftarrow \frac{1}{2}$ of $^{14}N^{16}O$ in the $^2\Pi_{\frac{1}{2}}$ ground state was observed some time ago[149, 150]. The NO molecule is considered an intermediate coupling case, slightly removed from Hund's case (a) and the theory of spin uncoupling with L uncoupling[226] was applied to analyse the later, more extensive, measurements[151, 152]. The nuclear hyperfine structure[227–230], the Zeeman effect[231] and the Stark effect[232, 233] were extensively investigated. Recently, French and Arnold[234] carried out line-width studies of the $J = \frac{3}{2} \leftarrow \frac{1}{2}$ rotational transition in NO using a Fabry–Perot cavity resonator. The results of these measurements give nitric oxide a collision diameter of 5.26 Å which exceeds the value obtained from kinetic theory predictions (3.53 Å) and which is attributed to dipole–dipole, dipole–quadrupole, quadrupole–quadrupole and dipole–induced dipole interactions. The astrophysically important fine-structure transitions in the radio-frequency region have been investigated by Neumann[235] by molecular-beam methods and predictions of other transitions were made. Meerts and Dymanus[236] have re-measured and extended these measurements and found some disagreement with Neumann's predictions.

(d) *Nitrogen sulphide* (NS) – Shortly after Carrington and co-workers[189, 237] in 1966 reported the electron paramagnetic resonance spectrum of NS in the $^2\Pi_{\frac{1}{2}}$ electronic state in the reaction products of hydrogen sulphide with active nitrogen, Uehara and Morino[238] produced this radical by a microwave discharge through a mixture of sulphur dichloride and nitrogen. Amano, Saito, Hirota and Morino[239] observed the rotational spectrum from 69 to 116 GHz and analysed the Λ doubling and the hyperfine interaction of the transitions $J = \frac{3}{2} \leftarrow \frac{1}{2}$ in the $^2\Pi_{\frac{1}{2}}$ state as well as $J = \frac{5}{2} \leftarrow \frac{3}{2}$ both in the $^2\Pi_{\frac{1}{2}}$ and $^2\Pi_{\frac{3}{2}}$ states for the two isotopic species $^{14}N^{32}S$ and $^{14}N^{34}S$. A negative sign of the Λ-doubling parameter α_p was obtained which is an indication that the dominant perturbing state is a $^2\Sigma^-$ state.

(e) *Chlorine monoxide* (ClO) – After the ClO radical was detected by electron paramagnetic resonance, the MW spectrum between 50 and 60 GHz in the $^2\Pi_{\frac{1}{2}}$ was reported[178] and later more extensive data on rotational transitions in $^2\Pi_{\frac{3}{2}}$ and $^2\Pi_{\frac{1}{2}}$ states of both isotopic species ^{35}ClO and ^{35}ClO appeared[179]. ClO is interesting in that it possesses a $^2\Pi$ ground state with an inverted spin doublet; the $^2\Pi_{\frac{3}{2}}$ level is the lowest lying electronic state.

(f) *Bromine monoxide* (BrO) – BrO appears to be the shortest lived of the five radicals SO, OH, NS, ClO and BrO for which MW spectra have been reported. Powell and Johnson[240] detected the rotational transitions $J = \frac{5}{2} \leftarrow \frac{3}{2}$ in the 60–70 GHz region for the isotopic species ^{79}BrO and ^{81}BrO in the $^2\Pi_{\frac{3}{2}}$ ground state.

7.3.2.3   Molecules with a $^1\Delta$ electronic state

The first and so far only MW observation of the absorption spectrum of a molecule in an excited electronic state was reported by Saito[181] in 1970 for SO. A single absorption at 127 770.47 MHz was observed and assigned to the $J = 3 \leftarrow 2$ rotational transition of SO in the $^1\Delta$ state. The basis for this assignment is that the observed transition is near the predicted frequency calculated by using the rotational constants obtained from the e.p.r. measurements of this state[241]. The observed Stark pattern could be analysed and the dipole moment was determined to be 1.336 D [181] in agreement with the gas-phase e.p.r. measurements[191].

7.3.3   Open-shell polyatomic molecules

The number of open-shell polyatomic molecules whose microwave absorption spectrum has been reported is so far limited to four triatomic species. The linear radical NCO possesses a $^2\Pi$ electronic ground state while the bent triatomic species can be placed in two groups: NO_2, with a $\tilde{X}^2A_1$ electronic ground state and NF_2 and ClO_2 with a $\tilde{X}^2B_1$ electronic ground state. A fifth molecule, the bent asymmetric HCO, has been studied using low-field magnetic resonance to obtain zero-field frequencies for several microwave transitions[301].

The effective Hamiltonian representing the model used to interpret the observed spectra must include all of the various dynamic and magnetic interactions to be found in polyatomic molecules in multiplet electronic states and in excited vibrational states. Theoretical treatments of such molecules are found in a series of papers[302–306]. The reader is also referred to the third volume of Herzberg's book '*Molecular Spectra and Molecular Structure*'[307].

7.3.3.1 Linear triatomic free radicals

The rotational spectra of linear triatomic radicals are in all essential features similar to those of diatomic molecules with the same type of electronic ground state. However, the presence and excitation of a doubly degenerate bending vibration, commonly denoted as v_2 in linear triatomic species, may give rise to a strong vibration–electronic interaction. The resulting energy states are labelled as Renner[302, 307] states. Two radicals, NCO and NCS [192], have been found experimentally in the ground vibrational state and in various Renner states by electronic paramagnetic resonance. In 1970 Saito and Amano[308] reported the MMW rotational spectrum of NCO in the ground vibrational state in the 50–90 GHz region. The observed rotational transitions $J = \frac{5}{2} \leftarrow \frac{3}{2}$ and $\frac{7}{2} \leftarrow \frac{5}{2}$ in the $^2\Pi_{\frac{3}{2}}$ state and $J = \frac{5}{2} \leftarrow \frac{3}{2}$ in the $^2\Pi_{\frac{1}{2}}$ state could be analysed by applying Hougen's formulation for linear molecules in $^2\Pi$ states. Since the NCO radical is important as a reaction intermediate in the reaction of oxygen with cyanogen (CN_2), and is probably also a constituent of interstellar space, it can be assumed that the MMW spectrum of this species will be pursued further.

7.3.3.2 Bent triatomic free radicals

The three molecules NO_2, ClO_2 and NF_2 are all asymmetric rotor molecules with unpaired electronic spins, and exhibit a complex magnetic fine and nuclear hyperfine structure due to the nature and the degree of coupling between the various angular momentum and spin vectors.

(a) *Nitrogen dioxide* (NO_2)—The slightly asymmetric rotor NO_2 with an asymmetry parameter $\kappa = -0.994$ has been the subject of spectroscopic investigations in almost all accessible spectral regions. The electronic angular momentum is 'quenched' so that the major contribution of the unpaired electron to the energy in the absence of external fields is a spin–rotation interaction. An extensive theoretical treatment of the effect of the magnetic interactions between the electronic spin, nuclear spin, and the molecular rotation on the rotational levels has been presented by Lin[157]. This treatment has been applied by Bird[155, 156] and Curl and co-workers[158–162] to several isotopic species including $^{14}N^{17}O^{16}O$ in which the complex hyperfine structure caused by two nuclei, ^{14}N and ^{17}O, both of which have magnetic and quadrupole coupling, has been analysed.

(b) *Chlorine dioxide* (ClO_2) — The complexity of the microwave spectrum of the ClO_2 radical has fascinated many investigators[163], and in a series of investigations Curl and co-workers[163-186] were able to evaluate a great many of the coefficients which describe the energy levels and to present an interpretation of the hyperfine coupling in terms of the electronic structure of the molecule[165]. The extension of the measurements into the MMW region made possible the determination of centrifugal distortion and potential constants[166].

(c) *Nitrogen difluoride* (NF_2) — A highly-sensitive K-band cavity spectrometer was employed by Hrubesh, Rinehart and Anderson[309] to detect the NF_2 radical. The radical was produced in adequate concentrations from the dissociation of the stable N_2F_4 molecule at elevated temperatures (90 °C) at a pressure of 0.1 torr. The analysis of the data is not yet completed.

The interpretation of the complex spectra of these molecules has provided an opportunity for testing the theory in its fine points. Although the number of species to which such a treatment is applicable is limited, we may find in future the opportunity to apply this knowledge to the rotational spectra of molecules in excited electronic states.

7.3.4 Linear molecules

The transition frequencies of OCS play a special role as they are universally used for tuning spectrometers operating in the MMW region. It is therefore

Table 7.6 Molecular constants of linear molecules measured in the MMW region

Molecule*	B_0/MHz	D_0/kHz	α_2/MHz	q_2/MHz	References
$^{16}O^{12}C^{32}S$	6 081.492 55(17)	1.301 92(4)	−10.56	6.344(18)	10, 349
$^{16}O^{12}C^{80}Se$	4 017.649	0.669 5	−6.89	3.172	347, 350
$^{14}N^{14}N^{16}O$	12 561.633 8(6)	5.282(4)	−16.9(2)	23.736	347, 331
^7LiOH	35 342.44(4)	–	–	295.810(5)	313
$H^{12}C^{14}N$	44 315.975 5(4)	87.22	−106.77(8)	224.476 6(4)	91, 324
DCN	36 207.462 7(2)	57.83	−126.9(3)	186.191 6(5)	91, 324
$^{35}ClCN$	5 970.831	1.663	−16.33	7.459	347
$^{79}BrCN$	4 120.221	0.884 4	−11.49	3.915	347, 351; 352
ICN	3 225.578(18)	0.88(9)	−9.50	2.643	351, 353, 354
HCP	19 976.004 8(88)	21.233(99)	−10.6(3)	48.6(1)	327
DCP	16 984.375(12)	14.49(21)	−32.85(4)	44.86(7)	327
HCCD	29 725.24(5)	33.6(3.2)			318
HCCF	9 706.19(1)	3.6(5)	$\alpha_4 = -27.65(5)$ $q_4 = 19.12(5)$ $\alpha_5 = -8.82(5)$ $q_5 = 12.57(5)$		312
HCCCN	4 549.06(1)	0.56	$\alpha_5 = -1.73(2)$ $q_5 = 2.56(2)$ $\alpha_6 = -9.24(2)$ $q_6 = 3.57(2)$ $\alpha_7 = -14.46(2)$ $q_7 = 6.54(2)$		312
HCNO	11 469.051 37(61)	4.261 9(30)	$\alpha_4 = -15.16(3)$ $q_4 = 23.673 2(11)$ $\alpha_5 = -38(16)$ $q_5 = 34.635 4(17)$		338, 325

* For other isotopic species see references.

understandable that the highest MMW frequency measured to date is the $J = 67 \leftarrow 66$ transition of $^{16}O^{12}C^{32}S$ (see Table 7.1). As a result, a very precise determination of the spectroscopic constants becomes possible, especially when the high-frequency measurements are combined with the Lamb-dip results of Winton and Gordy[310].

Other linear molecules whose spectra have been measured in the MMW region are listed with their ground-state rotational constants in Table 7.6. Recent newcomers to the list of linear molecules on the MW scene are those for which difficulties in chemical preparation and/or handling could success-fully be overcome. Those molecules are HCP[311], FCN[312], HCCF[312], LiOH, LiOD[313] and HCNO[314]. Because of their small dipole moments, HCCCl[312, 315], HCCBr[316, 317] and the almost non-polar HCCD[318, 319] con-stitute remarkable additions as well. The spectra of several dihalogeno-acetylenes and chlorodiacetylene have also been measured[320].

Because of the low-bending frequencies of many of these molecules, spectra of vibrationally-excited states have been observed. The interaction of the vibrational angular momentum, associated with excited states of the doubly degenerate bending mode, with the end-over-end rotation[321–323] increases with J. Its effects on the spectrum (l-type doubling and l-type resonance) therefore become more pronounced the higher one goes in frequency for a given molecule.

7.3.4.1 Triatomic linear molecules

Very precise spectroscopic constants have been derived by De Lucia and Gordy[91] from molecular-beam maser studies of the $J = 1 \leftarrow 0$ and $J = 2 \leftarrow 1$ rotational transitions of HCN and DCN (see Figure 7.4). Measurements of HCN and DCN have until recently been restricted to the ground vibrational state. However, rotational l-type doublet transitions $\Delta J = +1, l = 0$ in the vibrationally excited states $v_2 = 1$ and rotational transitions in $v_2 = 2$ of several isotopic species of HCN were recently reported by Winnewisser, Maki and Johnson[324]. This new data when combined with earlier microwave work[326] and existing infrared measurements gives a set of rovibrational

Table 7.7   Contributions to α_2 in HCN and HCP (from reference 327, by courtesy of Academic Press)

$$(\alpha_2 = (\alpha_2)_{h_1} + (\alpha_2)_{h_2} + (\alpha_2)_{anh})^*$$

Molecule	$(\alpha_2)_{h_1}$/MHz	$(\alpha_2)_{h_2}$/MHz	$f_{1\gamma\gamma}$/mdyn	$f_{3\gamma\gamma}$/mdyn	$(\alpha_2)_{anh}$/MHz	α_2/MHz
HCN	+112.1	−294.4	−0.134	−0.272	+75.9	−106.4†
DCN	+ 92.9	−270.4			+47.7	−129.8†
HCP	+ 24.3	− 80.9	−0.083	−0.299	+46.0	− 10.6‡
DCP	+ 22.4	− 85.4			+30.1	− 32.9‡

*Expressions for the various terms in this sum can be found in Lide and Matsumura[329].

†These are calculated values—to be compared with the observed values which are −106.77 ± 0.08 MHz and −126.9 ± 0.3 MHz respectively.

‡These are observed values—they were used to calculate the cubic force constants $f_{1\gamma\gamma}$ and $f_{3\gamma\gamma}$.

constants which are consistent with all available data. In general, these constants are defined by the equation:

$$B_v = B_e - \sum_i^3 \alpha_i (v_i + \frac{d_i}{2}) + \sum_{i=1}^3 \sum_{j=1}^3 \gamma_{ij}(v_i + \tfrac{1}{2}d_i)(v_j + \tfrac{1}{2}d_j) \quad (7.3)$$

In contrast to the definition of B_v given in Reference 324, we have followed here the example of Winnewisser and Winnewisser[325] by not including the term $l^2\gamma_{l_t l_t}$ in the defining equation. For a vibrational energy level such as $02^{0,2}0$ of HCN which has two components, we define B_v as above for the entire level. The effective unperturbed B value for the $l = 0$ and $l = 2$ levels may then be determined by $B_v + l^2\gamma_{l_t l_t}$. However, since l-type resonance mixes the $l = 0$ level with the symmetric member of the vibrationally degenerate $l = 2$ level[323], the definition of a separate B_v for the different l components of the $02^{0,2}0$ level is not strictly meaningful. A discussion of l-type doubling and l-type resonance is found in references 321–323.

Recent MMW measurements of methinophosphide, HCP [327], in the ground and $v_2 = 1$ states form an extension of earlier microwave work[311]. In comparing the interaction constants of HCP with those of HCN it is interesting to note that for both molecules the l-type doubling constant q_t is larger in the hydrides than it is in the deuterides. The apparently anomalous fact that α_2 (HCN) is smaller than α_2 (DCN) could be explained by Johns, Stone and Winnewisser[327] by applying the theory of curvilinear coordinates as developed by Pliva[328] to the standard expression for α_2 in the manner presented elegantly by Lide and Matsumura[329] for the alkali hydroxides CsOH and RbOH. The results of this analysis of α_2 for HCN and HCP are summarised in Table 7.7.

The vibrational dependence of the l-type doubling constant q_t and of the centrifugal distortion constant D_v were investigated in 1969 by Nakagawa and Morino[330], by calculating these higher-order contributions to the rovibrational energy for HCN and for OCS. Precise rotational and centrifugal distortion constants for several vibrational states 01^10, $02^{0,2}0$ and 100 of nitrous oxide, N_2O, in addition to the ground state, have been determined by Pearson, Sullivan and Frenkel[331] from rotational transitions between 125 and 302 GHz.

7.3.4.2 Four- and five-atomic linear molecules

A linear molecule with four or more atoms exhibits an interaction between the two or more degenerate bending modes in excited vibrational states. This interaction has been discussed by Amat and Nielsen[332] and is called 'vibrational l-type doubling'. Evidence of this effect has been observed in the infrared spectrum of acetylene[333, 334] but the spectrum of fulminic acid, HCNO, provided the first microwave data showing the effects of this vibrational l-type doubling[335, 336].

The only other four- or five-atomic molecules which have been investigated by MW techniques are acetylene halide derivatives with small dipole moments and large quadrupole splittings, so that the observation of excited state spectra is difficult. This is the case for HCCBr [316, 317] and HCCCl [312, 315] in particular. HCCF and HCCCN, which have more substantial dipole moments, have been measured in the MMW region, so that their rotational constants are well determined [312], but only for the ground state and the first excited state of each of the bending modes. The direct l-type doubling transitions in the MW region have been observed for $v_7 = 1$ and $v_7 = 3$ of HCCCN [337], but the corresponding rotational transitions have not yet been reported.

HCNO is rather exotic among four-atomic linear molecules whose rotational spectra have been studied. Earlier MW measurements [314, 335, 336] have been supplemented by Winnewisser and Winnewisser [325, 338] with extensive MMW data for all rotational transitions up to $J = 12 \leftarrow 11$ for the following vibrational states: 00000, 00001^1, $00002^{0,2}$, $00003^{1,3}$, 0001^10, $0002^{0,2}0$, and $000(11)^{0,2}$. From the investigation of the microwave spectrum of a combination state in which two degenerate bending modes are excited, it was found that the vibrational l-type doubling constant r_{45} has a J-dependent term which could be accurately determined, $r_{45}^{(1)} = 0.6565$ MHz. The transitions observed for $v_4 = 2$ and $v_5 = 2$ were analysed according to the Amat–Nielsen formulation [323] to obtain $g_{l_4l_4}$ and $g_{l_5l_5}$, the constants which determine the separation between the $l = 2$ and $l = 0$ vibrational energy level components of each of these states. It is found that not only the magnitude but also the sign of the quantity $g_{l_il_i}$ can be obtained. In the case of HCNO, $g_{l_4l_4}$ is small and positive (1.2 cm^{-1}) and $g_{l_5l_5}$ is large and negative (-24.2 cm^{-1}). The analogous analysis of the $v_5 = 3$ transitions yielded also a large negative $g_{l_5l_5}$, which however is a factor of four smaller than the value obtained for $v_5 = 2$. This large discrepancy is accompanied by other puzzling features of the constants obtained from the data.

If the r_s structure of HCNO is considered in comparison with that of other, related linear molecules, we find that the CH bond length is surprisingly short.

Molecule								
Methinophosphide[311]	H	—1.066 7(5)—	C	=1.542 1(5)=	P			
Hydrogen cyanide[324]	H	—1.063 16(10)—	C	=1.155 12(15)=	N			
Cyanoacetylene[312]	H	—1.058—	C	=1.205=	C	—1.378—	C	≡1.159≡ N
Fluoroacetylene[312]	H	—1.053—	C	=1.198=	C	—1.279—	F	
Fulminic acid[336]	H	—1.026 6(4)—	C	=1.167 9(4)=	N	—1.199 4(4)—	O	
Nitrous oxide[339]			N	=1.128 6(2)=	N	—1.187 6(3)—	O	

As Winnewisser, Maki and Johnson [324] have pointed out, the r_s distances or substitution bond lengths are still the most reliable and readily measurable quantities for structural comparisons between molecules. The structures given above reflect therefore the best r_s parameters available at present for the molecules listed. The numbers in parenthesis are the experimental errors in units of the last significant figures. It can be seen that the CH distance in HCNO is about 0.03 Å shorter than in any other molecule. Furthermore, this short CH bond length is found to be irreconcilable with Bernstein's rule[340], from which the CH bond length in HCNO is estimated to be 1.059 Å. One way of reconciling the short CH bond length in HCNO with the information quoted here is to interpret it as a projection on the CNO axis of a CH bond of a quasi-linear molecule[341]. A quasi-linear model[342] might also account for the unusual values of $g_{l_5l_5}$ obtained, as well as other anomalous effects. Thus, finally, after the available data has been analysed as though HCNO were strictly linear, the possibility of a non-linear model is now being seriously considered. One approach to the calculation of the energy levels of such a molecule is provided by the Hamiltonian developed by Hougen, Bunker and Johns[343] and its extension to rovibrational levels by Bunker and Stone[344] together with the application of this type of Hamiltonian to four-atomic molecules outlined by Sarka[345].

As of this writing, rotational transitions of the linear molecules HCN, HCCCN and OCS have been detected in interstellar space[346], providing new impulse to the effort to obtain precise information concerning the energy levels of these molecules.

7.3.5   Symmetric-top molecules

The very first MW absorption measurements were done on the inversion

Table 7.8   Symmetric rotors measured in the MMW region

Molecule	B_0/MHz	D_J/MHz	D_{JK}/MHz	References
Inorganic molecules				
$^{14}NH_3$	298 115.37	24.31	−45.27	356
ND_3	154 173.38	5.91	−10.49	356
NF_3	10 681.078(5)	0.014 534(68)	−0.022 694(86)	368
$^{31}PH_3$	133 480.15	3.95	−5.18	357
PD_3	69 471.09	1.02	−1.31	357
PF_3	7 819.99	0.007 84(5)	−0.117 7(4)	368
$^{75}AsH_3$	112 470.59(3)	2.925(3)	−3.718(4)	367
AsD_3	57 477.60(2)	0.741(2)	−0.928(3)	367
AsF_3	5 878.971(2)	0.463(3)	−0.617(5)	371
$^{121}SbH_3$	88 038.99(3)	1.884(4)	−2.394(15)	367
SbD_3	44 694.92(3)	0.473(4)	−0.598(9)	367
POF_3	4 594.262	0.001 02	+0.001 284	398
PSF_3	2 657.663	0.000 30(5)	+0.001 8(7)	399
$SiHF_3$	7 208.049	0.007 56	−0.012 5	398
Organic molecules				
CH_3F	25 536.146 6(15)	0.059 87(20)	+0.440 27(40)	310, 377
CHF_3	10 348.867(2)	0.011 33	−0.018 10(4)	377, 391
$CH_3{}^{35}Cl$	13 292.869(1)	0.018 0	+0.198 5(3)	377
$CH^{35}Cl_3$	3 302.083(3)	0.001 52(1)	−0.002 50(2)	400
$CD_3{}^{35}Cl$	10 841.91	0.010 84	+0.102 7	401
$CD^{35}Cl_3$	3 250.290(2)	0.001 826(7)	−0.002 31(2)	402
$CH_3{}^{79}Br$	9 568.192(5)	0.009 87	+0.128 6(1)	377, 397
CH_3I	7 501.276(5)	0.006 31	+0.098 7(1)	377
CH_3CN	9 198.897(6)	0.003 81(8)	+0.176 9(2)	387
CH_3NC	10 052.888	0.004 73	+0.227 1	403
CH_3CCH	8 545.869	0.002 875	+0.163 0	390
$(CH_2O)_3$	5 273.258(4)	1.35(1)	−2.03(1)	392

spectrum of ammonia in 1934 by Cleeton and Williams[355], and some of the most recent sub-MMW measurements reported in the literature are rotation–inversion transitions of the same molecule observed by Helminger, De Lucia and Gordy[356] including $J = 1 \leftarrow 0$, $K = 0$ of $^{14}NH_3$ at 572 498.15(15) MHz. The ground state spectral constants of symmetric rotors measured above 100 GHz are listed in Table 7.8.

7.3.5.1 Inorganic symmetric rotors

The present results of the rotation–inversion spectrum of ammonia can be summarised by noting that the $J = 1 \leftarrow 0$ transitions of $^{14}NH_3$ and $^{15}NH_3$ were reported recently[357], whereas the corresponding transitions of ND_3 have been known for some time[358]. The detection of the two inversion components of the $J = 2 \leftarrow 1$ transition of both $^{14}ND_3$ and $^{15}ND_3$ at a wavelength of 0.49 mm high-lights these investigations[356].

The $J = 1 \leftarrow 0$ rotation–vibration transition of NH_3 shows only one inversion component $(+ \leftarrow -)$ whereas for ND_3 both inversion components $(+ \leftarrow -)$ and $(- \leftarrow +)$ are observed with the expected intensity ratio of $1:10$. The B value is obtained from the measured rotation–vibration transition frequency in NH_3 by correcting for the inversion splitting of the $J = 0$, $K = 0$ and $J = 1$, $K = 0$ levels readily calculated by Costain's formula[359] recently extended[360] to include expansion coefficients up to $A_{20}J^5(J+1)^5$. Conversely in $^{14}ND_3$ the inversion doubling $\Delta v = \Delta_{00} + \Delta_{10} = 3172.78$ MHz is obtained directly from the $J = 1 \leftarrow 0$ transition, and is found to be in good agreement with the measurements of the direct inversion splittings[361].

With the detection of the (1,1) and (2,2) inversion lines of interstellar $^{14}NH_3$ at 1.26 cm wavelength in Sagittarius, the first radio emission lines of a polyatomic molecule were observed. Since that announcement[362] five additional inversion lines of NH_3 and various rotational transitions of CH_3CN [363] and CH_3CCH [364] have been detected in interstellar space.

For all pyramidal molecules, except NH_3, sufficient MW and MMW measurements are available to allow a complete determination of the molecular parameters. In the case of PH_3 the measurement of the $J = 2 \leftarrow 1$ transitions yielded accurate centrifugal stretching constants[357], but more important, the measurement of the $K = 1$ component established on the long-suspected unknown inversion doubling an experimental upper limit of 0.5 MHz in agreement with theoretical calculations[365]. Recent molecular-beam electric-resonance experiments reduce this experimental limit to less than 1 kHz for the ground state[366]. No observable inversion splitting is expected for other molecules like arsine, AsH_3, and stibine, SbH_3, whose MMW spectra have been re-investigated[367]. The nuclear quadrupole couplings of As and Sb are found to be larger in the deuterated species. Mirri and co-workers[368–371] have reported the MMW spectra of PF_3, PCl_3, AsF_3 and NF_3, determined the molecular constants and, by combining MMW and infrared data, the six constants of the general force field were obtained. The structure and force constant analysis of PF_3 has since been revised by Hirota and Morino[372, 373] on the basis of data from excited vibrational states.

To derive the equilibrium structure of a symmetric-top molecule, measurements in all $3n-6$ fundamental vibrational modes must be available for as many isotopic species as there are parameters. This has been achieved recently for NF_3 where at least two rotational transitions in all fundamental vibrational modes v_1, v_2, v_3 and v_4 have been measured, of which the last two are twofold degenerate. Otake, Matsumura and Morino[374] obtained $r_e(NF) = 1.365 \pm 0.002$ Å and $\angle(FNF)_e = 102° \, 22' \pm 2'$.

7.3.5.2   Organic symmetric rotors

Thomas, Cox and Gordy[375] have shown that for the methyl halides the vibrational frequencies can be simply related with the centrifugal distortion constants. This model, which considers the methyl group to be rigidly vibrating against the halogen, can be tested on the silyl halides as well once reliable experimentally determined centrifugal distortion parameters become available. The currently known molecular parameters of the silyl halides are found in a recent communication by Kewley et al.[376]. Extensive new measurements on the methyl halides have been reported by Sullivan and Frenkel[377] and their data for the excited state $v_6 = 1$ in $CH_3^{35}Cl$ showed irregularities which have not yet been fully explained. The substituted methanes in Table 7.8 are grouped so that the sign of the centrifugal distortion parameter D_{JK} can be compared for related molecules. Those which have a heavy symmetrical group (CF_3H, CCl_3H etc.) and a light hydrogen atom attached to the central carbon atom have a negative D_{JK}, whereas those with a light symmetrical group and a heavy atom or a linear group attached to the symmetrical group carbon (CH_3F, CH_3Cl etc.) have positive D_{JK}. Thus to a good approximation it is found that if the centre of mass lies inside (outside) the umbrella formed by the C_{3v}-group, D_{JK} is negative (positive). Oblate symmetric tops seem to have negative D_{JK} while prolate symmetric tops have positive D_{JK}.

7.3.5.3   Excited degenerate vibrational modes

The determination of the B_v value of symmetric tops in doubly degenerate vibrational modes can be complicated by the occurrence of K-type and l-type doubling, since the rovibrational energy levels are split into a number of components. In symmetric-top molecules with C_{3v} symmetry, first-order Coriolis coupling produces a splitting into an E level and an A_1A_2 pair for all $K \neq 3n$ where n is an integer, and two E levels for $K = 3n$. Higher order interactions produce a further splitting of the A_1A_2 pair into separate levels and this splitting is commonly referred to as l-type doubling. The magnitude of the splitting for $v_t = |l| = 1$ is given by $\Delta v = q_t J(J+1)$. Analytical expressions for the l-type doubling constant q_t are found in a number of studies. The theory of l-type doubling in symmetric tops was first discussed by Nielsen[322] and later more elaborately by Grenier-Besson[378, 379]. Recently, Oka[380] has elegantly derived equivalent results starting directly from the Wilson and Howard Hamiltonian[381].

  Energy expressions based on the approach of Nielsen and Grenier-Besson have been derived by Tarrago[382, 383] for states with $v_t = 1$ and $v_t = 2$ which include the effects of perturbations up to fourth order[382] and to sixth order[383].

Consideration of higher order effects has been applied in the analysis of the rotational spectrum of CH_3CN in the excited state $v_8 = 2$ by Bauer and Maes[384-386]. Transitions in the $v_8 = 1$ state had been reported earlier[387], and the study of the $v_8 = 2$ transitions revealed a previously unobserved resonance between the $K = \pm 4, l = 0$ and $K = \pm 2, l = 2$ levels. This resonance is analogous to that observed in fluoroform by Costain[388] and in PF_3 by Hirota and Morino[372, 373], and is explained by Maes and Amat[389].

Other molecules for which l-type doubling in $v_t = 1$ states has been studied in the MMW region are $CH_3^{35}Cl$ [377], CH_3CCH [390], CHF_3 [391] and $(CH_2O)_3$ [392]. The latter molecule, which is the cyclic trimer of formaldehyde, 1,3,5-trioxane, is an oblate top. Bellet, Colmont and Lemaire[392] report measurements for the ground state as well as three excited vibrational states, two of which are doubly degenerate. Several components of transitions from higher excited states, $v_t = 2$ and 3, were also observed. For the $v_E(2)$ state (notation after Oka et al.[393]), they show that the l-type doublet line positions of the higher J transitions (e.g. $J = 13 \leftarrow 12$ which fall into the MMW range) cannot be fitted any longer with the perturbation treatment given by Amat and Grenier-Besson[394], which is only valid under the assumption $|B_v - C_v + C_v \xi| \gg \frac{1}{2} a_J J(J+1)$, not met by the transitions indicated. The effects of some higher-order perturbations at high J values may also be analysed by measuring transitions in the MW region between levels connected by l-type resonance. Such measurements have been made for CHF_3 [395] and PF_3 [373]. The increasing use of MMW spectroscopy should lead to many more studies of spectra of this nature, providing an excellent meeting ground for experimental and theoretical work.

Combining infrared and MW data to obtain additional information is illustrated by recent work on CH_3I. The rotational constant A of a symmetric-top molecule is not readily measured and usually an estimate of A can be obtained from the determination of the Coriolis constants or from the Q-branch spacings of all the perpendicular bands and the ξ-sum rule. Matsumura, Nakagawa and Overend[396] measured the vibration–rotation transitions between the ground state and the $v_5 = 1$ and $v_3 = 1$, $v_6 = 1$ states and determined the energy difference between the $K = 2$ and $K = 5$ rotational levels in the vibrational ground state of CH_3I. This data together with the MMW measurements[375] yielding B_0, D_J and D_{JK}, led to the value $A_0 - 29D_K = 5.170\,82 \pm 0.000\,08$ cm^{-1} = $155\,017.31 \pm 2.4$ MHz.

7.3.6 Asymmetric-top molecules

Although all except the lightest asymmetric top rotors have an abundance of rotational transitions which occur in the centimetre wavelength region, measurements in the MMW region are desirable for a number of reasons. First, measurements over a wide range of frequencies are often necessary before a tentative assignment of the considerably more complex asymmetric rotor spectral transitions can be confirmed. Secondly, the theory of asymmetric rotors has advanced to a stage where it is advantageous to have precise MMW measurements available for an accurate evaluation of the theory in its finer details. Furthermore, sub-MMW measurements on various molecules may serve as highly accurate secondary frequency standards for infrared measurements. And finally the results from MMW radio-astronomy are strongly stimulating further activity in the investigation, particularly of those light asymmetric top rotors, whose spectra are located in the upper MMW region and which have not yet been fully investigated, if at all. Although most molecules are of the asymmetric-top class, the MMW spectra of only some 30 molecules which are listed in Table 7.9 have been investigated in the frequency range above 100 GHz.

Table 7.9 Asymmetric rotors, measured in the MMW region (from reference 356, by courtesy of Academic Press)

Molecule*	A/MHz	B/MHz	C/MHz	References
H_2O	835 783.30	435 044.51	278 446.89	478
D_2O	462 292.37(43)	217 979.95(27)	145 303.34(14)	456
HOD	701 931.50(22)	272 921.60(11)	192 055.25(10)	462
H_2S	316 304	276 512	147 536	482
D_2S†	164 571.118(45)	135 380.313(45)	73 244.068(71)	463
$H_2{}^{80}Se$	245 060(60)	231 772(60)	117 063(60)	421, 481
HOCl		15 117.50(15)	14 725.04(15)	448
HOF	590 000(6000)	26 758.00(10)	25 512.51(10)	447
NOF	95 189.77	11 844.12	10 508.31	433
$NO^{35}Cl$	87 650	5 737.69(3)	5 376.31(3)	422, 430
$NO^{79}Br$	83 340(104)	3 747.08(3)	3 586.14(6)	422, 430
S_2O†	41 915.440(28)	5 059.101(3)	4 507.163(3)	434
NO_2	239 898.3(11.0)	13 001.59(1.0)	12 304.72(3.0)	161
O_3	106 536.1813(72)	13 349.0465(84)	11 834.5493(10)	426
SO_2	60 778.553(7)	10 318.964(1)	8 799.806(1)	418
F_2O	58 782.630(50)	10 896.431(10)	9 167.412(10)	425
HNNN	609 679(899)	12 034.15(9)	11 781.51(9)	450, 451
HNCO	916 957(3000)	11 071.02(5)	10 910.58(5)	449, 451
HNCS	1 346 918(3000)	5 883.42(2)	5 845.62(2)	449, 451
H_2CO	281 972.3(1.3)	38 835.56(9)	34 002.94(15)	443
FHCO	91 156.56	11 760.23	10 396.72	479
H_2CS	291 710(42)	17 698.87(44)	16 652.98(48)	441
NO_2F	13 201.31	11 446.25	6 118.99	432
COF_2	11 813.550	11 753.049	5 880.896	431
$COCl_2$	7 918.84(4)	3 474.98(2)	2 412.22(2)	480
HSSH	146 858.170(32)	6 970.430(3)	6 967.689(3)	455
H_2C_2O	282 081(12)	10 293.9722(59)	9 915.2396(62)	444
HCOOH	77 512.23(2)	12 055.106(3)	10 416.156(3)	436
H_2CCHF	64 582.7	10 636.83	9 118.18	483
CH_3OH	A_1 674 437.00 A_2 157 328.96	24 680.03₃	23 769.67₂	474
CH_3SCN	15 787.077	4 155.613	3 354.103	475

* For other than the naturally most abundant isotopic species see original references.
†Reduced rotational constants A, B, C.

7.3.6.1 Centrifugal distortion treatments

Theoretical treatments of the rotational spectra of asymmetric rotors[404, 405] including centrifugal distortion effects[406-414] have been published. This theory, which has been applied with remarkable success to a large variety of molecules, is essentially based on the Hamiltonian given by Wilson and Howard[381] in 1936. A general first-order perturbation treatment was developed by Kivelson and Wilson[413, 414] and most applications follow this framework. The Hamiltonian is commonly written in the form:

$$H = H_2 + H_4 = AP_a^2 + BP_b^2 + CP_c^2 + \tfrac{1}{4} \sum_{\alpha\beta\beta} \tau'_{\alpha\alpha\beta\beta} P_\alpha^2 P_\beta^2 \qquad (7.4)$$

where the rotational constants A, B, C are related in the usual way to the reciprocal moments of inertia, and the τ' values may be expressed as functions of the harmonic force constants. Relatively later it was discovered by Watson[410] that the six angular momentum terms appearing in H_4 are not independent. Within the approximation discussed it is therefore not possible to evaluate simultaneously all nine constants appearing in equation (7.4). This indeterminacy was masked for planar molecules, since Dowling's planarity relationships were used to reduce the number of determinable centrifugal distortion constants from six to four (τ_{aaaa}, τ_{bbbb}, τ_{abab} and τ_{aabb})[414–417]. The effect of the indeterminacy of the six τ values and its removal has been discussed by Watson[410]. He has shown that with first-order perturbation theory five centrifugal distortion parameters can be determined in the general case of a non-planar molecule.

The τ constants may be expressed in Nielsen's notation as D_J, D_{JK}, D_K, D_1, D_2 and D_3 from the equations given by Allen and Olson[419]. The last three constants have sometimes in the literature been designated as δ_j, R_5, and R_6 but following Nielsen[420] and Herzberg[307] they are referred to here as D_1, D_2 and D_3 respectively.

Although the above-mentioned treatment is found to be sufficient for the bulk of asymmetric rotors, first-order correction terms in P^6 may have to be induced to obtain residuals of the same magnitude as the expected measurement error. The form of the Hamiltonian suitable for a first-order treatment of P^6 terms is that suggested by Watson

$$H_6 = H_J P^6 + H_{JK} P^4 P_a^2 + H_{KJ} P_a^4 P^2 + H_K P_a^6 + h_J P^4(P_b^2 - P_c^2)$$

$$+ h_{JK} P^2 [P_a^2(P_b^2 - P_c^2) + (P_b^2 - P_c^2)P_a^2] + h_K [P_a^4(P_b^2 - P_c^2) + (P_b^2 - P_c^2)P_a^4] \qquad (7.5)$$

Kirchhoff[418] has given a detailed description of the numerical evaluation of the centrifugal distortion parameters. He discusses specific procedures to be followed for a least-squares fit with accompanying error analysis, the question of determining the molecular force field from the centrifugal distortion constants alone, and finally how well microwave data can be fitted in the presence of arithmetical measurement and model errors. The procedures are illustrated by the specific examples, NSF, SiF_2, SO_2 and F_2O.

The accurate treatment of centrifugal distortion in SO_2 is a long standing problem[423]. Recently, Steenbeckeliers[424] reported that the rotational spectrum of SO_2 could not adequately be fitted if the planarity condition was invoked. This has been substantiated by the recent work of Kirchoff, who analysed more than 100 rotational transitions with a standard deviation of 0.260 MHz, but when the planarity constraint was released the standard deviation of the fit dropped to 0.083 MHz.

Since 1954 various similar careful distortion treatments of asymmetric-top molecules have been reported, such as F_2O [425], O_3 [426], Cl_2O [427] and NSF [428] which has now been re-analysed[418]. In particular, in the MMW region recent investigations are found for NOCl, NOBr[429, 430], COF_2 [331] and NO_2F [432] by Mirri and co-workers. In the latter molecule, which has C_{2v} symmetry, the general quadratic force field was determined. Using MMW measurements Cook presented a thorough centrifugal-distortion and force-constant analysis for NOF [433], while earlier measurements on S_2O [434] were extended by Cook, Lindsey and Winnewisser[435] into the MMW region to obtain reliable distortion constants. Bellet et al.[436] have also extended earlier measurements[437–440] to perform a complete centrifugal distortion analysis of c. 200 ground-state transitions of formic acid, HCOOH. They also report measurements on isotopic species which led to a complete r_s structure determination.

In some cases where MW and MMW data are available, a complete determination of the ground-state rotational and centrifugal distortion constants cannot be made. Examples in which this happens are molecules with C_{2v} symmetry, where the dipole moment is aligned along the a-axis. Hence the observed MW transitions have only a weak dependence on the A rotational constant. Molecules in which this occurs are formaldehyde, H_2CO, thioformaldehyde, H_2CS, ketene, H_2C_2O, and others. Although observation of the much weaker $\Delta K_a = 2$ transitions would alleviate this difficulty, these transitions are often too weak or too high in frequency to be easily observed. Refuge has then to be taken to infrared spectroscopy by which this difficulty can be circumvented, at least for some molecules. MW and infrared measurements on H_2CS by Johnson, Powell and Kirchoff [441] and Johns and Olson[442], as well as the analysis of new MMW data on H_2CO and its different isotopic species by Oka, Johns and Winnewisser[443, 445] and the MMW data on H_2C_2O by Johns, Stone and Winnewisser[444] serve as recent examples. For ketene the MW measurements of Johnson and Strand-berg[446] did not yield a value for the A constant at all, whereas the MMW data allow at least an estimate of $A = 282\,778 \pm 442$ MHz from the observed $K = 2$ asymmetry doublets. A simultaneous analysis of this data together with new infrared data lead to the value $A = 282\,081 \pm 12$ MHz [444].

7.3.6.2 Slightly asymmetric rotors

Many slightly asymmetric-top molecules with C_s symmetry present a similar problem: only a partial determination of their molecular constants is possible. Either the dipole moment component perpendicular to the a-axis is too small to allow observation of b or c transitions or they may occur at frequencies not accessible at present. The recent communications of the MMW spectra of hypofluorous acid, HOF, by Kim, Pearson and Appelman[447] and hypochloric acid, HOCl, by Mirri, Scappini and Cazzoli[448] serve as examples of this difficulty. While the rotational constants B and C are determined to an acceptable accuracy, the A constants can only be guessed from the MMW data. Similarly, the A constants of HNCS, HNCO and HNNN are not well determined at all, although MMW spectra and far-infrared spectra of these molecules are well known. HOF and HOCl have positive D_{JK} values (2.417(5) and 1.251(9) MHz, respectively) as was found in the cases of isocyanic[449] and hydrazoic acid[450]. The qualitative explanation given by Neely[499] for the latter molecules may be applied to HOF and HOCl as well; the decrease in bond angle formed by the off-axis H and the heavy-atom chain, which would increase the effective B value $(B - 2D_{JK}K^2)$

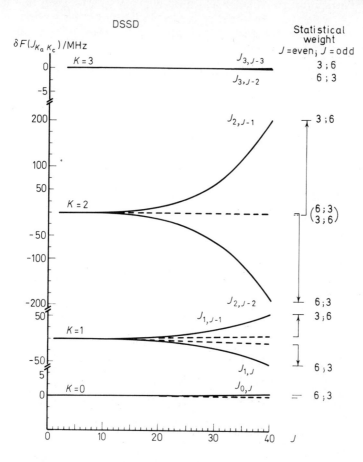

Figure 7.6 Rotational energy of DSSD as a function of J
(From Winnewisser[453], reproduced by courtesy of The American Institute of Physics)

for increasing K_a due to the rotation of the molecule around the a-axis (this would correspond to a negative D_{JK}), is thought to be overcompensated by a corresponding lengthening of the heavy-atom bonds. Indeed support for this interpretation is found in HOCl, where the ClO stretch–bend interaction constant has been determined to be $+0.5$ mdyn[448]. A positive sign for this constant is expected for the assumed model.

Spectra of especially slightly asymmetric-top molecules have been fitted often to a simplified energy expression in which the centrifugal distortion effects are accounted for only by the diagonal terms in P^4 and P^6. However, if accurate and sufficient data is available for a particular molecule, effects due to the neglected off-diagonal distortion terms become apparent. The wide span and high resolution of MMW spectroscopy are particularly useful for revealing these neglected terms.

An outstanding example for which off-diagonal quartic distortion terms cannot be neglected is hydrogen sulphide, HSSH, and its deuterated species DSSD[452–455]. The peculiar situation is met that the effect of the off-diagonal

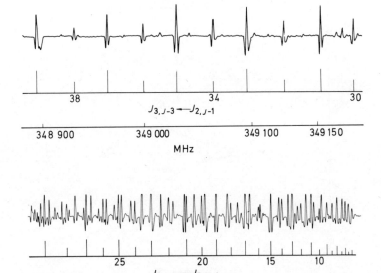

Figure 7.7 Centrifugal distortion splitting of the $^rQ_2$ branch of DSSD (From Winnewisser and Helminger[454], reproduced by courtesy of The American Institute of Physics)

centrifugal-distortion terms is as large or even larger than that of the diagonal terms, causing remarkable anomalies in the observed MMW spectra of these molecules. These anomalies are due to the additional K-doubling caused by the centrifugal distortion operator $(\hat{J}_+^4 + \hat{J}_-^4)$ in the rotational Hamiltonian[453]. This effect is referred to as centrifugal-distortion splitting.

The magnitude of the K-doubling in DSSD is illustrated by Figure 7.6 which shows the variation of the energy levels for different K values ($K = 0,1,2,3$) as a function of J. The diagonal terms in P^2 and P^4 have been subtracted from the energy, so that the plotted quantity δF is

$$\delta F(J_{K_aK_c}) = F(J_{K_aK_c}) - \{\tfrac{1}{2}(B+C)J(J+1) - D_K K^4 - D_{JK}K^2 J(J+1) - D_J J^2(J+1)^2\}$$ (7.6)

Therefore any deviations of δF from horizontal lines signify departures from symmetric-top behaviour. The observed K-doubling of the levels (solid lines) is the sum of two contributions; the asymmetry splitting (dashed lines) and the centrifugal-distortion splitting (indicated by arrows). For the $K = 1$ levels, asymmetry splitting and centrifugal-distortion splitting are of the same order of magnitude; however for the $K = 2$ levels the centrifugal distortion splitting is not only orders of magnitude larger ($c. \sim 10^4$ at $J = 40$) but is also opposite in sign to the asymmetry splitting. The DSSD molecule behaves strikingly like a symmetric top as far as the inertial asymmetry is concerned but is asymmetric with respect to the force field. For HSSH a similar reversal of the $K = 2$ levels has been found[455].

The effects of K-doubling are particularly well illustrated in the spectra of these Q branches[454]. The $^rQ_2$ branch of DSSD reproduced in Figure 7.7 illustrates centrifugal-distortion splitting in its purest form with no interference from other effects. The recorded spectrum can be divided into two regions, one with a simple pattern (top line) and one with a fairly complex pattern (bottom line). An analysis of the spectrum shows that the observed splitting is caused entirely by centrifugal distortion and that for the $J_{3,J-2} \leftarrow J_{2,J-2}$ transitions diagonal and off-diagonal centrifugal-distortion terms have opposite signs, which together with the different J dependence of these two contributions causes the formation of a second sub-band head for high J. This extraordinary spectrum may be fully appreciated if it is realised that without centrifugal distortion the entire $^rQ_2$ branch would coalesce into a single line at the position of the $J = 3$ line.

Unlike the asymmetry splitting the centrifugal-distortion splitting is not restricted to asymmetric rotors alone but can occur in symmetric-top molecules as well. K-doubling causes, in ammonia for example, a $K = 3$ and $K = 6$ splitting superposed on the inversion splitting[360] and recently a $K = 3$ splitting has been detected in the radio-frequency spectrum of phosphine[366].

7.3.6.3   Light asymmetric rotors

The spectra of light asymmetric rotors (H_2O, H_2S, etc.) can only be analysed by including higher-order terms in the angular momentum into the Hamiltonian. Each degree n in the angular momentum adds $(n+1)$ distortion coefficients to the reduced Hamiltonian. In the past such molecules have eluded MW study because of the occurrence of many, if not most, of their rotational transitions at frequencies beyond the reach of conventional MW spectrometers. In D_2S for example, only one transition ($1_{10}-1_{01}$) falls below 100 GHz and only six below 200 GHz, whereas most low-J transitions are above 200 GHz. As a result, most of the analyses of these molecules have been carried out by combining infrared data with available microwave data. Benedict et al. have analysed D_2O [456], Dowling H_2O [457] while Steenbeckeliers and Bellet[458-460] have analysed H_2O, HDO and D_2O using a somewhat limited set of MW data. In a recent series of papers DeLucia, Cook, Helminger and Gordy have measured in the frequency range up to 750 GHz the spectra of the molecules HDO (53 transitions)[462], HDS (46 transitions)[461], D_2S (66 transitions)[463] and H_2O (13 transitions)[478]. In the analysis of these extensive new results, they have employed a 22 parameter rotation–distortion Hamiltonian which included distortion effects as far as P^{10}, although only a partial set of terms in P^8 and P^{10} was necessary.

One should be cautioned about a direct comparison of molecular constants for one molecule given by various authors because of possible different formulations of the distortion theory used. An example of the increasing complexity of the rotation–distortion Hamiltonian is seen in a comparison of the rotation and distortion constants of $HD^{16}O$ determined by De Lucia et al.[461] and by Bellet and Steenbeckeliers[460].

7.3.6.4   Molecular structure of asymmetric rotors

The structures of many asymmetric top molecules have been determined from MW and MMW data[6, 18, 27]. From the four types of well defined bond lengths[6] r_o, r_s, r_z, r_e, usually only r_o and r_s structures are evaluated. Nevertheless, r_e structures have been obtained by Morino and co-workers[464–468] in SeO_2 and F_2O, O_3, SO_2 and H_2Se. For SO_2 the four structures were compared, distinctly revealing in its r_z structure the distortion of the molecule characteristic for each normal vibration[467]. The chain structures of the molecules H_2O_2 and H_2S_2 are well known[452]. Using the structural parameters (r_o) of HSSH and DSSD it was possible[469] to establish for HSSD a value of $0.003\,45 \pm 0.000\,30$ Å for the shortening of the SH bond caused by replacement of H by D.

7.3.6.5   Internal motion of asymmetric rotors

Although the study of internal rotation has been the subject of considerable activity within the last 20 years[6, 17, 20], almost all of the research has been confined to the region between 5 and 40 GHz. Only a very few molecules have been studied at higher frequencies. This is mainly due to experimental limitations including the difficulty of obtaining reliable assignments. The unambiguous identification of the informative P- and R-branch transitions presents the main problem in the MMW region.

We will mention only the four molecules CH_3OH, CH_3SCN, HOOH and HSSH which exhibit internal rotation and whose spectra have recently been studied in the MMW region. There exist, however, a number of excellent review articles to which the reader is referred[470, 471].

(a) *Methyl alcohol* (CH_3OH) – The understanding of the principal features of the methyl alcohol spectrum is a tribute to the theoretical investigations of Dennison and co-workers[472, 473]. Lees and Baker[474] recently reported the MMW spectrum of methyl alcohol. In the analysis of the *a*-type and b-type spectra, torsion–vibration–rotation interactions were represented by adjustable parameters in semi-empirical formulae. Lees and Baker showed that for any molecule the torsional barrier coefficients V_3 and V_6, the moments of inertia I_{a1} and I_{a2} about the near-symmetry axis and two of the adjustable interaction parameters k_6 and k_7 cannot be independently determined from the spectrum of a single isotopic species. The effective hindering potential barrier height V_3 for CH_3OH is between 372 and 373 cm^{-1}, and, ignoring all torsion-vibration-rotation interactions, V_3 was found to decrease slightly on deuteration.

(b) *Methyl thiocyanate* (CH_3SCN) – The microwave spectrum of CH_3SCN has also been the subject of many investigations. Dreizler and Mirri[475] extended the earlier measurements up to 99 GHz, observing both a-type and b-type transitions for $J = 13 \leftarrow 12$. The torsional fine structure observed for the ground vibrational state and for the first excited state of the CSC-bending vibration in CH_3SCN showed a marked dependence upon excitation of this vibration. The analysis was carried out with a Hamiltonian constructed from a molecular model having three rotational degrees of freedom, one torsional degree of freedom and one vibrational degree of freedom, and yielded the determination of a torsional potential function and a consistent set of rotational constants. Furthermore, it was found that the ground-state torsional splitting is nearly independent of the vibrational effects and yields therefore the most reliable torsional potential constants.

(c) *Hydrogen peroxide* ($HOOH$) *and hydrogen disulphide* ($HSSH$) – HOOH and HSSH are analogous molecules in so far as both possess a chain-like structure and consequently C_2 symmetry which allows internal rotation of the two OH or SH bars relative to each other. The barriers hindering the internal motion in the two molecules, however, are found to be surprisingly different. In hydrogen peroxide the effects of hindered internal rotation are observed in all regions of the spectrum[476, 477]. In HSSH an observable splitting due to barrier tunnelling is revealed only by the high resolution of MMW-spectroscopy[452]. By means of the saturation-dip method Winton and Winnewisser[73] measured this small doubling of only 116 kHz (see Figure 7.2) which gives a splitting of the rotational levels of only 58 kHz. The splitting of the ground state energy levels in HOOH has also been determined from MMW spectra and may be given as $342\ 885.0 \pm 2.0$ MHz[476], a factor of 6×10^6 larger than that for HSSH.

The numerical solution of the wave equation of the internal rotation problem for HOOH showed[477] that the energy level scheme derived from the infrared spectrum can be represented by a potential of the form $V_{cis} = 2460$ cm^{-1} and $V_{trans} = 386$ cm^{-1} and $\angle$ HOO $= 111.5$ degrees, indicating the position of the potential function minimum. These values together with spectroscopic constants allow an accurate prediction of the gross features of the microwave and MMW spectrum of HOOH. The locations of the various rotation–torsion sub-band centres could be calculated and were confirmed by the MMW measurements of Oelfke and Gordy[476].

Whereas in HOOH the barrier heights can be derived exclusively from infrared measurements, the barrier-height determination in HSSH is dependent predominantly upon the MMW data. A preliminary analysis indicates that $V_{trans} = 2370$ cm^{-1} and $V_{cis} = 2550$ cm^{-1} for HSSH.

7.4 SPECIAL TOPICS

7.4.1 Stark effect

Due to the relative ease of guiding a MMW beam between two parallel plates, which produce a very homogeneous electric field, accurate electric-dipole-moment measurements may be made in the MMW range by measuring the Stark shift of the rotational energy levels. Absolute measurements of the dipole moment of CH_3F at 102 GHz and of CD_3F at 82 GHz by Steiner and Gordy[484] in 1966 were more accurate than the measurement of the dipole moment of OCS which was accepted as a standard at that time[485]. These methyl fluoride measurements were found to be compatible with more recent MW[486] and electric-resonance measurements of the Stark effect in OCS [487] ($\mu = 0.714\,99 \pm 0.000\,05$ D) [488].

Stark-effect measurements in the MMW region have shown the effect of deuteration on the dipole moments of CH_3F [484], CH_3CN [484] and HNCO [52]. For HNCO, DNCO and HNNN, the variation of the dipole moment with the rotational quantum number K was determined by White and Cook[52]. Other molecules for which the dipole moment has been obtained with high accuracy from MMW transitions include D_2Se [489], HBr, HI [78] as well as DI, PH_3, CO, HCN and NO [6]. The values obtained for HBr and HI are particularly accurate because of the use of a molecular-beam absorption spectrometer[78]. Molecular-beam electric resonance at 89 GHz was used to determine the dipole moment of LiF in three excited states as well as the ground state, revealing a distinct non-linear variation with v (reference 81). The vibrational dependence of the dipole moment has been treated theoretically for diatomic[490] and triatomic linear molecules[491].

7.4.2 Molecular Zeeman effect

The high resolution of MW and MMW spectroscopy has proved to be effective in determining the small splitting in rotational transitions of diamagnetic molecules due to internal and external magnetic fields, as shown in the recent surveys by Flygare and Benson[492] and Gordy and Cook[6]. For many diatomic molecules the zero-field nuclear quadrupole and spin–rotation splitting has been measured with conventional MMW techniques (see Tables 7.2 and 7.3). Dymanus and co-workers have completed MMW molecular-beam absorption measurements on the magnetic hyperfine structure in H_2S [75], HBr [77] and HI [76]. The use of a MMW molecular-beam maser made possible a precise determination of the coupling constants in HCN [91]. In addition, maser measurements by DeLucia and Cederberg[493] on the $1_{10}–1_{01}$ transition of D_2S at 91 GHz, combined with similar results for HDS [494] and H_2S [75, 84], yielded the complete quadrupole coupling tensor and spin–rotation tensor for D_2S.

In the presence of an external magnetic field the energy levels of a diamagnetic molecule are further affected by the first-order rotational Zeeman effect, characterised by the g-tensor, and the second-order rotational Zeeman effect, characterised by the susceptibility tensor χ. Early work on DBr, DI,

$CO, PH_3, PD_3, SO_2, O_3, H_2S, HDS$ and D_2S was confined to the measurement of the first-order Zeeman effect[6]. The second-order rotational Zeeman effect, requiring high magnetic field strength, has been investigated only in the last few years[492]. In the MMW range, the magnetic susceptibility anisotropy of CO and CS has been reported by Gustafson and Gordy[130].

7.4.3 Microwave and laser — microwave double resonance

Two-photon absorption processes and their characteristics have been studied extensively in the MW and to a lesser extent in the MMW region. This method has been developed into a useful tool for the assignment of complicated rotational spectra of polyatomic molecules, for the measurement of dipole moments, the evaluation of rotational energy transfer and rotational relaxation. In the MMW region this method has been employed by Mirri et al.[431], to confirm the assignment of rotational transitions in COF_2. The application of high power pumping sources facilitates the observation of weak absorption signals. The sensitivity compares favourably with that achieved from Stark-effect modulated spectrometers as has been demonstrated, for example, through the observation of 'forbidden' eQq-induced $\Delta J = +3$ transitions in CH_3CH_2I [495]. These experiments have the potential of being applicable to observe weak $\Delta K \neq 0$ transitions in slightly asymmetric molecules. A second application of double resonance with high power pumping radiation is the study of non-equilibrium distributions in rotation and/or rotation–vibration energy levels in order to obtain quantitative information concerning collision-induced transitions, which in turn reveal the selection rules governing these transitions[496]. A MMW-laser double resonance experiment in CH_3Cl has been reported by Frenkel, Marantz and Sullivan[497] furnishing information on collisional energy transfer between rotational levels, vibrational levels and on the equalisation of the rotational temperature between adjacent J levels. This method has further potential for the study of collision processes and for high resolution MMW and infrared spectroscopy.

7.4.4 Spectroscopic determination of the velocity of light

Since 1950 a series of determinations of the velocity of light have been made by dividing the value of the rotational constant B_0 of a diatomic or linear molecule, as determined in frequency units, by the same constant as determined in reciprocal wavelength units. The molecules used have included $^{12}C^{16}O$, $H^{35}Cl$, $H^{79}Br$, $D^{35}Cl$ and $H^{12}C^{14}N$. In 1967 Pliva[498] obtained a new value of $c = 299\ 792.3(2)$ km s^{-1} from the data of the molecules $H^{35}Cl$, CO and N_2O. Since then, significant improvements, both in the quantity and accuracy of data available from MMW measurements, and certainly in the quantity of i.r. measurements, have taken place with respect to the molecules CO, $H^{35}Cl$, $D^{35}Cl$, N_2O, HCN and $^{16}O^{12}C^{32}S$. The importance of the recent measurements is indicated by the difference of 4 MHz between the first reported value[9] for the $J = 1 \leftarrow 0$ rotational transition of $H^{35}Cl$ and the more recent value[108]. A renewed effort to obtain a value for c from the new data should lead to a more reliable value, perhaps competitive with measurements upon which the currently accepted value of $c = 299\,792.50(10)$ km s^{-1} is based.

Acknowledgements

The authors have received help from many persons. We should like to thank especially Professor Wertheimer and Drs. De Lucia and Helminger for sending us preprints of unpublished results. One of us (M.W.) held a National Science Foundation Senior Fellowship when this review was being written.

References

1. Gordy, W., Smith, W V. and Trambarulo, R. F. (1953). *Microwave Spectroscopy*. (New York: Wiley. Republished in 1966)
2. Strandberg, M. W. P. (1954). *Microwave Spectroscopy*. (London: Methuen)
3. Townes, C. H. and Schawlow, A. L. (1955). *Microwave Spectroscopy*. (New York: McGraw-Hill)
4. Gordy, W. (1965). *Pure Appl. Chem.*, **11**, 403
5. Gordy, W. (1960). 'Millimeter and Submillimeter Waves in Physics', *Proceedings of the Symposium on Millimeter Waves*. New York: Polytechnic Press of the Polytechnic Institute of Brooklyn
6. Gordy, W. and Cook, R. L. (1970). *Microwave Molecular Spectra*, Volume IX, *Chemical Applications of Spectroscopy, Part II*. (West, W., editor) (New York: Wiley Interscience)
7. Burrus, C. A. and Gordy, W. (1954). *Phys. Rev.*, **93**, 897
8. Genzel, L. and Eckhardt, W. (1954). *Z. Physik*, **139**, 592
9. Jones, G. and Gordy, W. (1964). *Phys. Rev.*, **135**, A295
10. Helminger, P., De Lucia, F. C. and Gordy, W. (1970). *Phys. Rev. Lett.*, **25**, 1397
11. Tyler, J. K. (1963). *J. Molec. Spectrosc.*, **11**, 39
12. Möller, K. D. and Rothschild, W. G. (1971). *Far-Infrared Spectroscopy*, (Ballard, S., editor) (New York: Wiley-Interscience)
13. Kewley, R., Sastry, K. V. L. N., Winnewisser, M. and Gordy, W. (1963). *J. Chem. Phys.*, **39**, 2856
14. Costain, C. C. (1969). *Can. J. Phys.*, **47**, 2431
15. Winnewisser, M. (1971). *Z. Angew. Physik*, **30**, 359
16. Morino, Y. and Saito, S. (1972). 'Microwave Spectroscopy during the last Twenty-Five Years', *Molecular Spectroscopy: Modern Research*, (Rao, K. N. and Mathews, W. C., editors). (New York: Academic Press)
17. Rudolph, H. D. (1970). *Ann. Rev. Phys. Chem.*, **21**, 73
18. Lide, D. R. Jr. (1964). *Ann. Rev. Phys. Chem.*, **15**, 225
19. Flygare, W. H. (1967). *Ann. Rev. Phys. Chem.*, **18**, 325
20. Morino, Y. and Hirota, E. (1969). *Ann. Rev. Phys. Chem.*, **20**, 139
21. Wilson, E. B., Jr. (1968). *Science*, **162**, 59
22. Parkin, J. E. (1968). *Ann. Reports Chem. Soc.*, **65**, 111
23. Lovas, F. J. and Lide, D. R. Jr. (1971). *Advan. High Temp. Chem.*, **3**, 177
24. Sugden, T. M. and Kenney, C. N. (1965). *Microwave Spectroscopy of Gases*, (London: Van Nostrand)
25. Wollrab, J. E. (1967). *Rotational Spectra and Molecular Structure* (New York: Academic Press)
26. Martin, D. H., editor. (1967). *Spectroscopic Techniques for Far Infra-Red, Submillimetre and Millimetre Waves*. (Amsterdam: North Holland Publishing Co.)
27. Stark, B., Hüttner, W., Tischer, W. and Winnewisser, M. (1972). *Landolt-Börnstein, Numerical Data and Functional Relationships in Science and Technology*. Group II. Atomic and Molecular Physics, to be published. (Berlin: Springer-Verlag)

28. Starck, B. (1967). *Landolt-Börnstein,* Numerical Data and Functional Relationships in Science and Technology. Group II, Atomic and Molecular Physics, Vol. 4. Molecular Constants from Microwave Spectroscopy (Berlin: Springer Verlag)
29. National Bureau of Standards. *Microwave Spectral Tables,* Monograph 70, Vols. I-V (Washington, D.C.: U.S. Department of Commerce)
30. Guarnieri, A. and Favero, P. (1968). *Microwave Gas Spectroscopy Bibliography* 1954–1967, (Bologna: Instituto Chimico G. Ciamician Universita di Bologna)
31. Starck, B. (1970). *Bibliographie mikrowellenspectroskopischer Untersuchungen* III (1966–1969). Sektion Strukturdokumentation (Ulm: University of Ulm)
32. Palik, E. D. (1969). *Far-Infrared bibliography through* 1969. Up-to-date version of the original 'A Far Infrared Bibliography', published in *J. Opt. Soc. Amer.,* **50,** 1329 (1960). (Available from the office of Technical Services, Department of Commerce)
33. Gordy, W. and Kessler, M. (1947). *Phys. Rev.,* **12,** 644
34. Baker, J. G. (1967). Reference 26, Chapter 5
35. Fox, J., editor. (1964). *Proceedings of the Symposium on "Quasi-Optics" in Microwave Research Institute Symposia Series,* Vol. 14. (Brooklyn, N.Y.: Polytechnic Press of the Polytechnic Institute of Brooklyn) N.Y.
36. Strauch, R. G., Cupp, R. E., Lichtenstein, M. and Gallagher, J. J. (1964). in Reference 35, pp. 581–606
37. Fox, J., editor (1970). 'Proceedings of the Symposium on "Submillimeter Waves"', *Microwave Research Institute Symposia Series,* Vol. 20. (Brooklyn, N.Y.: Polytechnic Press of the Polytechnic Institute of Brooklyn)
38. King, W. C. and Gordy, W. (1953). *Phys. Rev.,* **90,** 319; (1954). *Phys. Rev.,* **93,** 407
39. Robson, P. N. (1967). Reference 26, Chapter 6
40. Martin, D. H. (1967). Reference 26, Chapter 1
41. Corcoran, V. J. and Smith, W. T. (1972). *Applied Optics,* **11,** 269
42. Wells, J. S. and Evenson, K. M. (1970). *Rev. Sci. Instr.,* **41,** 226
43. Dees, J. W. (1966). *Microwave J.,* **9,** 48
44. Wentworth, F. L., Dozier, J. W. and Rodgers, J. D. (1964). *Microwave J.,* **7,** 69
45. Ohl, R. S., Budenstein, P. P. and Burrus, C. A. (1959). *Rev. Sci. Instr.,* **30,** 765
46. Van Es, C. W., Gevers, M. and De Ronde, F. C. (1960). *Philips Tech. Rev.,* **22,** 115
47. Meredith, R. and Warner, F. L. (1963). *IEEE Trans. MTT.,* **11,** 397
48. Dees, J. W. and Sheppard, A. P. (1963). *Electronics,* **36,** 33
49. Bauer, R. J., Cohn, M., Cotton, J. M. and Packard, R. F. (1966). *Proc. IEEE,* **54,** 595
50. Veazey, S. E. and Gordy, W. (1965). *Phys. Rev.,* **138,** A1303
51. Winnewisser, G. and Cook, R. L. (1968). *J. Molec. Spectrosc.,* **28,** 266
52. White, K. J. and Cook, R. L. (1967). *J. Chem. Phys.,* **46,** 143
53. Mullard *Indium Antimonide Sub-millimeter wavelength infrared detector,* Industrial Electronics Division, Mullard Ltd., Mullard House, Torrington Place, London, W.C.1.
54. Advanced Kinetics, Inc., 1231 Victoria St., Costa Mesa, California.
55. Putley, E. H. (1965). *Applied Optics,* **4,** 649
56. Putley, E. H. and Martin, D. H. (1967). Reference 26, Chapter 4
57. Levinstein, H. (1965). *Applied Optics,* **4,** 639
58. Stahl, K. (1968). *Optik,* **27,** 11
59. Kneubühl, F. (1969). *Applied Optics,* **8,** 505
60. Willardson, R. K. and Beer, A. C. (1970). *Infrared Detectors in Semiconductors and Semimetals* Vol. 5. (London: Academic Press)
61. Gilbert, J. (1970). *Rev. Sci. Instr.,* **41,** 1050
62. Törring, T. and Schnabel, E. (1967). *Z. Physik,* **204,** 198
63. Gwinn, W. D., Luntz, A. C., Sederholm, C. H. and Millikan, R. (1968). *J. Comput. Phys.,* **2,** 439
64. Boyd, G. D. and Gordon, J. P. (1961). *Bell System Tech. J.,* **40,** 489
65. Culshaw, W. (1953). *Proc. Phys. Soc. (London),* **B66,** 597
66. Culshaw, W. (1959). *IRE Trans.,* **7,** 221; (1960). *IRE Trans.,* **8,** 182; (1961). *IRE Trans.,* **9,** 135
67. Lichtenstein, M., Gallagher, J. J. and Cupp, R. E. (1963). *Rev. Sci. Instr.,* **34,** 843
68. Valkenburg, E. P. and Derr, V. E. (1966). *Proc. IEEE,* **54,** 493
69. Bates, H. E., Gallagher, J. J. and Derr, V. E. (1968). *J. Appl. Phys.,* **39,** 3218
70. Lamb, W. E., Jr. (1964). *Phys. Rev.,* **134,** A1429
71. McFarlane, R. A., Bennett, W. R., Jr. and Lamb, W. E., Jr. (1963). *Appl. Phys. Lett.,* **2,** 189

72. Hanes, G. R. and Dahlstrom, C. E. (1969). *Appl. Phys. Lett.,* **14,** 362
73. Winton, R. S. and Winnewisser, G. (1970). *Symp. Molec. Struct. Spectrosc.,* Paper Q3, Columbus, Ohio, to be published
74. Rusk, J. R. and Gordy, W. (1962). *Phys. Rev.,* **127,** 817
75. Huiszoon, C. and Dymanus, A. (1966). *Phys. Lett.,* **21,** 164
76. Van Dijk, F. A. and Dymanus, A. (1968). *Chem. Phys. Lett.,* **2,** 235
77. Van Dijk, F. A. and Dymanus, A. (1969). *Chem. Phys. Lett.,* **4,** 170
78. Van Dijk, F. A. and Dymanus, A. (1970). *Chem. Phys. Lett.,* **5,** 387
79. Ramsey, N. F. (1965). *Molecular Beams.* (Oxford: Oxford University Press)
80. Kusch, P. and Hughes, V. W. (1959). 'Atomic and molecular beam spectroscopy', *Encyclopedia of Physics,* Vol. 37. (Flügge, S., editor) (Berlin: Springer-Verlag)
81. Wharton, L., Klemperer, W., Gold, L. P., Strauch, R., Gallagher, J. J. and Derr, V. E. (1963). *J. Chem. Phys.,* **38,** 1203
82. Ramsey, N. F. (1949). *Phys. Rev.,* **76,** 996
83. Strauch, R. G., Cupp, R. E., Derr, V. E. and Gallagher, J. J. (1966). *Proc. IEEE,* **54,** 506
84. Cupp, R. E., Kempf, R. A. and Gallagher, J. J. (1968). *Phys. Rev.,* **171,** 60
85. Marcuse, D. (1961). *J. Appl. Phys.,* **32,** 743
86. Marcuse, D. (1961). *Proc. IRE,* **49,** 1706
87. Marcuse, D. (1962). *IRE Trans. Inst.,* **11,** 187
88. Krupnov, A. F. and Skvortsov, V. A. (1964). *Soviet Physics JETP,* **18,** 74
89. Krupnov, A. F. and Skvortsov, V. A. (1964). *Soviet Physics JETP,* **18,** 1426
90. Krupnov, A. F. and Skvortsov, V. A. (1965). *Soviet Physics JETP,* **20,** 1079
91. De Lucia, F. and Gordy, W. (1969). *Phys. Rev.,* **187,** 58
92. De Lucia, F. and Gordy, W. (1970). Reference 37, pp. 99–114
93. De Lucia, F., Cederberg, J. W. and Gordy, W. (1970). *Bull. Amer. Phys. Soc.,* **15,** 562
94. Gordon, J. P., Zeiger, H. J. and Townes, C. H. (1955). *Phys. Rev.,* **99,** 1264
95. Weber, J. (1967). *Masers:* Selected Reprints with Editorial Comment Vol. 9, International Science Reviews Series. (New York: Gordon and Breach)
96. Röss, D. (1969). *Lasers Light Amplifiers and Oscillators.* (London and New York: Academic Press)
97. Button, K. J. and Lax, B. (1970). Reference 37, pp. 401–416
98. Button, K. J. (1971). 'Molecular Gas Lasers and their Applications', *Short Laser Pulses and Coherent Interactions.* (Haidemenakis, E. D., editor) (New York: Gordon and Breach)
99. Crocker, A., Gebbie, H. A., Kimmitt, M. F. and Mathias, L. E. S. (1964). *Nature (London),* **201,** 250
100. Gebbie, H. A., Stone, N. W. B., Findlay, F. D. and Robb, J. A. (1964). *Nature (London),* **202,** 169
101. Gebbie, H. A., Stone, N. W. B. and Findlay, F. D. (1964). *Nature (London),* **202,** 685
102. Duxbury, G., Jones, R. G., Burroughs, W. J., Bradley, C. C. and Stone, N. W. B. (1970). *Millimeter Waves.* pp. 79–83 in Reference 37
103. Corcoran, V. J., Gallagher, J. J. and Cupp, R. E. (1970). Reference 37, pp. 85–91
104. Cuppa, R. E., Corcoran, V. J. and Gallagher, J. J. (1970). *IEEE J. Quantum Electronics,* **QE-6,** 160; **QE-6.** 241
105. Corcoran, V. J., Cupp, R. E. and Gallagher, J. J. (1969). *IEEE J. Quantum Electronics,* **QE-5,** 424
106. Chang, T. Y. and Bridges, T. J. (1970). Reference 37, pp. 93–98
107. Evenson, K. M., Radford, H. E. and Moran, M. M., Jr. (1971). *Appl. Phys. Lett.,* **18,** 426
108. De Lucia, F. C., Helminger, P. and Gordy, W. (1971). *Phys. Rev.,* **3A,** 1849
109. Burrus, C. A., Gordy, W., Benjamin, B. and Livingston, R. (1955). *Phys. Rev.,* **97,** 1661
110. Rosenblum, B. and Nethercot, A. H. (1955). *Phys. Rev.,* **97,** 84
111. Dunham, J. L. (1932). *Phys. Rev.,* **41,** 721
112. Rosenblum, B., Nethercot, A. M. and Townes, C. H. (1958). *Phys. Rev.,* **109,** 400
113. Pearson, E. F. and Gordy, W. (1969). *Phys. Rev.,* **177,** 59
114. Van Vleck, J. H. (1936). *J. Chem. Phys.,* **4,** 327
115. Herman, R. M. and Asgharian, A. (1966). *J. Molec. Spectrosc.,* **19,** 305
116. Bunker, P. R. (1970). *J. Molec. Spectrosc.,* **35,** 306
117. Bunker, P. R. (1971). *J. Molec. Spectrosc.,* **39,** 90
118. Pearson, E. F. and Gordy, W. (1969). *Phys. Rev.,* **177,** 52
119. Clouser, P. L. and Gordy, W. (1964). *Phys. Rev.,* **134,** A863
120. Hoeft, J., Lovas, F. J., Tiemann, E. and Törring, T. (1970). *Z. Naturforsch.,* **25a,** 1750

121. Lovas, F. J., and Johnson, D. R. (1971). *J. Chem. Phys.*, **55**, 41
122. Nesbet, R. K. (1964). *J. Chem. Phys.*, **40**, 3619
123. Nesbet, R. K. (1965). *J. Chem. Phys.*, **43**, 4403
124. Wyse, F. C., Gordy, W. and Pearson, E. F. (1970). *J. Chem. Phys.*, **52**, 3887
125. Lide, D. R., Jr. (1965). *J. Chem. Phys.*, **42**, 1013
126. Hoeft, J., Lovas, F. J., Tiemann, E. and Törring, T. (1970). *Z. Naturforsch.*, **25a**, 1029
127. Yardley, J. T. (1970). *J. Molec. Spectrosc.*, **35**, 314
128. Mantz, A. W., Nichols, E. R., Alpert, B. D. and Rao, K. N. (1970). *J. Molec. Spectrosc.*, **35**, 325
129. Burrus, C. A. (1958). *J. Chem. Phys.*, **28**, 427
130. Gustafson, S. and Gordy, W. (1970). *J. Chem. Phys.*, **52**, 579
131. Wilson, R. W., Penzias, A. A., Jefferts, K. B., Kutner, M. and Thaddeus, P. (1971). *Astrophys. J.*, **167**, L97
132. Törring, T. (1968). *Z. Naturforsch.*, **23a**, 777
133. Raymonda, J. W., Muenter, J. S. and Klemperer, W. A. (1970). *J. Chem. Phys.*, **52**, 3458
134. Hoeft, J., Lovas, F. J., Tiemann, E. and Törring, T. (1969). *Z. Naturforsch.*, **24a**, 1217
135. Hoeft, J., Lovas, F. J., Tiemann, E. and Törring, T. (1970). *J. Chem. Phys.*, **53**, 2736
136. Hoeft, J., Lovas, F. J., Tiemann, E., and Törring, T. (1971). *Z. Naturforsch.*, **26a**, 240
137. Raymonda, J. and Klemperer, W. (1971). *J. Chem. Phys.*, **55**, 232
138. Mann, D. E., Thrush, B. A., Lide, D. R., Jr., Ball, J. J. and Acquista, N. (1961). *J. Chem. Phys.*, **34**, 420
139. Rank, D. H., Eastman, D. P., Rao, D. S. and Wiggins, T. A. (1962). *J. Opt. Soc. Amer.*, **52**, 1
140. Jones, G. and Gordy, W. (1964). *Phys. Rev.*, **136**, A1229
141. Plyler, E. K. (1960). *J. Res. Nat. Bur. Stand.*, **64A**, 377
142. Honig, A., Mandel, M., Stitch, M. L. and Townes, C. H. (1954). *Phys. Rev.*, **96**, 629
143. Fabricand, B. P., Carlson, R. O., Lee, C. A. and Rabi, I. I. (1953). *Phys. Rev.*, **91**, 1403
144. Herzberg, G. (1971). *The Spectra and Structures of Simple Free Radicals.* (Ithaca: Cornell University Press)
145. Burkhalter, J. H., Anderson, R. S., Smith, W. V. and Gordy, W. (1950). *Phys. Rev.*, **77**, 152; (1950) **79**, 651
146. Beringer, R. (1946). *Phys. Rev.*, **70**, 53
147. Lamont, H. R. L. (1948). *Phys. Rev.*, **74**, 353; *Proc. Phys. Soc.*, **61**, 562
148. Strandberg, M. W. P., Meng, C. Y. and Ingersoll, J. G. (1949). *Phys. Rev.*, **75**, 1524
149. Burrus, C. A. and Gordy, W. (1953). *Phys. Rev.*, **92**, 1437
150. Gallagher, J. J., Bedard, F. D. and Johnson, C. M. (1954). *Phys. Rev.*, **93**, 729
151. Gallagher, J. J. and Johnson, C. M. (1956). *Phys. Rev.*, **103**, 1727
152. Favero, P. G., Mirri, A. M. and Gordy, W. (1959). *Phys. Rev.*, **114**, 1534
153. McAfee, K. B., Jr. (1950). *Phys. Rev.*, **78**, 340A
154. McAfee, K. B., Jr. (1951). *Phys. Rev.*, **82**, 971L
155. Bird, G. R. (1955). *Phys. Rev.*, **98**, 1160A
156. Bird, G. R. (1956). *J. Chem. Phys.*, **25**, 1040
157. Lin, C. C. (1959). *Phys. Rev.*, **116**, 903
158. Hodgeson, J. A., Sibert, E. E. and Curl, R. F., Jr. (1963). *J. Phys. Chem.*, **67**, 2833
159. Esterowitz, L. and Rosenthal, J. (1964). *J. Chem. Phys.*, **40**, 1986
160. Bird, G. R., Baird, J. C., Jache, A. W., Hodgeson, J. A., Curl, R. F., Jr., Kunkle, A. C., Bransford, J. W., Rastrup-Andersen, J. and Rosenthal, J. (1964). *J. Chem. Phys.*, **40**, 3378
161. Lees, R. M., Curl, R. F., Jr. and Baker, J. G. (1966). *J. Chem. Phys.*, **45**, 2037
162. Foster, P. D., Hodgeson, J. A. and Curl, R. F., Jr. (1966). *J. Chem. Phys.*, **45**, 3760
163. Curl, R. F., Jr., Kinsey, J. L., Baker, J. G., Baird, J. C., Bird, G. R., Heidelberg, R. F., Sugden, T. M., Jenkins, D. R. and Kenney, C. N. (1961). *Phys. Rev.*, **121**, 1119
164. Curl, R. F., Jr., Heidelberg, R. F. and Kinsey, J. L. (1962). *Phys. Rev.*, **125**, 1993
165. Curl, R. F., Jr. (1962). *J. Chem. Phys.*, **37**, 779
166. Pillai, M. G. K. and Curl, R. F., Jr. (1962). *J. Chem. Phys.*, **37**, 2921
167. Tolles, W. M., Kinsey, J. L., Curl, R. F., Jr. and Heidelberg, R. F. (1962). *J. Chem. Phys.*, **37**, 927
168. Mariella, R. P., Jr. and Curl, R. F., Jr. (1970). *J. Chem. Phys.*, **52**, 757
169. Sanders, T. M., Schawlow, A. L., Dousmanis, G. C. and Townes, C. H. (1953). *Phys. Rev.*, **89**, 1158L
170. Dousmanis, G. C. (1954). *Phys. Rev.*, **94**, 789A

171. Sanders, T. M., Schawlow, A. L., Dousmanis, G. C. and Townes, C. H. (1954). *J. Chem. Phys.*, **22**, 245
172. Dousmanis, G. C., Sanders, T. M. and Townes, C. H. (1955). *Phys. Rev.*, **100**, 1735
173. Costain, C. C. (1957). *Can. J. Phys.*, **35**, 241
174. Powell, F. X., Fletcher, O. and Lippincott, E. R. (1963). *Rev. Sci. Instr.*, **34**, 36
175. Powell, F. X. and Lide, D. R., Jr. (1964). *J. Chem. Phys.*, **41**, 1413
176. Del Greco, F. P. and Kaufmann, F. (1962). *Discuss. Faraday Soc.*, **33**, 128
177. Winnewisser, M., Sastry, K. V. L. N., Cook, R. L. and Gordy, W. (1964). *J. Chem. Phys.*, **41**, 1687
178. Amano, T., Hirota, E. and Morino, Y. (1968). *J. Molec. Spectrosc.*, **27**, 257
179. Amano. T., Saito, S., Hirota, E., Morino, Y., Johnson, D. R. and Powell, F. X. (1969). *J. Molec. Spectrosc.*, **30**, 275
180. Uehara, H., Saito, S. and Morino, Y. (1970). *Symp. Molec. Struct. Spectrosc.*, Paper E3, Columbus, Ohio
181. Saito, S. (1970). *J. Chem. Phys.*, **53**, 2544
182. Baird, K. M. and Bredohl, H. (1971). *Astrophys. J.*, **189**, L83
183. Radford, H. E. (1959). *Nuovo Cimento*, **14**, 245
184. Radford, H. E. (1962). *Phys. Rev.*, **126**, 1035; (1961). ibid **122**, 144
185. Radford, H. E. (1964). *Phys. Rev. Lett.*, **13**, 534
186. Radford, H. E. and Linzer, M. (1963). *Phys. Rev. Lett.*, **10**, 443
187. Radford, H. E. (1964). *J. Chem. Phys.*, **40**, 2732
188. McDonald, C. C. (1963). *J. Chem. Phys.*, **39**, 2587
189. Carrington, A. and Levy, D. H. (1966). *J. Chem. Phys.*, **44**, 1298
190. Carrington, A., Dyer, P. N. and Levy, D. H. (1967). *J. Chem. Phys.*, **47**, 1756
191. Carrington, A., Levy, D. H. and Miller, T. A. (1966). *Proc. Roy. Soc. (London)*, **A293**, 108
192. Carrington, A., Fabris, A. R., Howard, B. J. and Lucas, N. J. D. (1971). *Molec. Phys.*, **20**, 961
193. Carrington, A., Levy, D. H. and Miller, T. A. (1970). *Advan. Chem. Phys.*, **18**, 149
194. Carrington, A. and Howard, B. J. (1970). *Molec. Phys.*, **18**, 225
195. Carrington, A., Fabris, A. R., Howard, B. J. and Lucas, N. J. D. (1970). 'Gas-phase electron resonance spectra of linear triatomic free radicals' in *Magnetic Resonance*', *Proc. Intern. Symp. Electron and NMR*, Melbourne, 1969
196. Carrington, A. (1970). 'Rotational Levels of Free Radicals', *Symp. Molec. Struct. Spectrosc.*, Paper E1, Columbus, Ohio. To be published in: *Molecular Spectroscopy: Modern Research*. (Narahari Rao, K. and Mathews, C. Weldon, editors). (1972). (New York: Academic Press)
197. Herzberg, G. (1964). *Molecular Spectra and Molecular Structure. I. Spectra of Diatomic Molecules*. (New York: D. Van Nostrand Company, Inc.)
198. Ter Meulen, J. and Dymanus, A. (1971). *Symp. Molec. Struct. Spectrosc.*, Paper R9, Columbus, Ohio
199. Jefferts, K. B., Penzias, A. A. and Wilson, R. W. (1970). *Astrophys. J.*, **161**, L87
200. Hougen, J. T. (1970). 'The Calculation of Rotational Energy Levels and Rotational Line Intensities in Diatomic Molecules', *National Bureau of Standards Monograph 115*. (Washington, D.C.: U.S. Government Printing Office)
201. Kovacs, I. (1969). *Rotational Structure in the Spectra of Diatomic Molecules*. (London: Adam Hilger Ltd.)
202. Mizushima, M. and Hill, R. M. (1954). *Phys. Rev.*, **93**, 745
203. Miller, S. L. and Townes, C. H. (1953). *Phys. Rev.*, **90**, 537
204. Zimmerer, E. W. and Mizushima, M. (1961). *Phys. Rev.*, **121**, 152
205. Tinkham, M. and Strandberg, M. W. P. (1955). *Phys. Rev.*, **97**, 937, 951
206. McKnight, J. S. and Gordy, W. (1968). *Phys. Rev. Lett.*, **21**, 1787
207. Evenson, K. M., Broida, H. P., Wells, J. S., Mahler, R. J. and Mizushima, M. (1968). *Phys. Rev. Lett.*, **21**, 1038
208. Babcock, H. D. and Herzberg, L. (1948). *Astrophys. J.*, **108**, 167
209. Amano, T., Hirota, E. and Morino, Y. (1967). *J. Phys. Soc. Jap.* **22**, 399
210. Solomon, J. E., Johnson, D. R. and Lin, C. C. (1968). *J. Molec. Spectrosc.*, **27**, 517
211. Daniels, J. M. and Dorain, P. B. (1966). *J. Chem. Phys.*, **45**, 26
212. Daniels, J. M. and Dorain, P. B. (1964). *J. Chem. Phys.*, **40**, 1160
213. Carrington, A., Levy, D. H. and Miller, T. A. (1967). *Proc. Roy. Soc.(London)*, **A298**, 340
214. Uehara, H. (1969). *Bull. Chem. Soc. Japan*, **42**, 886

215. Ehrenstein, G., Townes, C. H. and Stephenson, M. J. (1959). *Phys. Rev. Lett.*, **3**, 40
216. Weinreb, S., Barret, A. H., Meeks, M. L. and Henry, J. C. (1963). *Nature (London)*, **200**, 829
217. Ewen, H. I. and Purcell, E. M. (1951). *Phys. Rev.*, **83**, 881
218. Johnson, D. R. and Lin, C. C. (1967). *J. Molec. Spectrosc.*, **23**, 201
219. Powell, F. X. and Lide, D., Jr. (1965). *J. Chem. Phys.*, **42**, 4201
220. Robinson, B. J. and McGee, R. X. (1967). *Ann. Rev. Astron. Astrophys.*, **5**, 183
221. Cook, A. H. (1969). *Physica*, **41**, 1
222. Poynter, R. L. and Beaudet, R. A. (1968). *Phys. Rev. Lett.*, **21**, 305
223. Manchester, R. N. and Gordon, M. A. (1970). *Astrophys. Lett.*, **6**, 243
224. Ball, J. A., Gottlieb, C. A., Meeks, M. L. and Radford, H. E. (1971). *Astrophys. J.*, **163**, L33
225. Evenson, K. M., Wells, J. S. and Radford, H. E. (1970). *Phys. Rev. Lett.*, **25**, 199
226. Van Vleck, J. H. (1951). *Rev. Mod. Phys.*, **23**, 213
227. Dousmanis, G. C. (1955). *Phys. Rev.*, **97**, 967
228. Mizushima, M. (1954). *Phys. Rev.*, **94**, 569
229. Lin, C. C. and Mizushima, M. (1955). *Phys. Rev.*, **100**, 1726
230. Mizushima, M. (1957). *Phys. Rev.*, **105**, 1262
231. Mizushima, M., Cox, J. T. and Gordy, W. (1955). *Phys. Rev.*, **98**, 1034
232. Burrus, C. A. and Graybeal, J. D. (1958). *Phys. Rev.*, **109**, 1553
233. Mizushima, M. (1958). *Phys. Rev.*, **109**, 1557
234. French, I. P. and Arnold, T. E., Jr. (1968). *J. Chem. Phys.*, **48**, 5720
235. Neumann, R. M. (1970). *Astrophys. J.*, **161**, 779
236. Meerts, W. L. and Dymanus, A. (1971). *Symp. Molec. Structr. Spectr.*, Paper R11, Columbus, Ohio
237. Carrington, A., Howard, B. J., Levy, D. H. and Robertson, J. C. (1968). *Molec. Phys.*, **15**, 187
238. Uehara, H. and Morino, Y. (1969). *Molec. Phys.*, **17**, 239
239. Amano, T., Saito, S., Hirota, E. and Morino, Y. (1969). *J. Molec. Spectrosc.*, **32**, 97
240. Powell, F. X. and Johnson, D. R. (1969). *J. Chem. Phys.*, **50**, 4596
241. Carrington, A., Levy, D. H. and Miller, T. A. (1967). *J. Chem. Phys.*, **47**, 3801
242. Jefferts, K. B. (1969). *Phys. Rev. Lett.*, **23**, 1476
243. Jefferts, K. B., Penzias, A. A., Ball, J. A., Dickinson, D. F. and Lilley, A. E. (1970). *Astrophys. J.*, **159**, L15
244. Jefferts, K. B. (1968). *Phys. Rev. Lett.*, **20**, 39
245. Richardson, C. B., Jefferts, K. B. and Dehmelt, H. G. (1968). *Phys. Rev.*, **165**, 165
246. German, K. R. and Zare, R. N. (1969). *Phys. Rev. Lett.*, **23**, 1207
247. Evenson, K. M., Dunn, J. L. and Broida, H. P. (1964). *Phys. Rev.*, **136**, A1566
248. Evenson, K. M. (1969). *Phys. Rev.*, **178**, 1
249. Radford, H. E. (1964). *Phys. Rev.*, **136**, A1571
250. Crosley, D. R. and Zare, R. N. (1968). *J. Chem. Phys.*, **49**, 4231
251. Freund, R. S., Miller, T. A., De Santis, D. and Lurio, A. (1970). *J. Chem. Phys.*, **53**, 2290
252. West, B. G. and Mizushima, M. (1966). *Phys. Rev.*, **143**, 31
253. Gokhale, B. V. and Strandberg, M. W. P. (1951). *Phys. Rev.*, **82**, 327; **84**, 844
254. Miller, S. L., Javan, A. and Townes, C. H. (1951). *Phys. Rev.*, **82**, 454
255. Wilheit, T. T. and Barret, A. H. (1970). *Phys. Rev.*, **A1**, 213
256. Tischer, R. (1967). *Z. Naturforsch.*, **22A**, 1711
257. Hendrie, J. M. and Kusch, P. (1957). *Phys. Rev.*, **107**, 716
258. Bowers, K. D., Kamper, R. A. and Lustig, C. D. (1959). *Proc. Roy. Soc. London*, **A251**, 565
259. McDonald, C. C. and Goll, R. J. (1965). *J. Phys. Chem.*, **69**, 293
260. Carrington, A. and Levy, D. H. (1967). *J. Phys. Chem.*, **71**, 2
261. Carrington, A., Levy, D. H. and Miller, T. A. (1967). *Molec. Phys.*, **13**, 401
262. Carrington, A. (1968). *Proc. Roy. Soc. London*, **A302**, 291
263. Carrington, A. (1968). 'Molecular Spectroscopy', *Proc. IVth Int. Spectrosc. Conf. Brighton*, 157, (Hepple, editor) (Amsterdam: Elsevier)
264. Carrington, A., Currie, G. N., Levy, D. H. and Miller, T. A. (1969). *Molec. Phys.*, **17**, 535
265. Silvers, S. J., Bergeman, T. H. and Klemperer, W. (1970). *J. Chem. Phys.*, **52**, 4385
266. Radford, H. E. (1968). *Rev. Sci. Instr.*, **39**, 1687

267. Ball, J. A., Dickinson, D. F., Gottlieb, G. A. and Radford, H. E. (1970). *Astron. J.*, **75**, 762
268. Ehrenstein, G. (1963). *Phys. Rev.*, **130**, 669
269. Churg, A. and Levy, D. H. (1970). *Astrophys, J.*, **162**, L161
270. Clough, P. N., Curran, A. H. and Thrush, B. A. (1970). *Chem. Phys. Lett.*, **7**, 86
271. Carrington, A. and Lucas, N, J. D. (1970). *Proc. Roy. Soc. London*, **A314**, 567
272. Radford, H. E. and Linzer, M. (1963). *Phys. Rev. Lett.*, **10**, 443
273. Byfleet, C. R., Carrington, A. and Russel, D. K. (1971). *Molec. Phys.*, **20**, 271
274. Miller, T. A. (1971). *J. Chem. Phys.*, **54**, 1658
275. Radford, H. E. (1964). *J. Chem. Phys.*, **40**, 2732
276. Carrington, A., Currie, G. N. and Lucas, N. J. D. (1970). *Proc. Roy. Soc. London*, **A315**, 355
277. Evenson, K. M. (1968). *Appl. Phys. Lett.*, **12**, 253
278. Pratt, D. W. and Broida, H. P. (1969). *J. Chem. Phys.*, **50**, 2181
279. Carrington, A. and Howard, B. B. (1970). *Molec. Phys.*, **18**, 225
280. Brown, R. L. and Radford, H. E. (1966). *Phys. Rev.*, **147**, 6
281. Chardon, J. C. and Theobald, J. G. (1968). *Compt. Rend. Acad. Sci. Paris*, **266**, 602
282. Carrington, A., Levy, D. H. and Miller, T. A. (1966). *J. Chem. Phys.*, **45**, 3450
283. Uehara, H. and Morino, Y. (1970). *J. Molec. Spectrosc.*, **36**, 158
284. Carrington, A., Dyer, P. N. and Levy, D. H. (1970). *J. Chem. Phys.*, **52**, 309
285. Carrington, A., Currie, G. N., Dyer, P. N., Levy, D. H. and Miller, T. A. (1967). *Chem. Commun.*, 641
286. Carrington, A. (1970). *Chem. Brit.*, **6**, 71
287. Carrington, A., Currie, G. N., Miller, T. A. and Levy, D. H. (1969). *J. Chem. Phys.*, **50**, 2726
288. Lichten, W. (1962). *Phys. Rev.*, **126**, 1020
289. Lichten, W. (1971). *Phys. Rev.*, **A3**, 594
290. Chiu, L. Y. C. (1966). *Phys. Rev.*, **145**, 1
291. Marechal, M. A. and Jourdan, A. (1969). *Phys. Lett.*, **30A**, 31
292. Gammon, R. H., Stern, R. C., Lesk, M. E., Wicke, B. G. and Klemperer, W. (1971). *J. Chem. Phys.*, **54**, 2136
293. Stern, R. C., Gammon, R. H., Lesk, M. E., Freund, R. S. and Klemperer, W. (1970). *J. Chem. Phys.*, **52**, 3467
294. Gammon, R. H., Stern, R. C. and Klemperer, W. (1971). *J. Chem. Phys.*, **54**, 2151
295. Curran, A. H., McDonald, R. G., Stone, A. J. and Thrush, B. A. (1971). *Chem. Phys. Lett.*,**8**, 451
296. Falick, A. M., Mahan, B. H. and Meyers, R. J. (1965). *J. Chem. Phys.*, **42**, 1837
297. Miller, T. A. (1971). *J. Chem. Phys.*, **54**, 330
298. Uehara, H. and Morino, Y. (1970). *Bull. Chem. Soc. Jap.*, **43**, 2273
299. Miller, T. A. (1971). *J. Chem. Phys.*, **54**, 1658
300. Carrington, A., Currie, G. N., Levy, D. H. and Miller, T. A. (1969). *Molec. Phys.*, **17**, 535
301. Bowater, I. C., Brown, J. M. and Carrington, A. (1971). *J. Chem. Phys.*, **54**, 4957
302. Renner, R. (1934). *Z. Physik*, **92**, 172
303. Van Vleck, J. H. (1951). *Rev. Mod. Phys.*, **23**, 213
304. Creutzberg, F. and Hougen, J. T. (1967). *Can. J. Phys.*, **45**, 1363
305. Hougen, J. T. (1962). *J. Chem. Phys.*, **36**, 519
306. Raynes, W. T. (1964). *J. Chem. Phys.*, **41**, 3020
307. Herzberg, G. (1966). *Molecular Spectra and Molecular Structure III. Electronic Spectra and Electronic Structure of Polyatomic Molecules*, (New York: Van Nostrand Reinhold)
308. Saito, S. and Amano, T. (1970). *J. Molec. Spectrosc.*, **34**, 383
309. Hrubesh, L. W., Rinehart, E. A. and Anderson, R. E. (1970). *J. Molec. Spectrosc.*, **36**, 354
310. Winton, R. S. and Gordy, W. (1970). *Phys. Lett.*, **32A**, 219
311. Tyler, J. K. (1964). *J. Chem. Phys.*, **40**, 1170
312. Tyler, J. K. and Sheridan, J. (1963). *Trans. Faraday Soc.*, **59**, 2661
313. Freund, S. H., Godfrey, P. D. and Klemperer, W. *J. Chem. Phys.*, to be published
314. Winnewisser, M. and Bodenseh, H. K. (1967). *Z. Naturforsch.*, **22a**, 1724
315. Jones, H. and Sheridan, J., to be published
316. Jones, H., Owen, N. L. and Sheridan, J. (1967). *Nature (London)*, **213**, 175
317. Jones, H., Stiefvater, O. L. and Sheridan, J., to be published
318. McKnight, J. S. and Gordy, W. (1969). *Bull. Amer. Phys. Soc.*, **14**, 621

319. Muenter, J. S. and Laurie, V. W. (1964). *J. Amer. Chem. Soc.*, **86**, 3901
320. Bjørseth, A., Kloster-Jensen, E., Marstokk, K.-M. and Møllendal, H. (1970). *J. Molec. Struct.*, **6**, 181
321. Herzberg, G. (1942). *Rev. Mod. Phys.*, **14**, 219
322. Nielsen, H. H. (1949). *Phys. Rev.*, **75**, 1961; (1950). ibid., *Phys. Rev.*, **77**, 130
323. Amat, G. and Nielsen, H. H. (1958). *J. Molec. Spectrosc.*, **2**, 163
324. Winnewisser, G., Maki, A. G. and Johnson, D. R. (1971). *J. Molec. Spectrosc.*, **38**, 149
325. Winnewisser, M. and Winnewisser, B. P. (1972). *J. Molec. Spectrosc.*, **41**, 143
326. Maki, A. G. and Lide, D. R., Jr. (1967). *J. Chem. Phys.*, **47**, 3206
327. Johns, J. W. C., Stone, J. M. R. and Winnewisser, G. (1971). *J. Molec. Spectrosc.*, **38**, 437
328. Pliva, J. (1958). *Collect. Czech. Chem. Commun.*, **23**, 1846
329. Lide, D. R., Jr. and Matsumura, C. (1969). *J. Chem. Phys.*, **50**, 3080
330. Nakagawa, T. and Morino, Y. (1969). *J. Molec. Spectrosc.*, **31**, 208
331. Pearson, R., Sullivan, T. and Frenkel, L. (1970). *J. Molec. Spectrosc.*, **34**, 440
332. Amat, G. and Nielsen, H. H. (1958). *J. Molec. Spectrosc.*, **2**, 152
333. Scott, J. F. and Rao, K. N. (1965). *J. Molec. Spectrosc.*, **18**, 152, 451
334. Ghersetti, S., Pliva, J. and Rao, K. N. (1971). *J. Molec. Spectrosc.*, **38**, 53
335. Bodenseh, H. K. and Winnewisser, M. (1969). *Z. Naturforsch.*, **24a**, 1966
336. Bodenseh, H. K. and Winnewisser, M. (1969). *Z. Naturforsch.*, **24a**, 1973
337. Lafferty, W. J. (1968). *J. Molec. Spectrosc.*, **25**, 359
338. Winnewisser, M. and Winnewisser, B. P. (1971). *Z. Naturforsch.*, **26a**, 128
339. Costain, C. C. (1958). *J. Chem. Phys.*, **29**, 864
340. Bernstein, H. J. (1962). *Spectrochim. Acta.*, **18**, 161; Winnewisser, M. and Winnewisser, B. P., private communication
341. Winnewisser, B. P. and Winnewisser, M. (1971). *Symp. Molec. Struct. Spectr.*, Columbus, Ohio, Paper G7
342. Thorson, W. R. and Nakagawa, I. (1960). *J. Chem. Phys.*, **33**, 994
343. Hougen, J. T., Bunker, P. R. and Johns, J. W. C. (1970). *J. Molec. Spectrosc.*, **34**, 136
344. Bunker, P. R. and Stone, J. M. R. (1972). *J. Molec. Spectrosc.*, **41**, 310 (Paper G1, to be published)
345. Sarka, K. (1971). *J. Molec. Spectrosc.*, **38**, 545
346. Snyder, L. E. (1971). This Volume, Chapter 9
347. Burrus, C. A. and Gordy, W. (1956). *Phys. Rev.*, **101**, 599
348. McKnight, J. S. (1970). Private communication
349. Dorman, F. and Lin, C. C. (1964). *J. Molec. Spectrosc.*, **12**, 119
350. Low, W. (1955). *Phys. Rev.*, **97**, 1664
351. Simmons, J. W. and Anderson, W. E. (1950). *Phys. Rev.*, **80**, 338
352. Cowan, M. and Gordy, W. (1960). *Bull. Amer. Phys. Soc. Ser. II*, **5**, 241
353. Javan, A. (1955). *Phys. Rev.*, **99**, 1302
354. Townes, C. H., Holden, A. N. and Merritt, F. R. (1948). *Phys. Rev.*, **74**, 1113
355. Cleeton, C. E. and Williams, N. H. (1934). *Phys. Rev.*, **45**, 234
356. Helminger, P., De Lucia, F. C. and Gordy, W. (1971). *J. Molec. Spectrosc.*, **39**, 94
357. Helminger, P. and Gordy, W. (1969). *Phys. Rev.*, **188**, 100
358. Erlandsson, G. and Gordy, W. (1957). *Phys. Rev.*, **106**, 513
359. Costain, C. C. (1951). *Phys. Rev.*, **82**, 108
360. Schnabel, E., Törring, T. and Wilke, W. (1965). *Z. Physik.*, **188**, 167
361. Herrmann, G. (1958). *J. Chem. Phys.*, **29**, 875
362. Cheung, A. C., Rank, D. M., Townes, C. H., Thorton, D. D. and Welch, W. J. (1968). *Phys. Rev. Lett.*, **21**, 1701
363. Solomon, P. M., Jefferts, K. B., Penzias, A. A. and Wilson, R. W. (1971). *Astrophys. J.*, **168**, L107
364. Snyder, L. E. and Buhl, D. (1971). *B.A.A.S.*, **3**, 388
365. Costain, C. C. and Sutherland, G. B. B. M. (1952). *J. Phys. Chem.*, **56**, 321
366. Davies, P. B., Neumann, R. M., Wofsky, S. C. and Klemperer, W. (1971). *J. Chem. Phys.*, **55**, 3564
367. Helminger, P., Beeson, E. L., Jr. and Gordy, W. (1971). *Phys. Rev.*, **A3**, 122
368. Mirri, A. M., Scappini, F. and Favero, P. G. (1965). *Spectrochim. Acta*, **21**, 965
369. Mirri, A. M. and Damiani, D. (1966). *Atti. Accad. Nazl. Lincei. Rend.*, **40**, 635
370. Mirri, A. M. and Cazzoli, G. (1967). *J. Chem. Phys.*, **47**, 1197
371. Mirri, A. M. (1967). *J. Chem. Phys.*, **47**, 2823

372. Hirota, E. and Morino, Y. (1970). *J. Molec. Spectrosc.*, **33**, 460
373. Hirota, E. (1971). *J. Molec. Spectrosc.*, **37**, 20; (1971). ibid., **38**, 195
374. Otake, M., Matsumura, C. and Morino, Y. (1968). *J. Molec. Spectrosc.*, **28**, 316
375. Thomas, W. J. O., Cox, J. T. and Gordy, W. (1954). *J. Chem. Phys.*, **22**, 1718
376. Kewley, R., McKinney, P. M. and Robiette, A. G. (1970). *J. Molec. Spectrosc.*, **34**, 390
377. Sullivan, T. E. and Frenkel, L. (1971). *J. Molec. Spectrosc.*, **39**, 185
378. Grenier-Besson, M. L. (1960). *J. Phys. Radium*, **21**, 555
379. Grenier-Besson, M. L. (1964). *J. Phys. Radium*, **25**, 757
380. Oka, T. (1967). *J. Chem. Phys.*, **47**, 5410
381. Wilson, E. B., Jr. and Howard, J. B. (1936). *J. Chem. Phys.*, **4**, 260
382. Tarrago, G. (1965). *Cah. Phys.*, **19**, 149
383. Tarrago, G. (1970). *J. Molec. Spectrosc.*, **34**, 23
384. Bauer, A. and Maes, S. (1969). *Compt. Rend.*, **268B**, 1569
385. Bauer, A. (1971). *J. Molec. Spectrosc.*, **40**, 183
386. Bauer, A. and Maes, S. (1971). *J. Molec. Spectrosc.*, in the press
387. Bauer, A. and Maes, S. (1969). *J. Phys.*, **30**, 169
388. Costain, C. C. (1965). *Can. J. Phys.*, **43**, 244
389. Maes, S. and Amat, G. (1965). *Can. J. Phys.*, **43**, 321
390. Bauer, A. and Burie, J. (1969). *Compt. Rend.*, **268B**, 800
391. Costain, C. C. and Winnewisser, G., to be published
392. Bellet, J., Colmont, J.-M. and Lemaire, J. (1970). *J. Molec. Spectrosc.*, **34**, 190
393. Oka, T., Tsuchiya, K., Iwata, S. and Morino, Y. (1964). *Bull. Chem. Soc. Jap.*, **37**, 4
394. Amat, G. and Grenier-Besson, M. L. (1962). *J. Molec. Spectrosc.*, **8**, 22
395. Costain, C. C. (1967). *Symp. Molec. Struct. Spectrosc.*, Columbus, Ohio, Paper G5
396. Matsuura, H., Nakagawa, T. and Overend, J. (1970). *J. Chem. Phys.*, **53**, 2540
397. Winnewisser, G., Shimizu, F. and Shimizu, T., to be published
398. Burrus, C. A. and Gordy, W. (1957). *J. Chem. Phys.*, **26**, 391
399. Johnson, C. M., Trambarulo, R. and Gordy, W. (1951). *Phys. Rev.*, **84**, 1178
400. Favero, P. G. and Mirri, M. (1950). *Nuovo Cimento*, **30**, 502
401. Simmons, J. W., Garrison, A. K. and Alexander, C. (1965). *Bull. Amer. Phys. Soc.*, **10**, 492
402. McLay, D. B. and Winnewisser, G., to be published
403. Bauer, A. and Bogey, M. (1970). *Compt. Rend.*, **271B**, 892
404. King, G. W., Hainer, R. M. and Cross, P. C. (1943). *J. Chem. Phys.*, **11**, 27
405. Wilson, E. B., Jr. (1937). *J. Chem. Phys.*, **5**, 617; (1953). ibid., **21**, 1229
406. Chung, K. T. and Parker, P. M. (1963). *J. Chem. Phys.*, **38**, 8
407. Chung, K. T. and Parker, P. M. (1965). *J. Chem. Phys.*, **43**, 3865
408. Chung, K. T. and Parker, P. M. (1965). *J. Chem. Phys.*, **43**, 3869
409. Watson, J. K. G. (1966). *J. Chem. Phys.*, **45**, 1360
410. Watson, J. K. G. (1967). *J. Chem. Phys.*, **46**, 1935
411. Watson, J. K. G. (1968). *J. Chem. Phys.*, **48**, 181
412. Watson, J. K. G. (1968). *J. Chem. Phys.*, **48**, 4517
413. Kivelson, D. and Wilson, E. B., Jr. (1952). *J. Chem. Phys.*, **20**, 1575
414. Kivelson, D. and Wilson, E. B., Jr. (1953). *J. Chem. Phys.*, **21**, 1229
415. Olson, W. B. and Allen, H. C., Jr. (1963). *J. Res. Natl. Bur. Stand.*, **A67**, 359
416. Dowling, J. M. (1961). *J. Molec. Spectrosc.*, **6**, 550
417. Oka, T. and Morino, Y. (1961). *J. Phys. Soc. Japan*, **16**, 1235
418. Kirchhoff, W. H. (1972). *J. Molec. Spectrosc.*, **41**, 333
419. Allen, H. C. and Olson, W. B. (1962). *J. Chem. Phys.*, **37**, 212
420. Nielsen, H. H. (1959). *Encyclop. Phys.*, **37**, 1, 173
421. Oka, T. and Morino, Y. (1962). *J. Molec. Spectrosc.*, **8**, 300
422. Guarnieri, A. and Favero, P. G. (1965). *Nuovo Cimento*, **39**, 76
423. Kivelson, D. (1954). *J. Chem. Phys.*, **22**, 904
424. Steenbeckeliers, G. (1968). *Ann. Soc. Sci. Brux.*, **82**, 331
425. Pierce, L., Dicianni, N. and Jackson, R. H. (1963). *J. Chem. Phys.*, **38**, 730
426. Lichtenstein, M., Gallagher, J. J. and Clough, S. A. (1971). *J. Molec. Spectrosc.*, **40**, 10
427. Herberich, G. E., Jackson, R. H. and Millen, D. J. (1966). *J. Chem. Soc., A*, 336
428. Cook, R. L. and Kirchhoff, W. H. (1967). *J. Chem. Phys.*, **47**, 4521
429. Mirri, A. M. and Mazzariol, E. (1966). *Spectrochim. Acta*, **22**, 785
430. Mirri, A. M. and Mazzariol, E. (1967). *Spectrochim. Acta*, **23A**, 3035

431. Mirri, A. M., Scappini, F., Innamorati, L. and Favero, P. (1969). *Spectrochim. Acta,* **25A,** 1631
432. Mirri, A. M., Cazzoli, G. and Ferretti, L. (1968). *J. Chem. Phys.,* **49,** 2775
433. Cook, R. L. (1965). *J. Chem. Phys.,* **42,** 2927
434. Meschi, D. J. and Myers, R. J. (1959). *J. Molec. Spectrosc.,* **3,** 405
435. Cook, R. L., Lindsey, D. and Winnewisser, G. (1971). *J. Molec. Spectrosc.,* in the press
436. Bellet, J., Deldalle, A., Samson, C., Steenbeckeliers and Wertheimer, R. (1971). *J. Molec. Struct.,* **9,** 65
437. Erlandsson, G. (1958). *J. Chem. Phys.,* **28,** 71
438. Wertheimer, R. (1956). *Compt. Rend.,* **242,** 243, 1591
439. Mirri, A. M. (1960). *Nuovo Cimento,* **18,** 849
440. Kwei, G. H. and Curl, R. F. (1960). *J. Chem. Phys.,* **32,** 1592
441. Johnson, D. R., Powell, F. X. and Kirchhoff, W. H. (1971). *J. Molec. Spectrosc.,* **39,** 136
442. Johns, J. W. C. and Olson, W. B. (1971). *J. Molec. Spectrosc.,* **39,** 479
443. Oka, T., Johns, J. W. C. and Winnewisser, G. To be published
444. Johns, J. W. C., Stone, J. M. R. and Winnewisser, G. (1972). *J. Molec. Spectrosc.,* in the press
445. Oka, T. (1960). *J. Phys. Soc. Japan,* **15,** 2274
446. Johnson, M. R. and Strandberg, M. W. P. (1952). *J. Chem. Phys.,* **20,** 687
447. Kim, H., Pearson, E. F. and Appelman, E. H. (1972). *J. Chem. Phys.,* **56,** 1
448. Mirri, A. M., Scappini, F. and Cazzoli, G. (1971). *J. Molec. Spectrosc.,* **38,** 218
449. Kewley, R., Sastry, K. V. L. N. and Winnewisser, M. (1963). *J. Molec. Spectrosc.,* **10,** 418
450. Winnewisser, M. and Cook, R. L. (1964). *J. Chem. Phys.,* **41,** 999
451. Krakow, B., Lord, R. C. and Neely, G. O. (1968). *J. Molec. Spectrosc.,* **27,** 148
452. Winnewisser, G., Winnewisser, M. and Gordy, W. (1968). *J. Chem. Phys.,* **49,** 3465
453. Winnewisser, G. (1972). *J. Chem. Phys.,* **56,** 2944
454. Winnewisser, G. and Helminger, P. (1972). *J. Chem. Phys.,* **56,** 2954
455. Winnewisser, G. and Helminger, P. (1972). *J. Chem. Phys.,* **56,** 2967
456. Benedict, W. S., Clough, S. A., Frenkel, L. and Sullivan, T. E. (1970). *J. Chem. Phys.,* **53,** 2565
457. Dowling, J. M. (1968). *J. Molec. Spectrosc.,* **29,** 348
458. Steenbeckeliers, G. and Bellet, J. (1970). *Compt. Rend.,* **270B,** 1039
459. Steenbeckeliers, G. and Bellet, J. (1971). *J. Molec. Spectrosc.,* in the press
460. Bellet, J. and Steenbeckeliers, G. (1970). *Compt. Rend.,* **271,** 1208
461. Helminger, P., Cook, R. L. and De Lucia, F. C. (1971). *J. Molec. Spectrosc.,* **40,** 125
462. De Lucia, F. C., Cook, R. L., Helminger, P. and Gordy, W. (1971). *J. Chem. Phys.,* **55,** 5334
463. Cook, R. L., De Lucia, F. C. and Helminger, P. (1972). *J. Molec. Spectrosc.,* **41,** 123
464. Takeo, M., Hirota, E. and Morino, Y. (1971). *J. Molec. Spectrosc.,* in the press
465. Tanaka, T. and Morino, Y. (1970). *J. Molec. Spectrosc.,* **33,** 538
466. Morino, Y. and Saito, S. (1966). *J. Molec. Spectrosc.,* **19,** 435
467. Morino, Y., Kikuchi, Y., Saito, S. and Mirota, E. (1964). *J. Molec. Spectrosc.,* **13,** 95
468. Oka, T., Tomaya, M. and Morino, Y. (1962). *J. Molec. Spectrosc.,* **13,** 193
469. Winnewisser, G. (1972). *J. Molec. Spectrosc.,* **41,** 534
470. Dreizler, H. (1968). *Fortschritte der chemischen Forschung,* **10,** 59
471. Lin, C. C. and Swalen, J. D. (1959). *Rev. Mod. Phys.,* **31,** 841
472. Koehler, J. S. and Dennison, D. M. (1940). *Phys. Rev.,* **57,** 1006
473. Hecht, K. T. and Dennison, D. M. (1957). *Phys. Rev.,* **26,** 48
474. Lees, R. M. and Baker, J. G. (1968). *J. Chem. Phys.,* **48,** 5299
475. Dreizler, H. and Mirri, A. M. (1968). *Z. Naturforsch.,* **23a,** 1313
476. Oelfke, W. C. and Gordy, W. (1969). *J. Chem. Phys.,* **51,** 5336
477. Hunt, R. H., Leacock, R. A., Peters, C. W. and Hecht, K. T. (1965). *J. Chem. Phys.,* **42,** 1931
478. De Lucia, F. C., Helminger, P., Cook, R. L. and Gordy, W. (1972). *Phys. Rev.,* **5,** 487
479. Favero, P. G., Mirri, A. M. and Baker, J. G. (1960). *Nuovo Cimento,* **17,** 740
480. Mirri, A. M., Ferretti, L. and Forti, P. (1971). *Spectrochim. Acta,* **27A,** 937
481. Jache, A. W., Moser, P. W. and Gordy, W. (1956). *J. Chem. Phys.,* **25,** 209
482. Burrus, C. A., Jr. and Gordy, W. (1953). *Phys. Rev.,* **92,** 274
483. Mirri, A. M., Guarnieri, A. and Favero, P. (1961). *Nuovo Cimento,* **19,** 1189
484. Steiner, A. and Gordy, W. (1966). *J. Molec. Spectrosc.,* **21,** 291
485. Marshall, S. A. and Weber, J. (1957). *Phys. Rev.,* **105,** 1502

486. Dijkerman, H. A. and Ruitenberg, G. (1969). *Chem. Phys. Lett.,* **3,** 172
487. Muenter, J. S. (1968). *J. Chem. Phys.,* **48,** 4544
488. de Leeuw, F. H. and Dymanus, A. (1970). *Chem. Phys. Lett.,* **7,** 288
489. Mirri, A. M., Corbelli, G. and Forti, P. (1969). *J. Chem. Phys.,* **50,** 4118
490. Schlier, C. (1959). *Z. Physik,* **154,** 460
491. Braslawsky, J. and Ben-Aryeh, Y. (1969). *J. Molec. Spectrosc.,* **30,** 116
492. Flygare, W. H. and Benson, R. C. (1971). *Molec. Phys.,* **20,** 225
493. DeLucia, F. C. and Cederberg, J. W. (1971). *J. Molec. Spectrosc.,* **40,** 52
494. Thaddeus, P., Krishner, L. and Loubser, J. (1964). *J. Chem. Phys.,* **40,** 257
495. Oka, T. (1966). *J. Chem. Phys.,* **45,** 752
496. Shimizu, T. and Oka, T. (1970). *Phys. Rev.,* **A2,** 587, 1177
497. Frenkel, L., Marantz, H. and Sullivan, T. (1971). *Phys. Rev.,* **A3,** 1640
498. Pliva, J. (1967). *J. Molec. Spectrosc.,* **23,** 228
499. Neely, G. O. (1968). *J. Molec. Spectrosc.,* **27,** 177

8
Resonance Fluorescence and Non-radiative Relaxation in Polyatomic Gases

C. S. PARMENTER
Indiana University

8.1 INTRODUCTION

The material in this Chapter describes investigations of electronic fluorescence from polyatomics larger than triatomic at very low pressures. Most of the discussion concerns the comments which this fluorescence has made about electronic relaxation processes in isolated molecules, but some of the recent developments in the more spectroscopic aspects of this fluorescence are also included. Since the subject has not been periodically covered by reviews in the past, a number of references to older works have been included to provide necessary perspective to the discussions.

Gas-phase fluorescence has been a relatively minor area of molecular spectroscopy. When the studies of molecular luminescence are divided according to the phase of the experimental systems, we find only a small group concerned with gas-phase explorations, while on the other hand there is a vast accumulation of work with condensed-phase systems. Much of our present knowledge about the properties of excited electronic states has been derived from the condensed phase studies. They have provided information about the positions and assignments of upper electronic states (particularly triplet states), and it has given the basic description of electronic energy transfer and excited state relaxation mechanisms[1-3].

The past bias towards liquid and solid environments was in large part dictated by the issue of practical expediency. Gas-phase luminescence studies are in the main difficult to do, and of course with many materials of low vapour pressure, they are impossible. But unique aspects of the condensed phase have also contributed to the motivation for its use. It is generally the only path to low temperature studies, and until the advent of very short duration pulses, it provided the only opportunity for experiments with molecules in controlled orientations. Condensed phases are also natural media with which to probe the effect of environmental interactions on excited state properties.

It is just this last aspect of condensed-phase experiments that has provided impetus for much recent work in the gas phase. With condensed-phase experiments, one is always faced with the problem of disentangling the environment from the molecule. One such disentanglement, that of vibrational and rotational relaxation, is in fact generally impossible. With few exceptions[4], condensed-phase experiments allow examination of emission only from molecules that have been brought into thermal vibrational–rotational equilibrium prior to their radiative decay. Information about higher vibrational and rotational levels in the excited states is washed out. But furthermore, the extent to which the intimate environmental interactions have influenced the structure, the yields, and the lifetimes of condensed-phase molecular emission from those thermal levels is not clear. No way exists to deduce securely the *intrinsic* molecular properties from measured emission parameters.

Gas-phase emission spectroscopy provides an obvious circumvention of these environmental complications. However, formidable experimental problems precluded realisation of this potential, and until about 1960 the list of stable polyatomic species probed by gas-phase emission was rather limited. Since then technology has changed the picture. Great improvements have been made in the sensitivity and time response of photomultiplier tubes. New averaging and photon-counting circuits for these tubes have appeared. Concomitantly, better excitation sources have been developed. We have improved high-pressure xenon and xenon–mercury arcs, new nanosecond flash sources, and, of course, lasers.

With these tools, gas-phase emission is an awakening field. It is yielding much new information about excited states, particularly in molecules remote from environmental perturbation, so that the properties intrinsic to the isolated molecules are now being seen with improving detail. And in turn, this work is paralleled by a revival of theoretical interest in the problem of excited state behaviour in isolated polyatomics.

This Review will describe some of these advances derived from gas-phase emission, but it does not pretend to be a complete bibliography of all gas-phase studies in the past few years. Interesting areas will be selected and presented in some detail.

Frequent reference will be given to 'resonance emission'. This term is difficult to define in polyatomics, because emission exactly resonant with excitation in the sense of atomic resonance emission is of course not so important. Instead the term is often used to indicate emission in systems where the pressures are so low that collisions do not perturb the emitting species between the time of their preparation and the time of their radiative decay. But even this definition is not always satisfactory because short-lived radiative states prepared in excitation may undergo radiationless transitions to long-lived states, which also decay radiatively at low pressures. Thus ensembles of polyatomic molecules frequently display emission both from states radiating under collision-free or resonance conditions, and from other states which have been under some collisional interaction. For this reason, the term resonance emission must be used with charity when applied to polyatomics. Here, we simply take it to include experiments where the pressure is sufficiently low so that at least an interesting segment of the emission originates from excited molecules that have not been subject to significant collisional perturbation between excitation and their radiative decay.

8.2 RESONANCE EMISSION AND THE MECHANISM OF RADIATIONLESS TRANSITIONS IN ISOLATED MOLECULES

8.2.1 Intramolecular transitions in large molecules

In the past decade, resonance emission from vapours has excited considerable interest in the problem of non-radiative transitions between electronic states of polyatomic molecules. It has explored in a crucial way the basic physics of these phenomena, and it is now leading to elucidation of their finer details.

The questions discussed by resonance emission studies had their foundations in the pioneering work of Lewis and Kasha on molecular luminescence in low temperature glasses[5]. They established the generality of non-radiative processes by showing that the transitions $S_n \rightarrow S_1$ and $S_1 \rightarrow T$ are dominant excited state relaxation paths in a wide variety of polyatomic molecules. It is interesting to read in retrospect Coulson's introduction[6] to the 1950 Discussions of the Faraday Society where Kasha had summarised those findings[7]. Among several prophetic comments, Coulson remarked that the non-radiative transitions documented by the luminescence studies were among those problems that especially presented themselves for future resolution. A clear theoretical understanding of their mechanism was not apparent.

Coulson's observations were still largely true a decade later. For example, a second review by Kasha in 1960 includes no encompassing theory of these processes[8]. But a third review by Kasha (with Henry) appearing after passage of almost another decade finally revealed the beginning of the problem's resolution[9]. It described the theoretical discussions of Robinson and Frosch, of Ross and co-workers, of Siebrand and of Lin that had begun to put the subject in good order. Subsequently the theory has received extensive attention[10] which has brought it into close coincidence with experimental data. Even remarkably fine details of vibronic influences on non-radiative transitions can now be discussed with some quantitative success[11].

Resonance emission has played from the beginning an intimate role in this evolution. One of the early central issues concerned the nature – or perhaps

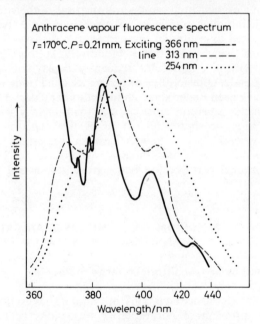

Figure 8.1 The resonance fluorescence spectrum of anthracene vapour at 0.29 torr and 170 °C excited by light pumping the S_1 and higher singlet states. Excitation is with Hg lines at 3660 Å (——), 3130 Å (- - - -) and 2537 Å (......).
(From Stevens and Hutton[13], by courtesy of Taylor and Francis)

the existence – of non-radiative transitions in isolated molecules. Would internal conversion and intersystem crossing be observable properties of molecules not subject to collisional interactions? What was the intrinsic molecular nature of those non-radiative processes which were so commonly observed in condensed media?

The answer to this question was in fact clearly evident in some beautiful resonance fluorescence spectra from anthracene vapour published by Pringsheim in 1938[12]. Those spectra showed that 'Kasha's rule' (fluorescence always originates from the first excited singlet state S_1) for condensed-phase studies could hold true in the vapour as well. Even at gas pressures so low that the

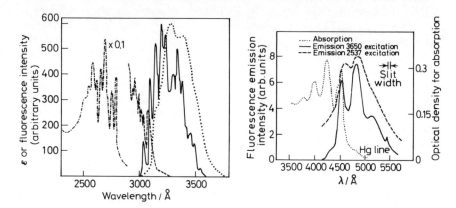

Figure 8.2 (Left) Resonance S_1 fluorescence from 0.1 torr of naphthalene vapour. The solid curve is fluorescence excited by pumping in the $S_1 \leftarrow S_0$ absorption system, and the dotted curve at wavelengths longer than 3000 Å is fluorescence excited by pumping in the S_2–S_0 system. The dashed-dotted lines superimposed on fluorescence and extending to higher energies are the extinction coefficients for absorption.
(From Watts and Strickler[17], by courtesy of the American Insitute of Physics)
(Right) The analogous S_1 fluorescence from tetracene. The solid curve is emission after excitation into the S_1 state and the dashed curve is fluorescence after excitation into a higher electronic state. The tetracene pressure is 0.033 torr.
(From Williams and Goldsmith[20], by courtesy of the American Institute of Physics)

hard-sphere collision interval greatly exceeded the relaxation lifetimes of the excited state pumped in the absorption act, anthracene fluorescence was observed only from the lowest excited singlet state. Emission was clearly from S_1 origins irrespective of whether S_1 or a higher singlet state was initially excited. This important finding has been confirmed by Stevens and Hutton[13] (see Figure 8.1) and by Gruzinskii and Borisevitch[14].

The significance of these results was not missed by Pringsheim (see p. 272 of reference 15) but Stevens and Hutton seem among the first to discuss the issue in detail when they made quantitative measurements of anthracene fluorescence to explore possible mechanisms of $S_2 \rightarrow S_1$ internal conversion in 'isolated' molecules. From their work and discussions it is clear that internal conversion, i.e. the non-radiative decay of the optically prepared electronic state into another electronic state of the same spin, is an intramolecular phenomenon independent of environmental inspiration.

A generality to the conclusions derived from anthracene fluorescence has been provided by similar comparisons of resonance fluorescence following pumping of S_1 or a higher singlet S_n in naphthalene[16–19] and tetracene[20]. Fluorescence spectra from both of these molecules are shown in Figure 8.2.

In each case the structure of fluorescence generated by pumping the higher singlet state approximates that following excitation of S_1. The electronic state carrying oscillator strength for radiative decay is clearly S_1 in every case. Watts and Strickler estimate that the rate of internal conversion exceeds that of radiative decay of S_2 in naphthalene by a factor of at least 500 [17]. These facts, combined with an estimate of the radiative lifetime of the S_2 state, indicate that the actual S_2 lifetime is exceedingly short relative to collision intervals. That of naphthalene is 5×10^{-11} s and that of the higher singlet in tetracene, with its very large extinction coefficient ($\varepsilon = 140\,000$)[2], must be equally short. These lifetimes, established at least in part, and probably almost entirely by $S_n \rightarrow S_1$ internal conversion, are considerably less than one thousandth of the gas-phase hard-sphere collision interval at the pressures used in the experiments and are of the same magnitude as those observed in solution. There is clearly little opportunity for collisional interactions, even with cross-sections considerably exceeding hard sphere, to influence significantly the rate of the S_n non-radiative decay in these systems.

Birks has described an alternate mechanism in which the collision-free generation of S_1 emission following S_n excitation occurs by radiationless intermolecular excitation migration through long-range dipole–dipole interactions[21]. A key aspect of such a mechanism would be partitioning of excess vibrational energy between the interacting molecules during exchange of electronic energy. In a sequence of excitation migration events, this would cause dissipation of vibrational energy in electronically excited molecules to establish a quasi-thermal distribution of vibrational energy in the S_1 .emission spectra. Their structure would be essentially independent of exciting wavelength at very low pressures. The spectra in Figures 8.1 and 8.2, however, are dependent on wavelength, and thermalisation of vibronic energy prior to emission does not occur in these cases. Nor are there any hints of such 'collision-free' vibrational thermalisation seen in S_1 fluorescence spectra from benzene[22–24], which have been generated by pumping single discrete $S_1 \leftarrow S_0$ absorption bands.

The preceding discussions relate how the structure of resonance fluorescence has provided an experimental basis for theoretical interpretation of internal conversion as an intramolecular phenomenon. Resonance fluorescence has also led to the experimental identification of intersystem crossing (non-radiative transitions between electronic states of different spin) as an intramolecular process in large molecules. In this case, however, measurement of kinetic parameters of radiative decay in addition to structure analysis has contributed to the identification.

Perhaps the first evidence came from a series of studies of the quantum yield of fluorescence from benzene excited with the 2537 Å Hg line at low pressures to vibronic levels about $2000\,\mathrm{cm}^{-1}$ above the S_1 zero-point energy[25–28]. The quantum yield of fluorescence at low benzene pressures was found to be constant and well below unity, indicating a persistent non-radiative decay. Indirect but convincing evidence indicated that at least some of this non-radiative decay resulted in formation of the triplet state, and hence intersystem crossing is identified as an intramolecular process from the $\pi\pi^*$ excited state.

Ware and Cunningham have also provided evidence supporting these conclusions from measurements of quantum yields and lifetimes of resonance fluorescence in anthracene[29] and perylene[30]. In each case, the observed lifetime is considerably shorter than that calculated for radiative decay alone, and fluorescence quantum yields are well below unity.

More direct evidence bearing on the intramolecular nature of intersystem crossing is now offered by resonance emission studies with other large molecules. Biacetyl[31] and pyrazine[32–35] exhibit both singlet and triplet emission following excitation of the singlet state at low pressure, and hence these systems allow the process to be characterised by direct observation of both the initial and final states.

The situation is perhaps most explicitly examined in biacetyl. In that system, the structure, yields and lifetimes of fluorescence and phosphorescence have been observed over a rather wide range of environmental situations. When biacetyl is excited to low S_1 vibrational levels by absorption of light near 4358 Å, both the lifetime and quantum yield of fluorescence are independent of pressure in the vapour, and, as well as one can tell within the experimental uncertainties, virtually the same in the vapour and in fluid solvent at room temperature[31, 36, 37]. These data indicate that the $S_1 \to T$ intersystem crossing which dominates the excited singlet relaxation[37] is not only an intramolecular phenomenon, but that it can be also nearly insensitive even to environmental interactions which establish significant perturbation of the static electronic energies.

While numerous examples of specific intermolecular effects on non-radiative transition rates can be cited, the insensitivity to external perturbation displayed in the $S_1 \to T$ transition in biacetyl is probably common for radiationless transitions in large molecules, as McClure suggested long ago[38] and as others have also noted. Ware and Cunningham's comparison of S_1 relaxation parameters in anthracene vapour and solution is an example[29]. Benzene itself offers another. We may compare the benzene fluorescence yield, which measures the competition between fluorescence and $S_1 \to T$ intersystem crossing, in the vapour and condensed phases *when relaxation in both phases proceeds only from the zero-point level of S_1*. In the gas phase, this yield can be obtained by monitoring fluorescence when the system is excited at very low pressures directly to the zero-point level with tuned narrow-band light[39]. In condensed phases relaxation from the zero-point level is secured by use of low temperature to effectively limit the Boltzmann distribution to that level. A 77 K solvent will place over 95% of the emitting molecules in that level. The quantum yields under these conditions are 0.22 in the gas phase[39] and 0.20 in condensed phase[40–42]. To within the accuracy of the data, these yields are identical. Surrounding benzene by an intimate collisional environment has little effect on the relaxation properties of this vibronic level.

It should be remarked here that a comparison of gas and condensed phase relaxation from Boltzmann S_1 vibronic distributions at 300 K in benzene shows a qualitative difference in non-radiative relaxation rates. The gas-phase yield (at high gas pressures to obtain a thermal fluorescence spectrum) is 0.18 while the condensed-phase yield is near 0.05 in a variety of solvents. The difference is probably due to fast collisional activation in condensed phase to vibronic levels above 2500 cm^{-1}, where non-radiative relaxation becomes extremely fast and fluorescence is non-competitive. A discussion of these data is given elsewhere[43].

8.2.2   Radiationless transitions in small molecules—the resonance and statistical limits

The preceding discussion describes some of the evidence from resonance emission which documents the intramolecular origins of non-radiative transitions in large molecules. Characteristic emission structure accompanied by fluorescence yields below unity and lifetimes shorter than calculated from oscillator strengths comprise the common denominators. But this behaviour is by no means universal. It is principally characteristic of larger molecules. Resonance emission from small polyatomics can be qualitatively different.

Douglas has been most influential in bringing this point forward through his discussions of resonance fluorescence lifetimes in triatomics[44]. Lifetimes in NO_2, SO_2 and CS_2 are *longer* (far longer in the first two) than radiative lifetimes calculated from integrated absorption coefficients. Douglas considered that the most probable origin of this 'anomalous' effect was the mixing of the excited singlet state carrying the radiative oscillator strength with a set of discrete but quasidegenerate vibronic levels of a lower electronic state carry little or no oscillator strength. Through this mechanism the upper state is split into many states and the absorption spectrum becomes richer in lines than in the absence of mixing. However, each line is of reduced strength, and the calculated radiative lifetime, which is related to a summation of these line intensities through the integrated absorption coefficient, will be unaffected by the presence or absence of mixing. The mixed state does not appreciably modify the total transition moment. But the situation is not reciprocal with respect to radiative decay. The radiative lifetime will be increased relative to that in the absence of mixing because each emitting state is 'diluted' with character of a non-emitting state, and the observed lifetime reflects a *mean* of the lifetimes of these individual 'diluted' states.

But such interactions had also been invoked in prior discussions of large molecule non-radiative transitions, where they were used to describe just the opposite behaviour[45–47]. In large molecules, fluorescent lifetimes are shorter rather than longer than those predicted from integrated absorption coefficients.

An explicit description of how this mixing could encompass both behaviours was subsequently given by Robinson[48]. The theory has been further developed in papers beginning with those of Bixon, Jortner, Berry, Chock and Rice[49–51]. These formulations show that the electronic coupling can span two limits defined as the 'small molecule' or resonance limit and the 'large molecule' or statistical limit. Further detailed discussion of the behaviour at and intermediate to these limits is given by Bixon and Jortner[52].

The partitioning towards one limit or the other reflects the density of interacting vibronic states in the lower (final) electronic manifold. In an extreme case, one has a continuum of states, and this is of course the situation in predissociation and auto-ionisation. In these cases the final states constitute a true continuum, at least as defined by the size of the 'box'. These non-radiative transitions have long been recognised as intramolecular phenomena. But an analogous situation is commonly met for non-radiative transitions between electronic states of a large molecule with many vibrational degrees of freedom. The final states of such a transition are high vibronic levels of the lower electronic state, and if the electronic energy gap between the two combining states is sufficiently large, the density of these vibronic levels will be high enough to generate a quasi-continuum of levels nearly degenerate with the initial state of the transition. This vibronic quasi-continuum provides a dissipative medium for electronic relaxation by internal conversion or intersystem crossing, just as the true continuum provides the dissipative medium for predissociation. In these cases, which are defined as the statistical (large molecule) limit, all of the elements required for an irreversible non-radiative transition (at least irreversible in the times accessible to the final states in the laboratory) are present within the molecule. The fluorescence lifetime for a molecule in or near this limit will be short relative to the dipole radiative lifetime, and the fluorescence quantum yield will be reduced below unity in the 'isolated' molecule. Medium effects on the excited state relaxation are expected to be minimal.

When the energy gap between electronic levels is small, or when the molecule itself is small with few vibrational degrees of freedom, the density of final vibronic levels may be too sparse to form a dissipative quasi-continuum, yet high enough so that many levels interact with the initial state. This situation is defined as near the resonance (small molecule) limit. The final state levels are far apart relative to their width, and in this case the manifestations of intramolecular non-radiative relaxation are no longer apparent. In time-dependent language the transitions are no longer irreversible on the experimental time scale and the characteristics described by Douglas are observed. Fluorescence lifetimes are longer than those calculated from the dipole-oscillator strengths, and collisions become crucial agents for non-radiative transitions. They impart irreversibility to the transitions, or, from an observational perspective, they induce the transitions. Medium effects in these cases are of course dominant.

Studies of non-radiative transitions in systems in the statistical limit are encompassed in the resonance fluorescence studies described in the previous Section. But explicit exploration of molecules expected to lie near the resonance limit are not yet abundant in the literature. The systems NO_2, SO_2 and CS_2 seem to be proper examples, and only a few others can be cited. Perhaps the most explicit among these is the six-atom molecule glyoxal (CHOCHO) in which resonance emission casts the $S_1 \rightarrow T$ intersystem crossing as an example of the resonance limit.

Glyoxal is well suited for study because both the initial singlet state (^1A_u) and the final triplet state (^3A_u) connected by the non-radiative transition can be observed directly in emission. Singlet and triplet emission from glyoxal excited with the 4358 Å Hg line by Anderson et al[53] is shown in Figure 8.3. Absorption of this line populates a vibronic level about 1200 cm^{-1} above the zero-point level in the ^1A_u (S_1) manifold, and in pure glyoxal at 0.2 torr this produces emission spread over a wide spectral region. In the 0.2 torr spectrum of Figure 5.3, the structure between 23 000 cm^{-1} and 19 500 cm^{-1} is emission from the singlet state pumped in absorption (with its 0,0 band near 22 000 cm^{-1}), and below about 19 500 cm^{-1} we observe triplet emission with its most prominent feature being the triplet 0,0 band near 19 200 cm^{-1}.

The structure seen at 0.2 torr, however, is not characteristic of the spectrum at other pressures. The structure is extremely sensitive to the involvement of excited glyoxal in collisions, and in that respect it differs grossly from the

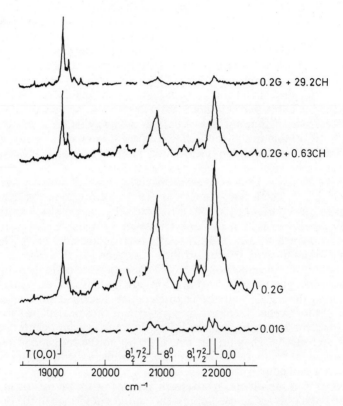

Figure 8.3 Emission from glyoxal excited with the 4358 Å Hg line. The lower spectra are from pure glyoxal with the pressure in torr specified to the right, and the upper spectra are from glyoxal at 0.2 torr with added cyclohexane (CH). Triplet emission ($^3A_u \rightarrow {}^1A_g$) is the structure centred around its 0,0 band near 19 200 cm^{-1}, and structure at high energies is singlet emission ($^1A_u \rightarrow {}^1A_g$) with its 0,0 band near 22 000 cm^{-1}. The exciting line is near 23 000 cm^{-1}.
(From Anderson *et al.*[53], by courtesy of North-Holland Publishing Co.)

imperturbable fluorescence often seen from large molecules. The story is told in just a few spectra. At the lowest pressure, only singlet emission can be observed. At intermediate pressures emission from both the singlet and triplet states is present. Finally, at the highest pressure in Figure 8.3 the spectrum has been transformed to one of only triplet phosphorescence. This change is consistent with collision-induced triplet formation from the singlet state reached by absorption.

These observations are entirely consistent with concepts about the resonance or statistical limits derived from estimates of the density of final triplet states quasi-degenerate with the singlet state excited with 4358 Å light. Calculations show the degenerate level density in the triplet ^3A_u manifold to be somewhat less than ten states per cm^{-1}. Those levels are thus much further apart than their width and the collision-induced appearance of the $S_1 \rightarrow T$ transition is expected. A wide variety of collision partners is effective, and from the fluorescence lifetime studies of Yardley, Holleman and Steinfeld[54], it is seen that cross-sections for 'collision-induced' intersystem crossing are generally on the order of one-tenth of hard sphere.

The constrast of $S_1 \rightarrow T$ intersystem crossing in glyoxal (HCOCOH) and biacetyl ($CH_3COCOCH_3$) offers an interesting illustration of the apparent sensitivity of the crossing mechanism to the density of final states. The structure of biacetyl emission is virtually insensitive to collisional perturbation whereas that of glyoxal emission is explicitly dependent on those perturbations.

These sensitivities document $S_1 \rightarrow T$ crossing, which are apparently in the statistical limit in biacetyl and, as has just been discussed, in the resonance limit in glyoxal. Since the electronic structure of the two molecules should be in close correspondence, the principal difference leading to those opposite behaviours would appear to be the additional 18 vibrational degrees of freedom in biacetyl. It is presumably the presence of this additional vibrational complexity that provides the density of states to bring biacetyl into the statistical limit. But the argument is clouded by the fact that calculations show the total density of triplet vibronic states quasi-degenerate with vibronic levels excited in the singlet state of biacetyl to be greater by only about a factor of 20 than the analogous density in glyoxal. It is not apparent that such a modest increase could have such profound consequences on the relaxation characteristics of these states. However, an unknown parameter remains to be evaluated in these systems, and that is the location of the ^1B_g and ^3B_g electronic states. These states must lie close to the ^1A_u and ^3A_u excited states that are detected in fluorescence and phosphorescence, and they could have a decisive influence on the characteristics of non-radiative decay in both molecules.

8.2.3 The statistical limit in small molecules

It is perhaps unfortunate that the language 'large and small molecule limit' has entered the literature to describe the statistical and resonance cases of non-radiative transitions. 'Small' molecules can of course display 'large molecule' behaviour and vice versa. The essential parameter is the vibrational 'degeneracy' of the final state of the transition. The vibrational levels in the final state of a small molecule will be very densely packed if the electronic energy gap between the combining states is high, or if relaxation proceeds from a high vibronic level of the initial state. Thus the relaxation in this small molecule can lie in the statistical limit. Similarly, a large molecule may display the collision-sensitivity of the resonance limit if the electronic energy gap between the combining states is small.

These situations are expected to be a common occurrence. The non-radiative transitions $S_1 \rightarrow S_0$ and $T_1 \rightarrow S_0$ in almost all polyatomics are expected to be in the statistical limit because of the large electronic energy gap between the initial and final states. Conversely, the intersystem crossing $S_1 \rightarrow T$ or the internal conversion $S_2 \rightarrow S_1$ may be near the resonance limit in molecules of even rather severe vibrational complexity because the energy gap can be small and of the order of vibrational frequencies.

These situations yet remain to be extensively documented, but there are indications of each from vapour luminescence studies. In fact glyoxal may document the case of statistical behaviour in a 'small' molecule by the marked difference in collision-sensitivity between its singlet and triplet relaxation. The singlet relaxation discussed before is an example of $S_1 \rightarrow T$ crossing near the resonance limit. The fluorescence is sensitive to collisions with a great variety of partners. On the other hand, the $T_1 \rightarrow S_0$ crossing that may account for non-radiative relaxation of the triplet state should display just the opposite behaviour. Calculations[53] indicate that the density of final ground-state vibrational levels quasi-degenerate with the triplet zero-point level is of the order of 10^6 states per cm^{-1}. This is certainly high enough to create the dissipative array of final states required for crossing in the statistical limit. The triplet emission is consistent with these indications. At high collision frequencies, where virtually all singlets or at least a constant fraction of singlets are induced to cross to the triplet state, we find that triplet phosphorescence becomes independent of pressure (see Figure 8.3). Phosphorescence is insensitive to further increases of either cyclohexane or helium pressures. While both of these gases are effective in inducing the formation of the triplet state, neither seems to effect its decay.

Lifetime studies of glyoxal's triplet emission confirm the insensitivity of triplet relaxation to collisional interactions[53]. Except for a few specific partners, the triplet lifetime is not appreciably influenced by external perturbations. Triplet relaxation must indeed be in the statistical limit as expected. However, absolute quantum yields of phosphorescence have yet to be reported, so that this is at present an imperfect illustration of $T_1 \rightarrow S_0$ intersystem crossing. Appreciable photochemistry occurs in this system[55], and the extent to which it, rather than $T_1 \rightarrow S_0$ crossing, controls triplet relaxation under the conditions of the phosphorescence studies is not known.

8.2.4   The resonance limit in large molecules

Indications of non-radiative transitions near the resonance limit in very large molecules can be found in several fluorescence studies, and even in a study of condensed phase systems. Nitzan, Jortner and Rentzepis[56] have argued that an observed lengthening of the lifetime of S_1 benzophenone molecules, as higher vibronic levels are excited in solution, is a consequence of a near-resonant $S_1 \rightarrow T$ intersystem crossing. Observations of $S_2 \rightarrow S_0$ resonance fluorescence in pyrene[57, 58], 3,4-benzpyrene[57, 59] and naphthalene[60] have been associated with S_2–S_1 coupling near the resonance limit, and a reinvestigation of biacetyl fluorescence by Nitzan, Jortner, Kommandeur and Drent[61] has revealed indications of near-resonance behaviour in an $S_1 \to T$ crossing in even this 'well explored' example of a molecule whose crossing, at least under certain conditions, seemed to be in the statistical limit.

The fruits of improved experimental technique are nicely revealed in the biacetyl study. By using an aluminium integrating sphere coupled with sensitive fluorescence detection circuits, Drent and Kommandeur have found two novel aspects of biacetyl fluorescence. The first was the observation of infrared fluorescence from *vibration–vibration* transitions in the C—O stretching mode of the triplet state of biacetyl[62]. This excited-state infrared fluorescence is interesting not only because of its uniqueness, but also because it comments on aspects of the $S_1 \to T$ intersystem crossing which populates that triplet state. It reveals by an independent method of observation the presence of a triplet state, formed under collision-free conditions.

The second discovery of Drent and Kommandeur is a marked sensitivity of $S_1 \to S_0$ fluorescence to small pressures of added helium when biacetyl is excited with the 4047 Å Hg line[61]. This effect is highly dependent upon exciting wavelength, and it is not observed when lower vibrational levels of the S_1 state are excited with the 4358 Å line.

As described before, an intramolecular $S_1 \to T$ intersystem crossing dominates the S_1 decay after 4358 Å excitation and reduces the fluorescence yield to about 10^{-3}. No hint of collision sensitivity exists. Neither the fluorescence nor the phosphorescence acknowledges the perturbations of added gases after 4358 Å excitation. But a subtle and highly significant difference in behaviour was revealed by Drent and Kommandeur when they were able to monitor the $S_1 \to S_0$ fluorescence after excitation with the 4047 Å line. The major parameters are the same. An intramolecular intersystem crossing again dominates the relaxation to reduce the fluorescence yield to a small value. However, in this case the fluorescence is sensitive in a peculiar manner to very modest perturbations from added gas. Addition of 0.6 torr of helium to 0.050 torr of biacetyl reduces the fluorescence intensity by about a factor of 2, but further addition restores the intensity. It returns to its original gas-free value upon addition of 3–4 torr of helium. Not only is the occurrence of a minimum in the fluorescence rather unique, but so is the fact that the collision sensitivity appears and disappears in a pressure region where the hard-sphere collision interval exceeds the lifetime of the excited singlet state (as calculated from the fluorescence yield and S_1–S_0 oscillator strength) by more than an order of magnitude.

The authors have shown how this behaviour can be consistent with two coincident intersystem crossings. One is in the near-resonance limit and the other is in the statistical limit. The latter, involving weak coupling of the singlet state to a dissipative array of triplet levels, establishes the dominant intramolecular intersystem crossing that is seen with excitation at both 4047 Å and 4358 Å. But by accident, the 4047 Å line can also excite a singlet level that is strongly coupled to only one or only a few nearly degenerate triplet levels. This establishes opportunity for a (small) fraction of singlet decay to be in the near-resonant limit. Analysis of the problem shows that if the zero-order energy separation of the pair, the S–T interaction matrix element of the pair, and the singlet line-width are all of similar magnitude, the stage is properly set for reproducing the observed behaviour. The triplet level (but not the singlet) is extremely sensitive to collision broadening by coupling to the sea of nearly degenerate triplet vibronic levels which surround it. Because

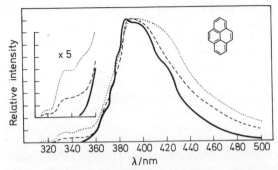

Figure 8.4 Resonance fluorescence spectra of pyrene obtained by excitation (at 3230 Å) into the $S_1 \leftarrow S_0$ absorption system (solid line), and into higher singlet states with shorter wavelength excitation: dashed line excited at 2680 Å; dotted line excited at 2360 Å.
(From Nakajima and Baba[58], by courtesy of the Japanese Chemical Society)

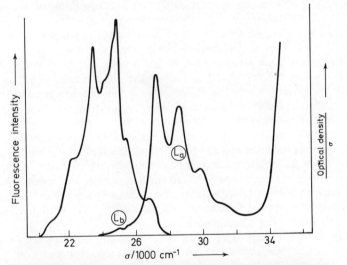

Figure 8.5 The resonance fluorescence spectrum of 3,4-benzpyrene vapour (left) observed after excitation into the $S_2 \leftarrow S_0$ (L_a) absorption system (right). The $S_1 \leftarrow S_0$ absorption system (L_b) is relatively weak.
(From Geldof et al.[57], by courtesy of North-Holland Publishing Co.)

of this, collision broadening of the triplet level can at first bring it into significant interaction with the singlet to cause crossing and reduce the fluorescence. However, a continued increase in collisional perturbation will eventually reduce this S–T interaction by broadening the triplet level appreciably beyond the energy regions of strong singlet interaction. Crossing will then be reduced and the fluorescence yield will asymptotically return to its original low pressure value.

Unfortunately the vibrational complexity of biacetyl with respect to these experiments is too severe to discuss the validity of the mechanism by identification of any specific levels that could so couple. One can, as the authors have done, only provide evidence for the plausibility of the situation.

S_2–S_0 fluorescence in pyrene, 3,4-benzpyrene, and naphthalene, comprise, with the special case of azulene[63] and a few azulene derivatives[64,65], the exceptional instances of fluorescence from a singlet state above S_1. The fluorescence spectra in these cases are given in Figures 8.4, 8.5 and 8.6. S_2

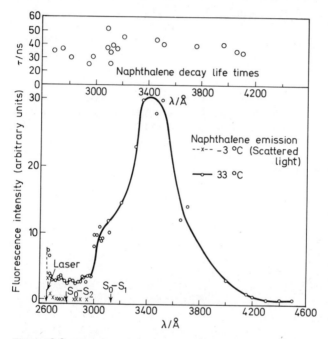

Figure 8.6 The resonance emission spectrum and the fluorescence decay times monitored at various fluorescence wavelengths when naphthalene is pumped in the $S_2 \leftarrow S_0$ absorption system by picosecond pulses from the fourth harmonic (2650 Å) of a Nd^{3+} laser. The fluorescence structure at wavelengths shorter than about 3000 Å is attributed to S_2 fluorescence. The positions of the excitation and the 0,0 bands of the S_2–S_0 and S_1–S_0 transitions are marked.
(From Wannier et al.[60], by courtesy of North-Holland Publishing Co.)

emission from 3,4-benzpyrene and 1,2-benzanthracene has also been seen in solution[66]. In each molecule, S_1 fluorescence dominates the emission intensity, but additional structure of low intensity is observed on the high-energy edge of S_1 emission when excitation populates the higher singlet states. Since this structure lies outside the region of 'normal' S_1 emission and in the region appropriate for S_2 emission, and since this structure appears only when excitation reaches into S_2 or higher states, it has been assigned as $S_2 \rightarrow S_0$ fluorescence.

Before theoreticians had so elegantly revealed the possible subtleties associated with internal conversion, the appearance of this S_2 fluorescence would have been easily 'understood' by a simple kinetic argument. It would have been the (rarely) successful competition between a normal $S_2 \rightarrow S_0$ radiative decay and the fast $S_2 \rightarrow S_1$ internal conversion that dominates S_2 relaxation. This argument would be put forward even currently if all $S_2 \rightarrow S_1$ internal conversion was considered to be in the statistical limit, and in each molecule, a priori expectations of the statistical limit are justified. While the S_2–S_1 electronic energy gaps are relatively small (1500–4000 cm^{-1}), the vibrational complexities are sufficiently high to establish the required large density of states (there are 72, 90 and 48 vibrational degrees of freedom in pyrene, 3,4-benzpyrene and naphthalene, respectively). In fact, the intense S_1 fluorescence accompanying the S_2 emission is presumably a consequence of such an internal conversion, and the existence of an ordinary S_2 radiative decay which is just fast enough to be observed in competition with this internal conversion would not be too surprising.

However, there is additional information about S_2 fluorescence that shows the situation is not so simple. For example, Geldof et al.[57] find that the intensity of S_2 emission in both pyrene and 3,4-benzpyrene is reduced by the presence of just 9 torr of n-hexane. The lifetime of those molecules producing S_2 fluorescence must be far longer than the value 10^{-11}–10^{-12} s predicted by the relationship

$$\tau \approx Q_f \tau_{rad}$$

which would be appropriate if all S_2 decay is in the statistical limit. (Q_f is the fluorescence quantum yield and τ_{rad} is the 'pure' radiative lifetime calculated from the S_2–S_0 oscillator strength.) The collision interval in 9 torr of gas is of the order of 10^{-8} s, and the lifetime of the molecules emitting in the S_2 region must at least approach this value.

These indications are confirmed by the direct lifetime measurements of Wannier et al. on 3,4-benzpyrene[59] and naphthalene[60]. Using a method ingenious in its simplicity for truncating a chain of picosecond pulses from the mode-locker laser used for excitation, they observed that the lifetimes of S_2 fluorescence are indeed much longer than expected from the simple picture of statistical internal conversion. The lifetimes are in fact almost identical to those of S_1 emission. The S_2 fluorescence lifetime in 3,4-benzpyrene is about 65 ns compared to 70 ns for that of fluorescence monitored in the S_1 region, and it seems hard to decide from the data whether this small difference exceeds the fluctuations in the measurements themselves. The S_2 and S_1 lifetimes in naphthalene are also close. They are 30–35 ns and 40–45 ns, respectively (see Figure 8.6).

The two groups of workers have provided an essentially similar accounting of these long lifetimes and the origin of S_2 fluorescence, although their discussions are couched in different languages. They require that the fluorescing S_2 molecules are strongly coupled to a comparatively small number of the available quasi-degenerate S_1 vibronic levels. Such restrictive strong coupling to a small subset of S_1 levels in molecules with these narrow electronic energy gaps is plausible because of coupling restrictions imposed by symmetry and Franck–Condon arguments. This restrictive coupling places $S_2 \rightarrow S_1$ internal conversion *for these molecules* near the resonance limit. Thus one observes that S_2 fluorescence is quenched by collisions which 'induce' the $S_2 \rightarrow S_1$ internal conversion, and one observes S_2 fluorescence lifetimes that are long relative to the radiative lifetime calculated from S_2–S_0 oscillator strengths.

The authors arguments are carefully developed and fit the facts well. But a case might also be offered that the 'S_2 fluorescence' is in fact ordinary $S_1 \rightarrow S_0$ fluorescence from high S_1 vibrational levels. As higher S_1 levels are excited (in these instances by the fast statistical $S_2 \rightarrow S_1$ internal conversion), their resonance fluorescence would extend over an increasingly wide spectral range, and some of the extension will occur in the high energy side of 'ordinary' (low level) S_1 emission, coincident with the position of S_2 fluorescence.

The extension arises from additional progression members possible only from higher S_1 vibrational levels. At low excitation energies, the few progressions responsible for suggestions of vibronic structure in these spectra contain essentially only the transitions $0 \rightarrow 0, 0 \rightarrow 1, 0 \rightarrow 2 \dots$ in vibrational quanta of those few modes active in forming progressions. However, when higher excitation energies populate several quanta of a mode active in progressions, one will observe transitions analogous to those plus additional ones. If, say, three quanta are excited, the transitions $3 \rightarrow 3, 3 \rightarrow 4, 3 \rightarrow 5 \dots$ will occur very near the position $0 \rightarrow 0, 0 \rightarrow 1, 0 \rightarrow 2 \dots$ transitions, being displaced from them only by the slight red-shifting, resulting from the (generally) small difference between excited and ground-state vibrational frequencies. But now, opportunity for additional progression members $3 \rightarrow 2, 3 \rightarrow 1$ and $3 \rightarrow 0$ occurs for which no analogous transitions were possible from the lower level. These progression members must exist and they will generate vibronic structure to higher energy from the 'normal' S_1 progression structure.

This S_1 structure in the S_2 region will display kinetic parameters consistent with those reported for S_2 fluorescence. It would be quenched as added gases cause fast vibrational relaxation of the high S_1 levels. It would increase in intensity as the excitation energy increases. It would also increase in intensity with increasing temperature since the energy distribution of vibrational levels initially excited by light absorption reflects the ground-state level distribution. It could also be present in gas and solution spectra[66] as a result of thermal activation of higher modes in the S_1 state. It will of course display a lifetime identical to that of fluorescence monitored in the 'normal' S_1 region.

Perhaps the best criteria for distinguishing S_2 fluorescence from high vibronic level S_1 fluorescence are found in the emission structure. While both S_2 and S_1 emission may show a progression in the same ground-state frequency, it is only by fortuitous coincidence that a real S_2 structure could fit smoothly as a continuation of the progression in the S_1 region. In addition, the progression intensity of S_2 emission and S_1 high vibronic level emission should be qualitatively different. That from S_2 emission is expected to mirror $S_2 \leftarrow S_0$ absorption with maximum intensity in the first few (highest energy) progression members. That from S_1 emission is expected to be opposite. It is controlled by the relative populations of higher S_1 vibronic levels, and these populations are set either by Boltzmann factors (thermal activation) or by Franck–Condon factors ($S_2 \rightarrow S_1$ internal conversion). In either case there is strong population bias to lower levels, so that the progression members will *decrease* in intensity at increasing energies from the $S_1 \rightarrow S_0$ 0,0 band.

Unfortunately, it is difficult to decide the issue from the data so far published. The intensity patterns tend to support the case for S_1 emission, but rather less can be said from a study of the positions of vibronic structure. The vibronic structure is most clearly developed in 3,4-benzpyrene. Its resonance fluorescence in Figure 8.5 has two separate but overlapping progressions in a ground-state frequency of 1200–1400 cm^{-1}. The structure assigned to S_2 fluorescence is clearly not associated with the progression most prominent in the S_1 region, and this may support the case for true S_2 fluorescence. It would be most useful to see this spectrum as exitation probes S_1 levels just below the S_2 state. Easterly *et al.*[66] give evidence from their solution work to support also the S_2 case. The vibronic structure in pyrene under conditions producing S_2 fluorescence is too poorly developed to yield much helpful information. A combination of the S_2 structure of naphthalene reported by Wannier *et al.*[60] with the S_1 structure observed by Stockburger[19] when pumping the 0,0 band of the $S_2 \leftarrow S_0$ system suggests that a single long progression with a ground-state frequency 1200–1400 cm^{-1} runs through both regions, but if this is to be the case, we must assume that one band between the S_1 and S_2 regions is unreported. The S_2 structure in naphthalene is so difficult to see, however, that judgements based on it would most charitably be considered as tenuous.

8.3   OPTICAL SELECTION STUDIES AND THE VIBRONIC STRUCTURE OF EXCITED STATE RELAXATION PROCESSES

The resonance fluorescence studies so far discussed comment on the basic physics of non-radiative transitions. The emission has demonstrated the intramolecular course of these transitions when the final states constitute a quasi-continuous array of levels, and with close stimulation from developing theory it has also revealed the essential role of collisions to 'induce' or to impart 'irreversibility' to these transitions when the final states are more sparsely positioned. The collaborations of theory and experiment have resulted in a new understanding of these transitions between electronic states in polyatomic molecules.

More detailed studies of these processes are now just emerging to carry the picture forward. They are providing a description of the dependence of these transitions on vibrational (and rotational) energy in the initial states. The goal of these experiments is to map out a vibronic 'spectrum' of excited state relaxation.

The approach has been to explore first those systems which display relaxation processes in the statistical limit. This enables collisional interactions to be removed from the problem. One can then retreat to the gas phase at low pressure so that excited states with selected vibrational energies can be pumped and then preserved for electronic relaxation after their initial preparation. Pumping of selected vibrational domains in the excited state (usually S_1) has been achieved by absorption of tuned narrow-band exciting light. Preservation of those selected states for relaxation is generally not difficult because gas pressures need only be kept below about 0.1 torr to avoid significant collision scrambling of vibrational or rotational energies before electronic decay. The collision-free lifetimes of fluorescing states usually do not exceed 500 ns.

The most successful experiment would derive from initial selection of a single vibronic level in excitation, but this places demands on the molecular system which nature finds contradictory. To display collision-free radiation-less transitions, the molecule must have considerable vibrational complexity, which is generally achieved by size. At the same time it must also have a vibrationally simple absorption system so that tuned light can pump a single band not overlapped by others.

A compromise is indicated, and it has generally been at the expense of clean preparation of the excited state. Most of the 'single vibronic level' relaxation studies so far reported concern molecules with very crowded absorption spectra. In some of those molecules, even line sources could not prepare individual upper levels. They are single vibronic level studies in spirit only. But nevertheless they do represent preparation of excited states with a relatively high degree of vibronic selection, and they are extremely valuable for their comments about the vibronic influence on excited state decay.

Fluorescence studies resolve excited state relaxation into the two channels of radiative and non-radiative decay. By themselves they do not define the nature of the non-radiative channel, and further experiments involving parameters other than fluorescence are required for this resolution. This is frequently a problem in the optical selection experiments. It is difficult to tell exactly what non-radiative processes are being described.

Two fluorescence measurements are required to uniquely resolve relaxation into the radiative and non-radiative channels. We may characterise radiative decay by a first-order rate constant k_r and the non-radiative channel by a first-order constant k_{nr}. The fluorescence quantum yield Q_f, observed as the ratio of photons emitted to photons absorbed, is a function of the ratio of these constants

$$Q_f = \frac{k_r}{k_r + k_{nr}}$$

The fluorescence lifetime τ_{obs}, observed as the exponential decay of fluorescence in a time-modulated experiment, is a function of their sum

$$\tau_{obs} = (k_r + k_{nr})^{-1}$$

Measurements of one, or in fewer cases, both of these parameters at just a few exciting wavelengths are available for numerous molecules. Together with a good deal of other data, they indicate undeniably the general sensitivity of relaxation channels to vibrational excitation in the upper state. But such measurements from resonance fluorescence excited with narrow-band light tuned to sample many vibronic regions in a single molecule are not yet so extensive. On systems larger than triatomics, reports have been published on β-naphthylamine[67, 68], naphthalene-h_8[18, 69, 70], naphthalene-d_8[18], aniline[71], glyoxal[54], fluorobenzene[72], benzene[22, 39, 73–75] and benzene-d_8[74, 75]. These reports display the sensitivity of excited state relaxation to vibrational excitation in considerable detail. They are the vanguard of an experimental area that will become increasingly important as more intense and mono-chromatic exciting sources are available.

8.3.1   Relaxation from selected vibronic energy domains

Tuned excitation has so far succeeded in exciting single vibronic levels only in benzene and glyoxal. The room-temperature absorption spectra of the others are so crowded that excitation can only produce a distribution of excited vibronic states. Because of this common feature, we will discuss these cases together in this Section. Benzene and glyoxal are the subject of comments in the next.

There is a uniform behaviour in the fluorescence from those systems excited to vibronic domains. The fluorescence lifetimes become generally shorter and the fluorescence yields become generally less as the energy of the monitored vibronic domains increases (this is true also in benzene). In those few cases where lifetimes and yields are simultaneously available, the drift is seen to be dominated by an increase in non-radiative decay. This is surely true of all systems, however, because the rate of radiative decay is not expected to show a very great dependence on vibrational excitation. In those cases where the radiative transition is symmetry-allowed, its rate should be virtually independent of vibronic energy, and when the fluorescence is vibronically induced (as in benzene), its rate will be only a very modest function of vibronic energy.

It is probably fair to say that there has not been a uniformly exhaustive exploration of the nature of the non-radiative channel(s) being monitored in these studies. It is presumed in every case to be principally $S_1 \rightarrow T$ inter-system crossing, and in some instances there is evidence to support this. As far as is known at present, those presumptions are correct, and as such, we see that these experiments all show a boost in the intersystem crossing rate as the energy gap between the combining electronic states increase.

This is a direct exploration of the 'energy gap law' derived from data on condensed-phase systems in which the variation in electronic energy gap between the combining states was established by variation in molecular structure, i.e. by studying a variety of molecules with different T–S gaps. Theoretical formulations of the law are found in Siebrand's discussions[76] and in papers by Englman and Jortner[77], and by Fisher[78]. An explicit discussion of the clean exponential dependence of non-radiative decay on excitation energy seen in Schlag and von Weyssenhoff's pioneering study of β-naphthylamine[68] has been given by Fisher and Schlag[79].

In two cases, β-naphthylamine[68] and naphthalene[69, 70], fluorescence data are available for excitation which spans both S_1 and higher singlets. The fluorescence lifetimes in each case show a smooth energy dependence as excitation climbs from S_1 through the position of the S_2 state. No hints of the presence of that state are seen in the data. This is consistent with the prejudice of prior experience which would require that (a) fast $S_2 \to S_1$ internal conversion precludes any contributions from S_2 fluorescence to that being timed, and that (b) $S_2 \to S_1$ internal conversion forming the emitting S_1 levels is on a time scale much faster than that for relaxation of those S_1 states. Thus, as excitation leaves the S_1 region and climbs into the S_2 region, fluorescence is still observed only from S_1 molecules and in their relaxation they have no memory of the path by which they were formed. Lifetime measurements see them only as S_1 molecules with high vibrational energies.

Complementary quantum yield measurements in these cases would have been useful because quantum yields are not so parochial. They can com-

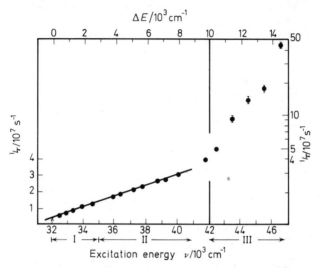

Figure 8.7 Experimental fluorescence decay rate of naphthalene vapour as a function of excitation energy. Regions I, II, and III correspond to excitation into the first, second, and third excited singlet state respectively. ΔE represents the excess of excitation energy above the vibrationless first excited singlet state. The ordinate for values of $1/\tau$ larger than 4×10^7 s^{-1} is continued on a logarithmic scale (right-hand side) for ease of representation of the data.
(From Laor and Ludwig[69], by courtesy of the American Institute of Physics)

ment on events associated with the formation as well as decay of those high S_1 states. For example, a second S_2 relaxation channel competitive with $S_2 \to S_1$ internal conversion would be quickly revealed as a discontinuity (reduction) in the fluorescence yield as the exciting light climbed from S_1 into S_2.

In naphthalene, lifetimes are even available as excitation climbs from S_1 through S_2 and well into S_3 [69]. They are shown in Figure 8.7. An interesting change occurs in the energy dependence of the lifetime. Throughout the S_1 and S_2 regions, the dependence is linear and shallow, but it changes into a steep and nearly exponential dependence near the position of S_3. There is no evidence that fluorescence is other than S_1 [18], and the break indicates that either a new S_1 relaxation channel has become accessible or that a previous S_1 channel such as $S_1 \to T$ crossing has suddenly become accelerated.

There is at present no specific information concerning the new relaxation channel. Because of the broad exciting band pass, it is also difficult to locate the position of the onset to see if its location near the S_3 state is more than coincidence. But it remains a clear illustration that channels for relaxation of *an* excited electronic state in large molecules can have a rather sharp energy threshold and a persistent presence above it. Such thresholds have long been the central hypothesis in discussions of unimolecular reactions from high vibrational levels, and fluorescence studies using atomic line sources for excitation have also hinted at their presence in excited states in large molecules. It is expected that low pressure fluorescence studies with

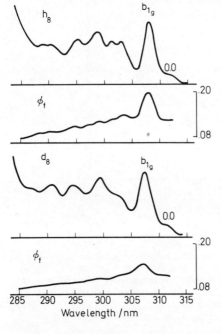

Figure 8.8   A comparison of the $S_1 \leftarrow S_0$ absorption spectrum and the excitation energy dependence of the resonance fluorescence yields ϕ_f from naphthalene and naphthalene-d_8. The $S_2 \leftarrow S_0$ 0,0 band is near 285 nm.
(From Uy and Lim[18], by courtesy of North-Holland Publishing Co.)

tuned excitation will become increasingly successful in their characterisation. Progress has been made in two other cases (benzene and cyclobutanone) and these will be described later.

We now turn to consideration of the structural details in the vibronic energy dependence seen in these fluorescence experiments. No structure can be seen in the broad-band excitation experiments on β-naphthylamine (30 Å) or naphthalene-h_8 (50 Å). These show only a smooth dependence of relaxation parameters on the excitation energy. The more narrow-band experiments have considerable fluctuations in lifetimes and yields which are superimposed on the overall drifts of these quantities as excitation energies change. Figure 8.8 from Uy and Lim[18] shows the structure in fluorescence

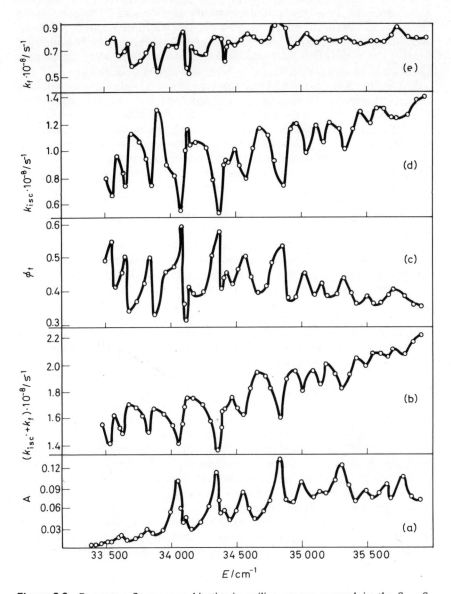

Figure 8.9 Resonance fluorescence kinetics in aniline vapour pumped in the $S_1 \leftarrow S_0$ absorption system with 2.7 Å tuned excitation: (a) the absorbance of the $S_1 \leftarrow S_0$ transition; (b) the reciprocal fluorescence lifetime τ_{obs}^{-1}; (c) the fluorescence quantum yield; (d) the non-radiative relaxation constant $k_{nr} = k_{isc}$; (e) the radiative relaxation constant $k_r = k_f$. These are each plotted as a function of the energy E of the exciting light

(From von Weyssenhoff and Kraus[71], by courtesy of the American Institute of Physics)

yields from naphthalenes excited in $S_1 \leftarrow S_0$ absorption with a 9 Å band pass. Figure 8.9 shows the structure in both τ and Q_f from 2.7 Å excitation of the $S_1 \leftarrow S_0$ system of aniline by von Weyssenhoff and Kraus[71]. While no fluctuations are seen in the 3 Å exploration of fluorobenzene by Nakamura[72], a rather sharp cut-off in fluorescence is seen in that study when the exciting light climbs to energies in the $S_1 \leftarrow S_0$ system above c. 2550 Å.

The fluctuations in fluorescence parameters of naphthalene and aniline mirror structure in the absorption spectrum, and the task is to understand why this should be so. Consider first the yields in Figure 8.8, which span the first $3000 \, \text{cm}^{-1}$ of vibronic energy in the S_1 states of naphthalene and naphthalene-d_8. The yields in each case display maxima corresponding to the position of absorption maxima. The latter are due in part to excitation of a b_{1g} vibration in the S_1 state. Uy and Lim have proposed that this vibration has a crucial effect on the $S_1 \rightarrow T$ intersystem crossing which in turn is reflected in the fluorescence yields. They argue that when the zero-point level or only totally symmetric vibrations are dominantly excited in the S_1 state, coupling to the triplet states T_1 and T_2 can occur through first-order spin–orbit interaction. When b_{1g} modes become dominant among excited vibrations, however, the $S_1 \rightarrow T$ interactions can proceed through additional vibronically induced paths as well. For this reason, $S_1 \rightarrow T$ coupling will have appreciable sensitivity to the vibronic states initially excited, and this will appear in the fluorescence parameters when measured as a function of exciting wavelength. (The maximum in yields at the position of b_{1g} excitation in Figure 8.8 is deceiving without further information. Uy and Lim point out that both non-radiative and radiative decay are accelerated at that position.)

Such qualitative symmetry arguments for vibronic transitions in large molecules must in the general case be put forward with care because they are to a large extent reciprocal. If, say, a b_{1g} vibration is to allow vibronically induced coupling to a triplet state, it must do so by combining with a totally symmetric vibronic level in that triplet state. But a nearly equal opportunity usually exists for that vibronically induced path to operate when totally symmetric levels are excited in the initial state. In this case, the required b_{1g} symmetry is simply present in the level reached in the final state. Since these transitions occur in the statistical limit, the degeneracy of the final vibronic states is so high that there is always ample opportunity for this to occur. And, in the statistical limit, any energy restrictions on the presence of degenerate final levels of the correct symmetry for one case must usually apply with nearly equal severity for the other. For these reasons the rate of non-radiative decay may have little sensitivity to excitation of vibration with specific symmetries.

Recently a lifetime study by Laor and Ludwig[69] using 10 Å and 3 Å excitation band-widths has become available to complement the quantum yield measurements. The 3 Å data are shown in Figure 8.10. The inverse lifetimes are plotted as a function of the excitation wavelength in the upper part of the Figure. The marked fluctuations with the excitation are reminiscent of the 9 Å yields in Figure 8.8. However, those fluctuations disappear when the data is plotted as an approximate function of the vibrational energy in the S_1 state of the fluorescing molecules. This relationship cannot be made with great accuracy because the spectrum is too crowded with sequence structure to avoid pumping distributions, but this correlation of relaxation data with excited state energies strongly suggests that the vibronic energies are dominant in controlling the relaxation rates. Individual symmetries indeed seem not so important.

Laor and Ludwig propose that the structure seen when relaxation measurements are plotted against exciting wavelength is caused by passage of a

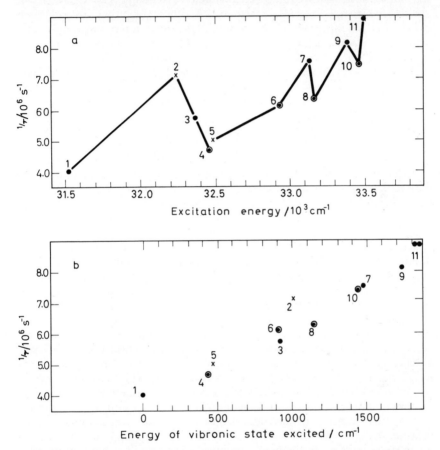

Figure 8.10 The reciprocal lifetimes observed for S_1 fluorescence decay in naphthalene vapour at 0.1 torr and 25 °C with a 3 Å excitation band-pass. The lifetimes are plotted as a function of the energy of the exciting light in the upper display, and as a function of the energy of the dominant emitting vibronic level in the lower plot. All the data concern excitation only in the $S_1 \leftarrow S_0$ absorption band
(From Hsieh *et al.*[70], by courtesy of North-Holland Publ. Co.)

swept exciting source through regions of the spectrum which alternately pump cold and hot absorption bands. This imparts a sequential *energy* fluctuation in the excited vibronic states which mirrors the intensity patterns in the absorption spectrum, since the cold and hot regions roughly correspond to maxima and minima respectively in a crowded absorption spectrum.

The resourceful 2.7 Å exploration of relaxation from the 0–2500 cm^{-1} vibronic region of the S$_1$ state of aniline shown in Figure 8.9 accomplishes the synthesis of lifetimes and yields which are so far lacking in most other studies. The radiative rates (k_f in Figure 8.9) describe the appearance of an allowed transition, and as expected, the rate does not show a general dependence on vibronic energy. The rate constant k_r (their k_f) averaged through the fluctuations is essentially independent of energy. The cause of the fluctuations is not currently assignable. To the extent they are real, they suggest that an appreciable contribution to the transition moment may come from a vibronically induced mechanism. Examination of the structure of the resonance fluorescence itself may yield information on this point.

Von Weyssenhoff and Kraus[71] attribute the fine structure of fluctuations in the non-radiative rate constant to the consequence of sequentially pumping with swept excitation S$_1$ levels dominated by vibrations of the wrong symmetry to efficiently couple the singlet and triplet manifolds. This may indeed be true, and as the authors describe there are a variety of coupling routes to support their arguments. However, these data are the result of fluorescence from distributions of upper levels excited by pumping in a crowded spectrum, and the problem with hot sequence bands will be quite analogous to the naphthalene case. All the comments made there apply here. It will be most valuable to see how much vibronic structure remains in the rate constant k_{nr} when these extensive data are projected onto a scale more accurately describing the vibronic energy of the fluorescing molecules.

Hemminger and Lee[80] have reported initial data from a combined study of both resonance fluorescence and photochemical decomposition in cyclobutanone excited with dominant populations in single vibronic levels by tuned 0.5–0.75 Å light. The resonance fluorescence yield drops by a factor of five as the energy of the emitting S$_1$ vibronic level increases from 1440 to 1540 cm^{-1} above the zero point level. The decrease is persistent as levels above this threshold are probed, and the yield quickly becomes too low to observe. Extensive chemical relaxation yielding C$_2$ and C$_3$ hydrocarbons, CO and ketene accompanies the fluorescence, and it also displays structured dependence on the vibronic level reached in excitation. The ratio of C$_2$ to C$_3$ products undergoes a marked change as various members of a progression exciting quanta of an O-atom wagging vibration are pumped. The authors propose that this may be the consequence of opening a new channel (a bond cleavage to give the diradical CH_2—CH_2—CH_2—C=O) as the vibronic energy in S$_1$ climbs. While much remains to be done if a clear picture of S$_1$ relaxation is to emerge from this system, it represents an encouraging initiation of photochemical studies using tuned optical selection of the excited state. Combined studies of resonance fluoresence and photochemical activity from single vibronic levels hold high promise for yielding a major advance in the physics of large molecule chemical relaxation.

8.3.2   Relaxation from single vibronic levels

Tuned optical-selection studies of resonance fluorescence in molecules whose absorption spectra are sufficiently open to truly pump single vibronic levels have been reported in only a few cases. Of these the benzenes have been most extensively explored, and comments about this work will be given in the last parts of this section. The study of cyclobutanone discussed in the preceding paragraphs may qualify, and we can also include studies of glyoxal and formaldehyde. The latter, however, probably represent somewhat special cases because the $S_1 \rightarrow T$ non-radiative crossing in each is near the resonance limit and hence may be absent in low pressure studies (see Section 8.2.2 for a discussion of glyoxal). Both of these cases may also be complicated by the presence of chemical relaxation from these singlet states, but this in fact should provide even more compelling motivation for study of their resonance fluorescence with tuned excitation.

There has so far been little work reported on resonance emission in glyoxal, but because of the extremely well structured (and well assigned[81, 82]) absorption spectrum, several studies with selected vibronic excitation are now in progress. The only published report, however, is that concerning resonance fluorescence lifetimes by Yardley, Holleman and Steinfeld[54] who swept a 1 Å band from a tuned dye laser through the region 4510–4585 Å of the $S_1 \leftarrow S_0$ absorption system. The fluorescence lifetime under collision-free conditions is about 2 µs. This is a factor of four less than the radiative lifetime calculated from the absorption coefficient[83], but with such a sharply structured absorption spectrum it is not possible to claim an accuracy in the calculated lifetime that is better than that difference. The observed lifetime is also invariant with exciting wavelength, even though the tuned laser sweeps over the rotational contours of at least five identified absorption bands[81, 82]. Thus none of the free-molecule S_1 relaxation channels have appreciable sensitivity to (modest) changes in the S_1 vibronic and rotational levels. In the absence of fluorescence yield data, however, one is unable to know if either of the possible statistical non-radiative channels (photochemistry or $S_1 \rightarrow S_0$ internal conversion) are coupled to the states reached in this preliminary study.

Formaldehyde is expected to display S_1 relaxation processes which in many respects are analogous to those of glyoxal. While there have been numerous probes of formaldehyde relaxation using atomic line excitation, only one report is currently available in which efforts have been made to reach single vibronic levels. Yeung and Moore[84] have measured the resonance fluorescence lifetime from each of two levels excited in the S_1 state of D_2CO with a 1 Å band from a tuned ultraviolet laser. In contrast to glyoxal, the lifetimes from the two levels differ by a factor of 50 (5 v. 0.1 µs). No information is yet available, however, to assess the significance of these data. Neither the nature nor even the existence of non-radiative relaxation channels has been established. The extension of this work beyond this preliminary stage will be extremely fruitful.

The exploration of single vibronic level fluorescence in the S_1 state of the benzenes C_6H_6, C_6D_6 and C_6H_5F is now at a comparatively advanced stage. The data are most extensive on C_6H_6. Resonance-fluorescence spectra, quantum yields and lifetimes are available from over twenty securely identified levels in this molecule. These data provide an unambiguous examination of the effect of varied vibrational energies, modes and symmetries on fluorescence and $S_1 \rightarrow T$ intersystem crossing.

A review of the modest amount of data from single vibronic level C_6H_6

fluorescence so far published[22, 23, 39, 73–75] has been recently given by the author[43] together with a discussion of the more conventional gas-phase fluorescence and relaxation studies of benzene that are essential for interpretation of the single vibronic level experiments. That material will not be repeated here. The bulk of the data, however, has yet to appear. There now remains to be published a particularly fine set of quantum yields and lifetimes on C_6H_6, C_6D_6, and C_6H_5F by Abramson, Spears and Rice[85], a set of quantum yields and resonance-fluorescence spectra by Parmenter, Schuh and Schuyler[86], a probe of quantum yields from selected rotational distributions in some C_6H_6 vibronic levels[87], an exploration of the lifetimes of high non-fluorescing C_6H_6 vibronic levels through line-width measurements in the absorption spectrum[88], and a detailed theoretical account of the vibronic level dependence of the $S_1 \rightarrow T$ intersystem crossing[11]. When these reports are available they will comprise the most intimate account of excited state relaxation yet offered. But because so much yet remains to be published on this system, it will not be reviewed in detail here. However, a few of the important findings that have come forward in C_6H_6 are offered below.

(a) Fluorescence and intersystem crossing have little sensitivity to relaxation from varying rotational distributions within a vibronic level.

(b) The radiative relaxation rates depend on the vibronic level from which fluorescence originates. As far as they have been tested, they are in fair accord with the relative rates predicted by the Herzberg–Teller picture of vibronically-induced transitions.

(c) Non-radiative relaxation from S_1 levels not more than $2500\,\text{cm}^{-1}$ above the zero-point level are governed by $S_1 \rightarrow T$ intersystem crossing with only a shallow dependence on the initial vibronic level. The crossing rates increase slowly and rather smoothly with increasing vibronic energy. Severe fluctuations in the crossing rate from different levels with similar energies do not occur, but some sensitivity to modes can be seen. Correlation can be made between the rates and the number of vibrational quanta initially excited. The relative crossing rates are consistent with a theoretical description of Heller, Freed and Gelbart[11].

(d) An accelerated non-radiative decay occurs in C_6H_6 from most initial levels with energies more than $3300\,\text{cm}^{-1}$ above the zero-point level. The onset of this decay is exceedingly sharp as excitation climbs through levels of increasing energy. The non-radiative decay rate increases by a factor of 10^3 over a vibronic energy interval of $400\,\text{cm}^{-1}$. The nature of this decay has not yet been identified.

8.4 RESONANCE FLUORESCENCE SPECTRA FROM LARGE MOLECULES

Much of the recent activity in resonance fluorescence spectroscopy has centred on the structure of fluorescence from diatomics and a few triatomics such as NO_2 and SO_2. But new explorations of the fluorescence spectra from larger molecules have also begun to come forward. The structure in these spectra will be the subject of some comments in this Section.

8.4.1   Glyoxal

The availability of lasers has inspired much of the fluorescence work in small molecules. The laser's combined assets of monochromaticity and high power which can be easily imaged into a fluorescence cell fit precisely the needs of fluorescence spectroscopy with dilute gases. But so far not many laser lines are readily available in the ultraviolet, and those that do exist are subject to the same caprice of nature which governs conventional atomic line sources. They generally do not coincide with the most interesting regions of molecular absorption for the purposes of generating resonance fluorescence in large molecules. The tunable ultraviolet laser now under development[84] will obviously have an enormous impact on large molecule fluorescence.

In the meanwhile there have been few laser-excited probes of fluorescence spectroscopy in large polyatomics. However, that of Holzer and Ramsay[89] using a CW argon-ion laser to excite fluorescence in glyoxal is a most intriguing example of what is to come. They have found a remarkable coincidence of a fixed laser line with a unique aspect of the glyoxal absorption spectrum. Laser excitation of fluorescence has revealed the previously unobserved *cis* isomer of glyoxal by pumping directly the exceedingly weak 0,0 band in the $S_1 \leftarrow S_0$ transition of that isomer.

In a sequence of separate experiments, emission was excited by each of the argon-ion lines at 4765, 4880, 4965, 5017 and 5145 Å. The higher energy lines pump hot bands reaching low vibronic levels in the S_1 (^1A_u) state of the dominant *trans* isomer

$$\underset{H}{\overset{O}{\diagdown}} \underset{}{\overset{}{C}} \text{-} \underset{}{\overset{}{C}} \underset{O}{\overset{H}{\diagup}}$$

while the lines at 5017 and 5145 Å may also pump directly the triplet ^3A_u state of glyoxal.

Since the glyoxal pressure (50 torr) was high enough to cause vibrational equilibration in the excited states before emission, many bands common to every exciting line were observed. They were readily assigned to predictable $^1A_u-^1A_g$ and $^3A_u-^1A_g$ structure in *trans*-glyoxal. Some of that structure had been previously reported by Brand[81] using conventional fluorescence excitation. However, a group of seven bands became particularly prominent upon excitation with the 4880 Å argon ion line, and these were inconsistent with those expected or elsewhere observed from the $^{1,3}A_u$ states of the normal *trans* isomer.

Following the clue that these bands were associated with 4880 Å excitation, Currie and Ramsay[90] then examined at high resolution the 4875 Å absorption band whose rotational envelope overlapped the 4880 Å laser line. The rotational structure and constants in that band were found to be markedly different from those of the many other glyoxal bands analysed by Ramsay and co-workers[82]. However the constants were precisely those expected of a *cis* isomer, and the band at 4875 Å is reasonably assigned as the 0,0 band of the $^1B_1 \leftarrow {}^1A_1$ absorption system in *cis*-glyoxal. The temperature sensitivity of this band indicates that the zero-point energy of the ground state of *cis*-glyoxal is $1125 \pm 100 \text{ cm}^{-1}$ above that of *trans*-glyoxal.

8.4.2 Anthracene

Another interesting coincidence of line excitation, but this time not involving lasers, occurs between a Cd line and the 0,0 band of the $S_1 \rightarrow S_0$ absorption of anthracene.

Anthracene is typical of all large molecules whose absorption spectra are continuous due to vibronic congestion in that its resonance fluorescence spectrum mirrors that congestion. Even when excited with a sharp atomic line, the fluorescence is generally continuous and without sharp structure. Vibronic structure appears only as broad maxima, and even these tend to disappear as the excitation energy becomes increasingly larger. Pringsheim[12] long ago noted that an exception occurs in anthracene. While the 3660 Å Hg lines produce continuous fluorescence, the higher energy Cd line at 3610 Å (which also pumps a region of absorption lacking discrete structure) excites fluorescence with many sharp bands superimposed on a continuous background.

Haebig[91] has recently explored this further, and he has shown that the structure is related to that of fluorescence from the zero-point S_1 level of anthracene in a 4 K heptane matrix. The two sets of structure differ only by a uniform solvent shift. It is apparent that the Cd line is accidently coincident with the $0,0 \ S_1 \leftarrow S_0$ absorption band[92] which, in the vapour at 160 °C, is buried in a wealth of structure from sequence bands. A sufficiently large population of that unique level is selected by the Cd line to produce in this exceptional case a structured fluorescence spectrum after excitation into a diffuse region of absorption.

An instructive parallel occurs in naphthalene fluorescence excited by the 3130 Å Hg lines, because those lines fall very near the 0,0 band of the $S_1 \leftarrow S_0$ naphthalene system. However, the fluorescence in naphthalene differs markedly from that of anthracene with respect to congestion; that of naphthalene is fairly open and cleanly structured[93], while the anthracene zero-point structure is superimposed on an intense background of congestion continuum. The additional absorption congestion of anthracene arising from its increased vibrational complexity is clearly reflected in comparison of these two cases of fluorescence excitation near the 0,0 bands.

8.4.3 Nitrogen heterocyclic molecules

Logan and Ross[32, 33] have extended observations of gas-phase emission to an additional class of compounds. With Hg line exciting sources, they have been successful in recording resonance emission from (1) pyrazine, (2) pyrimidine, (3) isoquinoline, and (4) indazole.

No emission could be generated in (5) pyridine or (6) pyridazine.

(5)   (6)

The emission of pyrazine is perhaps the most interesting because it consists of both fluorescence and phosphorescence following excitation of the $n\pi^*$ S_1 state by the 3130 Å Hg line. Both emissions are rich in structure, but the fluorescence is somewhat diffuse despite an exceedingly sharp $S_1 \leftarrow S_0$ absorption spectrum. The addition of 550 torr of added isopentane fails to change the structure of the resonance fluorescence which indicates that the S_1 emitting state has a lifetime probably somewhat less than a nanosecond. In contrast, the triplet phosphorescence is vibrationally equilibrated even at low gas pressures, and this long-lived emission is sensitive to small added amounts of known triplet quenchers[35]. Nakamura[35] has explored the relaxation kinetics of pyrazine using these emissions and has also extended the wavelengths of exciting light to 3267, 3000 and 2900 Å. He could observe no emission using the 2537 Å Hg line to pump the S_2 state of pyrazine.

The other luminescent heterocyclics display only fluorescence, although the simultaneous presence of phosphorescence in pyrimidine might be difficult to detect. They each have a highly structured fluorescence spectrum, but this structure is not always free of background congestion. Logan and Ross have offered partial analysis of some of this fluorescence[33].

The fluorescence of isoquiniline is particularly sharp with little congestion underlying the structure. This is probably the consequence of still another fortuitous coincidence reminiscent of anthracene. The 3130 Å Hg lines used in excitation coincide closely to the position of the 0,0 band in the iso-quinoline $S_1 \leftarrow S_0$ absorption system. It would be interesting to compare this vapour zero-point spectrum with zero-point fluorescence more conventionally obtained in a low temperature matrix.

8.4.4 Fluorescence spectra generated with tuned excitation

One of the new developments in large molecule electronic fluorescence has been the use of tunable narrow-band excitation to pump selected excited state vibronic domains or, in some cases, single vibronic levels. Lifetime and yield measurements of fluorescence excited by these sources have been discussed in the previous Sections. In this Section, we discuss the fluorescence spectra from these selected vibronic domains. The spectra so far reported represent the vanguard of a novel excursion into large molecule fluorescence spectroscopy.

For convenience, these spectra will be called SVL fluorescence spectra (single vibronic level) to distinguish them from conventional resonance fluorescence excited by fixed line sources. In many cases, however, the adjectives SVL constitute an optimistic appraisal. The spectra are often only dominated by contributions from a single vibronic level. True SVL fluorescence spectra are difficult to obtain with the tunable exciting sources at present in use.

To date, reports of SVL fluorescence spectra encompass only a few larger molecules. Blondeau and Stockburger have published spectra from naphthalene[19, 94], the xylenes[94], toluene[94], aniline[94] and benzene[94]. Further exploration of benzene and benzene-d_6 has been reported by Atkinson, Parmenter and Schuyler[22–24, 39]. The list is expected to grow rapidly because numerous other laboratories are now also studying SVL fluorescence. All of the work has so far used tuned exciting light obtained by isolating a 15–50 cm^{-1} band from the continuum of a xenon arc.

8.4.4.1 Naphthalene, toluene and aniline

Blondeau and Stockburger[19] have obtained (approximately) SVL fluorescence spectra from the level v'_8 (b_{3g}) $= 438$ cm^{-1} and the level $v'_8 + v'_7$ (a_g) $= 1425$ cm^{-1} of the S_1 state of naphthalene. These are shown in Figure 8.11 together with the fluorescence spectrum from a Boltzmann vibronic distribution established by a high pressure of added gas. Both of the SVL spectra show considerable structure. The assignment of structure from the upper levels is consistent with the vibronic activity seen in absorption and in the Boltzmann fluorescence.

The structure in the two SVL spectra, and particularly that from the lower level, is open relative to the congestion present in the Boltzmann fluorescence. This is testimony to the success of the workers in their attempt to pump single vibronic levels. While Blondeau and Stockburger point out that sequence bands near the main transition are also pumped, they have obviously been successful in limiting excitation to a narrow selection of upper vibronic states.

The remaining naphthalene spectrum in Figure 8.11 is that of fluorescence generated by pumping the 0,0 band of the $S_2 \leftarrow S_1$ absorption system. As discussed before, it shows emission only in the region of S_1 fluorescence. (For reasons unknown, the S_2 fluorescence seen by Wannier et al.[60] is not visible here.) The transformation from the structured fluorescence from the lower levels 438 and 1425 cm^{-1} above the zero-point level to this diffuse S_1 emission from levels with about 3400 cm^{-1} of vibronic energy is rather striking.

Blondeau and Stockburger[94] have also pumped some narrow vibronic domains in toluene, and these, as in the case of naphthalene, show a marked decrease in the amount of sharp vibronic structure in fluorescence as the energies of the emitting vibronic states increase. One can compare in Figure 8.12 the transformation in structure as the dominant S_1 fluorescence levels change from the zero-point levels to levels 528 and 932 cm^{-1} above it. Those authors further report that all evidence of discrete structure is gone when the dominant vibronic energies rise to about 1200 cm^{-1}.

The increasing diffuseness of these fluorescence spectra invites comment on the problem of intramolecular vibrational 'redistribution' in large molecules. As one ascends to successively higher vibronic levels in an electronic state, the density of levels will eventually become so great that many levels overlap the level width, and intramolecular vibrational coupling can lead to the vibrational energy 'migration' or 'vibrational redistribution' among levels which is so extensively discussed in the literature of unimolecular reactions. SVL fluorescence is in principle an extremely sensitive probe of this effect.

Operation of the SVL fluorescence probe is perhaps best seen by first considering the familiar phenomenon of $S_1 \rightarrow T$ intersystem crossing. In that case, absorption in a *sharp* band optically selects an S_1 vibronic level that is weakly coupled to many vibronic levels in the triplet state. These

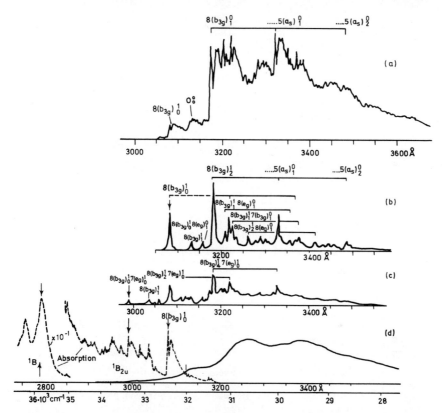

Figure 8.11 Fluorescence spectra from naphthalene vapour excited with tuned and narrow-band light. Spectrum (a) is 'equilibrated' fluorescence from naphthalene in the presence of 50 torr of hexane. Spectra (b) and (c) are resonance fluorescence dominated by emission from the S_1 levels ν_8 (438 cm^{-1}) and $\nu_8 + \nu_7$ (1425 cm^{-1}) respectively. Spectrum (d) is resonance fluorescence obtained by pumping the 0,0 band of the $S_2 \leftarrow S_0$ absorption system
(From Blondeau and Stockburger[19], by courtesy of North-Holland Publishing Co.)

triplet levels form a dissipative array of coupled states, and appear as the final states in the non-radiative intersystem crossing that subsequently destroys the S_1 level selected in absorption. Their presence is easily detected in a fluorescence quantum yield or lifetime experiment or in some experiment that probes the build-up of final triplet states.

Vibrational redistribution within the excited single state is quite analogous.

An initial level among a coupled set of levels (but in this case, levels in the same electronic state) is optically selected by absorption in a single *sharp* band. Subsequent to its preparation, relaxation into the dissipative vibrational array will destroy the initial state. If the time scale of such relaxation is competitive with radiative decay, that relaxation cannot fail to have important consequences in resonance fluorescence. The fluorescence spectrum will be poorly structured and congested. It will be a compendium of emission from many of those coupled states rather than from the single level reached

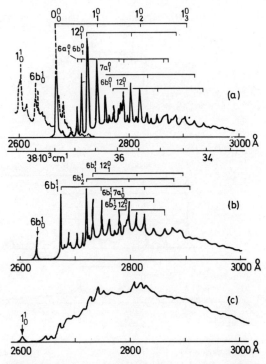

Figure 8.12 Resonance fluorescence from narrow-band tuned excitation of toluene vapour. The dominant S_1 emitting levels are the zero-point level in spectrum (a), the level v_{6b} (528 cm^{-1}) in spectrum (b), and the level v_1 (932 cm^{-1}) in spectrum (c). The dashed line in spectrum (a) indicates part of the $S_1 \leftarrow S_0$ absorption.
(From Blondeau and Stockburger[94], by courtesy of Verlag Chemie Gmbh)

in absorption. It will not have the sharp and open appearance of single level fluorescence.

The transformation from sharp to congested structure as SVL fluorescence experiments climb to higher vibrational levels in excitation offers a tool that far exceeds the sensitivity of absorption spectra as a probe of intramolecular coupling. Weak interactions which have profound effects on excited state behaviour may be totally invisible in absorption. For example, one may again consider $S_1 \rightarrow T$ intersystem crossing. The level v_6 in the

$^1B_{2u}$ state of benzene (reached by absorption of tuned light in the A_0^0 absorption band), is coupled with a large array of vibronic levels in the triplet ($^3B_{1u}$) state that cause about 75 per cent of the molecules in the initial level to decay by intersystem crossing rather than by fluorescence[39]. But high resolution examination of the $S_1 \leftarrow S_0$ absorption band reaching v_6 reveals no trace of this coupling. The band is sharp and highly structured[95]. The dissipative broadening is several orders of magnitude less than Doppler-widths and hence not visible in absorption. However, this coupling dominates the non-stationary character of the initial vibronic level.

Unfortunately, the detection of dissipative intramolecular vibronic coupling by diffuseness and congestion in resonance fluorescence is complicated by several factors. In some molecules, including benzene, fast non-radiative decay cuts off fluorescence from the higher vibronic levels. The probe is lost for the interesting energy region. A second problem occurs because present exciting sources are far from being monochromatic. Their $15 \, cm^{-1}$ widths place a similar limit on the resolution in fluorescence spectra. This will tend to obscure subtle onsets of spectral congestion due to intramolecular vibronic coupling, but it still provides adequate resolution to see the effect long before it would be apparent as diffuseness in an absorption band.

The principal and most general barrier, however, is the problem of distinguishing diffuseness caused by vibronic coupling from that caused by the trivial mechanism of exciting many upper vibronic levels directly in the absorption act. For example, this trivial mechanism is certainly responsible for the underlying fluorescence congestion in anthracene when the 0,0 band is pumped by the Cd line, and it can be expected to be a prime contributor to diffuseness in most fluorescence generated with the present $15 \, cm^{-1}$ tunable sources. An exceedingly careful comparison of the absorption spectrum and the exciting power profile must always be made (even with line sources) if much is to be learned about the cause of diffuseness in a fluorescence spectrum.

Diffuseness in toluene and naphthalene fluorescence are plausibly assigned to mechanisms other than intramolecular vibrational coupling, although the latter may also be a contributor. In each case the vibronic structure in the absorption spectrum is crowded relative to the excitation band-pass. Thus the trivial mechanism of congestion created in excitation is important, and it can be expected to become more so as higher levels are excited because the absorption spectrum itself increases in congestion at shorter wavelengths. In fact, it is unlikely that vibrational redistribution could make any contribution to congestion in the toluene spectra shown in Figure 8.12. The density of S_1 vibronic levels at 528 and $932 \, cm^{-1}$ is surely too low to establish opportunity for dissipative coupling. On the other hand, the very diffuse S_1 fluorescence spectrum observed in Figure 8.11, when excitation pumps the $S_2 \leftarrow S_0$ 0,0 band in naphthalene, is consistent with the concept of $S_2 \rightarrow S_1$ internal conversion into a dissipative array of many S_1 levels quasi-degenerate with the S_2 state. Watts and Strickler[17] have calculated that the density of those S_1 levels is about 10^5 per cm^{-1}.

Blondeau and Stockburger have also published two SVL spectra from aniline. Both of these have extensive sharp structure, and continued exploration of this fluorescence should provide helpful comments on interpretation of the kinetic data concerning S_1 relaxation in aniline which was discussed in Section 8.3.1.

8.4.4.2 Single vibronic level fluorescence from benzene

Among the molecules from which fluorescence spectra have been obtained with tuned excitation, benzene is the only one whose absorption spectrum is sufficiently open to limit excitation to truly a single vibronic level with the sources currently available. Fluorescence spectra have now been recorded and analysed from over 20 levels lying within 2900 cm^{-1} of the zero-point S_1 level of C_6H_6 and C_6D_6 [96], but not all of these have been published.

Three spectra from C_6H_6 are shown together with an equilibrated fluorescence spectrum in Figure 8.13. These spectra are in many aspects typical of all SVL fluorescence spectra so far recorded from benzene. First, they are

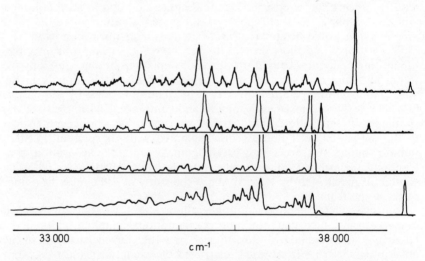

33 000   38 000

cm^{-1}

Figure 8.13   SVL fluorescence spectra from various levels in the $^1B_{2u}$ (S_1) state of benzene. The position of the exciting light is in every case the structure at highest energy. The originating levels are, from the top, $v_6 + v_1$ (1446 cm^{-1}), v_6 (523 cm^{-1}) and the zero-point level. The bottom spectrum is from a Boltzmann distribution of vibronic levels.
(From Parmenter and Schuyler[39], by courtesy of North-Holland Publ. Co.)

each unique. Spectra from different levels have marked variations in structure. Furthermore, as the energy of the emitting level increases, structure appears at shorter wavelengths than that in the equilibrated fluorescence. In addition, the SVL fluorescence structure is more open and simple than that of the equilibrated fluorescence. The assignments are in every case consistent with that from the upper level reached in absorption. No indications of intra-molecular vibrational redistribution exist in these spectra, nor are any expected. The low densities of vibronic levels in the energy regions accessible for fluorescence experiments in benzene would preclude it.

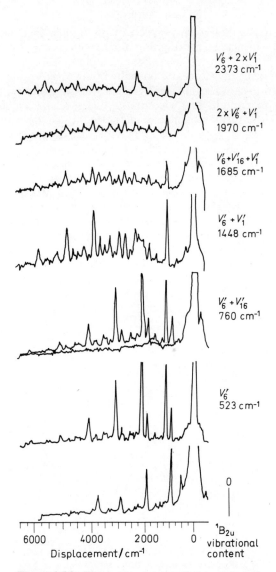

$V'_6 + 2 \times V'_1$
2373 cm^{-1}

$2 \times V'_6 + V'_1$
1970 cm^{-1}

$V'_6 + V'_{16} + V'_1$
1685 cm^{-1}

$V'_6 + V'_1$
1448 cm^{-1}

$V'_6 + V'_{16}$
760 cm^{-1}

V'_6
523 cm^{-1}

0

$^1B_{2u}$
vibrational
content

6000   4000   2000   0

Displacement / cm^{-1}

Figure 8.14 Fluorescence spectra from seven single vibronic levels in benzene at 0.2 torr pressure are shown with the exciting wavelengths aligned. The amount of vibrational energy and the particular vibrational modes excited in the $^1B_{2u}$ state are shown in the column to the right. Since these spectra are linear in wavelength, the scale shown for displacement in cm^{-1} from the exciting band (located at 0 cm^{-1} in each case) is approximate and fits well only the lower-most spectrum. Wavelength increases to the left in the spectra. The large band at 3800 cm^{-1} in the $\varepsilon_{vib} = 0$ cm^{-1} spectrum only is present in the scattered exciting light.
(From Atkinson et al.[24], by courtesy of the American Chemical Society)

While each SVL fluorescence spectrum is unique, their common heritage is readily displayed by aligning them with respect to the position of the exciting light. This has been done in Figure 8.14 (where the exciting lines are prominent due to scattered light in an early version of the fluorescence apparatus). With this arrangement, the structure from diverse S_1 levels is seen to contain many common energy displacements from the exciting line. These displacements are of course the ground-state vibrational energies which control the position of vibronic structure relative to the exciting line in all resonance fluorescence spectra.

The assignment of this fluorescence is in general rather straightforward, because the frequencies of the ground-state vibrations are all known from independent sources. The assignments have been discussed in several reports[22-24, 43]. It is simply mentioned here that in every case they are consistent with the Herzberg–Teller selection rules so successfully used in the classic work on the absorption spectrum of benzene[97].

Those assignments have also been of considerable help in refining the vibronic assignment of the absorption spectrum. The presence of assigned

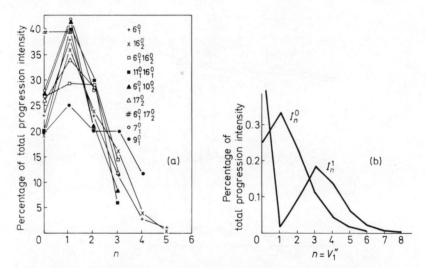

Figure 8.15 (a) A plot of the relative intensities of progression members in nine progressions seen in the vapour-fluorescence spectrum shown in Figure 8.16. The progressions are each built by transitions $v' = 0 \rightarrow v''_1 = n$ with $n = 0, 1, 2 \ldots$ occurring in combination with other vibrational changes as indicated to induce the transition
(From Parmenter and Schuyler[23], by courtesy of the American Institute of Physics)
(b) Calculated intensities in the above progressions and in the progression $v'_1 = 1 \rightarrow v''_1 = n$
(From Atkinson et al.[24], by courtesy of the American Chemical Society)

fluorescence from so many levels gives a clearer indication of the activity of vibrations in the S_1–S_0 transition of benzene than had previously been available. This provides a new criterion to use in assigning both fluorescence and absorption, because one can now check for consistency in the activity of certain vibrations in the two types of spectra. A second comment from resonance fluorescence on absorption assignments is more direct. In some cases one can simply inquire whether SVL fluorescence spectra are consistent with the upper state proposed in the absorption assignment.

Examples of the interplay of absorption and fluorescence assignments are given in two papers discussing the assignment of resonance fluorescence excited in benzene by absorption of the 2537 Å Hg line[24, 98]. Further discus-

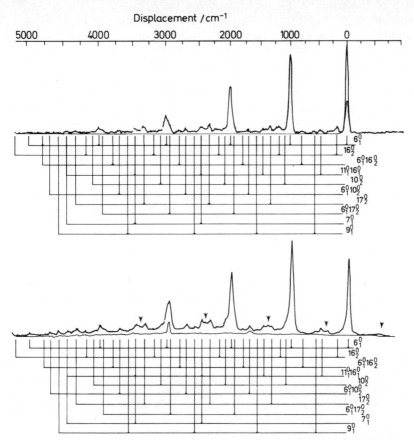

Figure 8.16 A comparison of fluorescence from the zero-point level of the $^1B_{2u}$ state of benzene vapour (top) with that from the zero-point level of benzene in a 77 K cyclohexane matrix (bottom). The assignment scale of the vapour spectrum has been placed under the 77 K spectrum to facilitate comparison. The triangular indices in the latter mark a forbidden progression in the symmetric (a_{1g}) vibration v_1 that begins with the symmetry-forbidden 0,0 band. This progression does not appear in vapour spectra (From Parmenter and Schuyler[23], by courtesy of the American Institute of Physics)

sion of the SVL fluorescence and its use in refining the absorption spectrum is given in some detail in a recent review of radiative and non-radiative processes in the S_1 state of benzene vapour[43].

Benzene SVL fluorescence spectra have also yielded an improved picture of Franck–Condon factors governing the intensities of the members of a progression in the symmetric ring breathing vibration v_1 ($v_1'' = 992 \text{ cm}^{-1}$; $v_1' = 920 \text{ cm}^{-1}$). The relative intensities of the members $v_1' = 0 \rightarrow v_1'' =$

$0, 1, 2 \ldots$ from nine distinct progression origins in the zero-point C_6H_6 fluorescence spectrum are shown in Figure 8.15 together with the calculated relative intensities for that progression. The correspondence is satisfactory. These comparisons have been made before by observations of the corresponding $v_1' = 0, 1, 2 \ldots \leftarrow v_1'' = 0$ members in the absorption spectrum[99]. Calculation of intensities in the emission progression $v_1' = 1 \rightarrow v_1'' = 0, 1, 2 \ldots$ is also shown in Figure 8.15. This case is most interesting because the 1–1 member is of almost vanishing intensity[100]. Observation of the progression containing that member is difficult (but possible[100]) in absorption because it can occur only in a hot band with very low intensity. In SVL fluorescence, however, $v_1' = 1 \rightarrow v_1'' = 0, 1, 2 \ldots$ progressions are easily seen by simply pumping an upper state with $v_1' = 1$. The predicted minimum is characteristic of such SVL fluorescence, and in fact it is chiefly responsible for the pronounced difference in structure between the v_6 and $(v_6 + v_1)$ spectra and between the $(v_6 + v_{16})$ and $(v_6 + v_{16} + v_1)$ spectra in Figure 8.14. SVL fluorescence also allows exploration of the $v_1' = 2 \rightarrow v_1'' = 0, 1, 2 \ldots$ progressions in benzene, and they, as the others, are consistent with the calculated intensities.

We finally remark on an interesting comparison of gas and condensed (77 K) phase fluorescence from the zero-point level in benzene. The two spectra are shown in Figure 8.16. While the entire condensed phase spectrum is shifted in energy, the structure of the two is remarkably similar. But one vital difference is readily observed. The forbidden 0,0 band and its accompanying $v_1' = 0 \rightarrow v_1'' = 0,1,2 \ldots$ progression becomes prominent among the minor structure in condensed phase. That progression is symmetry-forbidden and entirely absent in the vapour spectrum. It is quite obvious that this and other forbidden structure which is so extensively documented from condensed phase spectra are induced by site interactions of benzene with its solvent.

References

1. McGlynn, S. P., Azumi, T. and Kinoshita, M. (1969). *Molecular Spectroscopy of the Triplet State.* (Englewood Cliffs, N.J.: Prentice Hall)
2. Birks, J. B. (1970). *Photophysics of Aromatic Molecules.* (London: Wiley-Interscience)
3. Calvert, J. G. and Pitts, J. N. (1966). *Photochemistry.* (New York: Wiley)
4. Tinti, D. S. and Robinson, G. W. (1968). *J. Chem. Phys.,* **49**, 3229
5. Lewis, G. N. and Kasha, M. (1944). *J. Amer. Chem. Soc.,* **66**, 2100; (1945). **67**, 994
6. Coulson, C. A. (1950). *Discuss. Faraday Soc.,* **9**, 1
7. Kasha, M. (1950). *Discuss. Faraday Soc.,* **9**, 14
8. Kasha, M. (1960). *Radiation Research Supplement 2,* 243. (New York: Academic Press)
9. Henry, B. and Kasha, M. (1968). *Ann. Rev. Phys. Chem.,* **19**, 161
10. See for example Jortner, J., Rice, S. A. and Hochstrasser, R. M. (1969). *Advan. Photochem.,* **7**, 149
11. Heller, D. F., Freed, K. F. and Gelbart, W. M. (1972). *J. Chem. Phys.* (in press)
12. Pringsheim, P. (1938). *Ann. Acad. Sci. Tech. Varsovie,* **5**, 29
13. Stevens, B. and Hutton, E. (1960). *Molec. Phys.,* **3**, 71
14. Gruzinskii, W. V. and Borisevitch, N. A. (1963). *Opt. Spectrosc.,* **15**, 246 [*Opt. Specktrosk.,* **15**, 457]
15. Pringsheim, P. (1949). *Fluorescence and Phosphorescence.* (New York: Interscience)
16. Stockburger, M. (1962). *Z. Physik. Chem.,* **35**, 179
17. Watts, R. J. and Strickler, S. J. (1966). *J. Chem. Phys.,* **44**, 2423
18. Uy, J. O. and Lim, E. C. (1970). *Chem. Phys. Lett.,* **7**, 306
19. Blondeau, J. M. and Stockburger, M. (1971). *Chem. Phys. Lett.,* **8**, 436

20. Williams, R. and Goldsmith, G. J. (1963). *J. Chem. Phys.*, **39**, 2008
21. Reference 2, p. 112
22. Parmenter, C. S. and Schuyler, M. W. (1969). *Transitions Nonradiat. Mol., Reunion Soc. Chim. Phys.*, 20th, 92. (Suppl. *J. Chim. Phys.*, 1969–1970)
23. Parmenter, C. S. and Schuyler, M. W. (1970). *J. Chem. Phys.*, **52**, 5366
24. Atkinson, G. H., Parmenter, C. S. and Schuyler, M. W. (1971). *J. Phys. Chem.*, **75**, 1572
25. Kistiakowsky, G. B. and Parmenter, C. S. (1965). *J. Chem. Phys.*, **42**, 2942
26. Anderson, E. M. and Kistiakowsky, G. B. (1969). *J. Chem. Phys.*, **51**, 182; (1968). **48**, 4787
27. Douglas, A. E. and Mathews, C. W. (1968). *J. Chem. Phys.*, **48**, 4788
28. Parmenter, C. S. and White, A. H. (1969). *J. Chem. Phys.*, **50**, 1631
29. Ware, W. R. and Cunningham, P. T. (1965). *J. Chem. Phys.*, **43**, 3826
30. Ware, W. R. and Cunningham, P. T. (1966). *J. Chem. Phys.*, **44**, 4364
31. Parmenter, C. S. and Poland, H. M. (1969). *J. Chem. Phys.*, **51**, 1551
32. Logan, L. M. and Ross, I. G. (1965). *J. Chem. Phys.*, **43**, 2903
33. Logan, L. M. and Ross, I. G. (1968). *Acta Physica Polonica*, **34**, 721
34. Lahmani, F., Ivanov, N. and Magat, M. (1966). *C.R. Acad. Sci., Paris, Ser. C*, **283**, 1005
35. Nakamura, K. (1971). *J. Amer. Chem. Soc.*, **93**, 3138
36. Anderson, L. G. and Parmenter, C. S. (1970). *J. Chem. Phys.*, **52**, 466
37. Bäckström, H. L. J. and Sandros, K. (1960). *Acta. Chem. Scand.*, **14**, 48
38. McClure, D. S. (1949). *J. Chem. Phys.*, **17**, 905
39. Parmenter, C. S. and Schuyler, M. W. (1970). *Chem. Phys. Lett.*, **6**, 339
40. Lim, E. C. (1962). *J. Chem. Phys.*, **36**, 3497
41. Eastman, J. W. (1968). *J. Chem. Phys.*, **49**, 4617; (1970). *Z. Naturforsch.* **A25**, 949
42. Parker, C. A. and Hatchard, C. G. (1962). *Analyst (London)*, **87**, 664
43. Parmenter, C. S. (1972). *Advan. Chem. Phys.*, **22** (in press)
44. Douglas, A. E. (1966). *J. Chem. Phys.*, **45**, 1007
45. Robinson, G. W. and Frosch, R. P. (1962). *J. Chem. Phys.*, **37**, 1962; (1963). **38**, 1187
46. Hunt, G. R., McCoy, E. F. and Ross, I. G. (1962). *Austr. J. Chem.*, **15**, 591
47. Byrne, J. P., McCoy, E. F. and Ross, I. G. (1965). *Austr. J. Chem.*, **18**, 1589
48. Robinson, G. W. (1967). *J. Chem. Phys.*, **47**, 1967
49. Bixon, M. and Jortner, J. (1968). *J. Chem. Phys.*, **48**, 715
50. Jortner, J. and Berry, R. S. (1968). *J. Chem. Phys.*, **48**, 2757
51. Chock, D. P., Jortner, J. and Rice, S. A. (1968). *J. Chem. Phys.*, **49**, 610
52. Bixon, M. and Jortner, J. (1969). *J. Chem. Phys.*, **50**, 3284, 4061
53. Anderson, L. G., Parmenter, C. S., Poland, H. M. and Rau, J. D. (1971). *Chem. Phys. Lett.*, **8**, 232
54. Yardley, J. T., Holleman, G. W. and Steinfeld, J. I. (1971). *Chem. Phys. Lett.*, **10**, 266
55. Parmenter, C. S. (1964). *J. Chem. Phys.*, **41**, 658
56. Nitzan, A., Jortner, J. and Rentzepis, P. M. (1971). *Chem. Phys. Lett.*, **8**, 445
57. Geldof, P. A., Rettschnick, R. P. H. and Hoytink, G. J. (1969). *Chem. Phys. Lett.*, **4**, 59
58. Nakajima, A. and Baba, H. (1970). *Bull. Chem. Soc. Japan*, **43**, 967
59. Wannier, P., Rentzepis, P. M. and Jortner, J. (1971). *Chem. Phys. Lett.*, **10**, 102
60. Wannier, P., Rentzepis, P. M. and Jortner, J. (1971). *Chem. Phys. Lett.*, **10**, 193
61. Nitzan, A., Jortner, J., Kommandeur, J. and Drent, E. (1971). *Chem. Phys. Lett.*, **9**, 273
62. Drent, E. and Kommandeur, J. (1971). *Chem. Phys. Lett.*, **8**, 303
63. Beer, M. and Longuet-Higgins, H. C. (1955). *J. Chem. Phys.*, **23**, 1390
64. Viswanath, G. and Kasha, M. (1956). *J. Chem. Phys.*, **24**, 574
65. Binsch, G., Heilbronner, E., Jankow, R. and Schmidt, D. (1967). *Chem. Phys. Lett.*, **1**, 136
66. Easterly, C. E., Christophorou, L. G., Blaunstein, R. P. and Carter, R. P. (1970). *Chem. Phys. Lett.*, **6**, 579
67. Schlag, E. W. and von Weyssenhoff, H. (1968). *Ber. Bunsinges. Physik. Chem.*, **72**, 153
68. Schlag, E. W. and von Weyssenhoff, H. (1969). *J. Chem. Phys.*, **51**, 2508
69. Laor, U. and Ludwig, P. K. (1971). *J. Chem. Phys.*, **54**, 1054
70. Hsieh, J. C., Laor, U. and Ludwig, P. K. (1971). *Chem. Phys. Lett.*, **10**, 412
71. von Weyssenhoff, H. and Kraus, F. (1971). *J. Chem. Phys.*, **54**, 2887
72. Nakamura, K. (1970). *J. Chem. Phys.*, **53**, 998
73. Ware, W. R., Selinger, B. K., Parmenter, C. S. and Schuyler, M. W. (1970). *Chem. Phys. Lett.*, **6**, 342

74. Gelbart, W., Spears, K. G., Freed, K. F., Jortner, J. and Rice, S. A. (1970). *Chem. Phys. Lett.*, **6**, 345
75. Selinger, B. K. and Ware, W. R. (1970). *J. Chem. Phys.*, **53**, 3160; (1970). **52**, 5482
76. Siebrand, W. (1966). *J. Chem. Phys.*, **44**, 4055; (1967). **47**, 2411
77. Englman, R. and Jortner, J. (1970). *Mol. Phys.*, **18**, 145
78. Fisher, S. F. (1970). *J. Chem. Phys.*, **53**, 3195; (1969). *Chem. Phys. Lett.*, **4**, 333
79. Fisher, S. F. and Schlag, E. W. (1969). *Chem. Phys. Lett.*, **4**, 393
80. Hemminger, J. C. and Lee, E. K. C. (1971). *J. Chem. Phys.*, **54**, 1405
81. Brand, J. C. D. (1954). *Trans. Faraday Soc.*, **50**, 431
82. Birss, F. W., Brown, J. M., Cole, A. R. H., Lofthus, A., Krishnamachari, S. L. N. G., Osborne, G. A., Paldus, J., Ramsay, D. A. and Watmann, L. (1970). *Can. J. Phys.*, **48**, 1230
83. Anderson, L. G., private communication
84. Yeung, E. S. and Moore, C. B. (1971). *J. Amer. Chem. Soc.*, **93**, 2059
85. Spears, K. G. and Rice, S. A. (1972). *J. Chem. Phys.* (in press); Abramson, A. S., Spears, K. G. and Rice S. A. (to be published)
86. Parmenter, C. S., Schuh, M. D. and Schuyler, M. W. (to be published)
87. Parmenter, C. S. and Schuh, M. D. (1972). *Chem. Phys. Lett.* (in press)
88. Callomon, J. H., Lopez-Delgado, R. and Parkin, J. E. (1972). *Chem. Phys. Lett.* (in press)
89. Holzer, W. and Ramsay, D. A. (1970). *Can. J. Phys.*, **48**, 1759
90. Currie, G. N. and Ramsay, D. A. (1971). *Can. J. Phys.*, **49**, 317
91. Haebig, J. E. (1968). *J. Molec. Spectrosc.*, **25**, 117
92. Byrne, J. P. and Ross, I. G. (1965). *Can. J. Chem.*, **43**, 3253
93. Hollas, J. M. (1962). *J. Molec. Spectrosc.*, **9**, 138
94. Blondeau, J. M. and Stockburger, M. (1971). *Ber. Bunsenges. Physik. Chem.*, **75**, 450
95. Callomon, J. H., Dunn, T. M. and Mills, I. M. (1966). *Phil. Trans. Roy. Soc. (London)*, *Ser. A*, **259**, 499
96. Schuyler, M. W. (1970). *Ph.D. Thesis*, Indiana University
97. Herzberg, G. (1966). *Electronic Spectra of Polyatomic Molecules*, 555–561, 665–6. (Princeton, N.J.: D. Van Nostrand)
98. Atkinson, G. H. and Parmenter, C. S. (1971). *J. Phys. Chem.*, **75**, 1564
99. Craig, D. P. (1950). *J. Chem. Soc.*, 2146
100. Smith, W. L. (1968). *J. Phys. (Proc. Phys. Soc.)*, **B1**, 89